Geometry, Groups and Mathematical Philosophy

CONTEMPORARY MATHEMATICS

811

Geometry, Groups and Mathematical Philosophy

International Conference in Honor of
Ravi S. Kulkarni's 80th Birthday
Geometry, Groups and Mathematical Philosophy
May 21–25, 2022
Bhaskaracharya Pratishthana, Pune, India

Krishnendu Gongopadhyay
Shashikant A. Katre
Editors

AMS AMERICAN MATHEMATICAL SOCIETY

2020 *Mathematics Subject Classification.* Primary 51M10, 53Axx, 22Exx, 20Fxx, 05Cxx.

Library of Congress Cataloging-in-Publication Data

Names: Geometry, Groups and Mathematical Philosophy (Conference) (2022 : Pune, India) | Kulkarni, Ravi S., honoree. | Gongopadhyay, Krishnendu, editor. | Katre, S. A. (Shashikant A.), editor.

Title: Geometry, groups and mathematical philosophy : international conference on geometry, groups and mathematical philosophy in honor of Ravi Kulkarni's eightieth birthday, May 21-25, 2022, Bhaskaracharya Pratishthana, Pune, India / Krishnendu Gongopadhyay, Shashikant A. Katre, editors.

Description: Providence, Rhode Island : American Mathematical Society, [2025] | Series: Contemporary mathematics, 0271-4132 ; volume 811 | Includes bibliographical references.

Identifiers: LCCN 2024036323 | ISBN 9781470472153 (paperback) | ISBN 9781470478612 (ebook)

Subjects: LCSH: Geometry, Differential–Congresses. | Symmetry (Mathematics)–Congresses. | Mathematics–Philosophy–Congresses. | AMS: Geometry – Real and complex geometry – Hyperbolic and elliptic geometries (general) and generalizations. | Differential geometry – Classical differential geometry. | Topological groups, Lie groups – Lie groups. | Group theory and generalizations – Special aspects of infinite or finite groups. | Combinatorics – Graph theory.

Classification: LCC QA641 .G4453 2022 | DDC 516.3/6–dc23/eng/20241220
LC record available at https://lccn.loc.gov/2024036323
Contemporary Mathematics ISSN: 0271-4132 (print); ISSN: 1098-3627 (online)
DOI: https://doi.org/10.1090/conm/811

∞ The paper used in this book is acid-free and falls within the guidelines
established to ensure permanence and durability.
Visit the AMS home page at `https://www.ams.org/`

10 9 8 7 6 5 4 3 2 1 30 29 28 27 26 25

This volume is dedicated to Ravi S. Kulkarni on his eightieth birthday

Contents

Preface ix

My pursuit of truth through mathematics
RAVINDRA S. KULKARNI 1

Philosophy of mathematics – four issues
RAVINDRA S. KULKARNI 13

Counting cusp excursions of reciprocal geodesics
ARA BASMAJIAN and ROBERT SUZZI VALLI 21

Elementary groups in PSL$(3, \mathbb{C})$
W. BARRERA, A. CANO, J. P. NAVARRETE, and J. SEADE 31

Quasilocal mass and angular momentum
PO-NING CHEN, MU-TAO WANG, YE-KAI WANG, and SHING-TUNG YAU 49

Geodesics on the modular surface and continued fraction expansions of their endpoints
S. G. DANI 61

Finite group actions on 1-complexes and homology
ALLAN L. EDMONDS 75

Affine structures on surfaces and the twisted cubic cone
WILLIAM M. GOLDMAN 83

Computing mixed multiplicities, mixed volumes and sectional Milnor numbers
KRITI GOEL, VIVEK MUKUNDAN, SUDESHNA ROY, and J. K. VERMA 93

Differential central simple algebras
PARUL GUPTA, YASHPREET KAUR, and ANUPAM SINGH 113

z-Classes and rational conjugacy classes
VIKAS S. JADHAV 123

On Jacobian group and complexity of the Δ-graph
A. D. MEDNYKH, I. A. MEDNYKH, and I. N. YUDIN 133

Groups, communication and coordination
ROHIT PARIKH 145

Curvature and torsion via quadrilateral gaps
NITIN NITSURE 157

Schur index and extensions of Witt-Berman's theorems
SOHAM SWADHIN PRADHAN 163

Nonlinear retracts and the geometry of Banach spaces
M. A. SOFI 183

Computations for invariant bilinear forms under an invertible linear transformation: an expository article
JAGMOHAN TANTI 201

A comparative overview of Riemannian and Finsler geometry
BANKTESHWAR TIWARI 209

Revisiting Kulkarni's topological proof of Millington's theorem
DEVENDRA TIWARI and HARSHAVARDHAN REDDY 241

Kulkarni-Nomizu tensor fields
MUKUT MANI TRIPATHI 259

Preface

This volume is an outcome of the international conference: Geometry, Groups and Mathematical Philosophy in honor of Professor Ravindra S. Kulkarni, also known as Ravi S. Kulkarni, turning eighty! The conference took place in hybrid (both online and offline) mode at the Bhaskaracharya Pratishthana (BP), an Institute of Mathematics in Pune, India during 21–25 May, 2022. Ravi S. Kulkarni has been a distinguished professor at residence at this Institute since 2008. The main aim of the conference was to celebrate the influence of Ravi Kulkarni in Mathematics. Kulkarni's pioneering work from 1968 until 1998 falls in the broad area of differential geometry. From 2000 onwards, Kulkarni has influenced some unexpected directions in Group theory with ideas coming from geometry. Kulkarni has also been interested in general mathematics, mathematical philosophy, and Indian knowledge system.

In the conference, there were talks by mathematicians from India and abroad, especially USA, giving flavours from all these areas that intrigued Prof. Ravi Kulkarni over the years. Accordingly, the conference was divided into three parts: Differential geometry (May 21 & 24), Symmetries from a geometric viewpoint (May 23), and general mathematics, mathematical philosophy and Indian mathematics (May 25). A Felicitation program was held on Ravi Kulkarni's Birthday (May 22) from 4.30 p.m. to 7 p.m. (IST) presided by the eminent computer scientist Padma Bhushan Professor Vijay Bhatkar. Professors Karmeshu, Vinod Saxena, Satya Deo, Sudesh K. Khanduja, J. K. Verma, A. R. Shastri, Dilip Patil, S. A. Katre and various mathematicians expressed their admirations to Prof. Kulkarni on this occasion.

The conference speakers included many friends, colleagues and former students of Prof. Ravi Kulkarni, who had interacted with him on these topics over the years. The timetable, especially on May 21 & 24, was prepared keeping in mind the time difference of the overseas locations of the speakers. Conference talks are available at the YouTube channel of Bhaskaracharya Pratishthana. Most of the speakers have given articles based on their talks for these proceedings. The articles will give some glimpses of the work of Prof. Ravi Kulkarni and some future directions. The list of speakers in the conference is given in the following.

Speakers.

Kulkarni@80: Differential geometry. • Ara Basmajian (CUNY) • Sagun Chanillo (Rutgers Uni., New Brunswick) • S. G. Dani (CEBS, Mumbai) • Allan Edmonds (Indiana) • William Goldman (Maryland) • Yunping Jiang (CUNY) • Alexander Mednykh (Novosibirsk) • Ulrich Pinkall (TU Berlin) • Nitin Nitsure

(TIFR, Mumbai) • José Seade (UNAM, Mexico) • Dragomir Saric (CUNY) • S. -T. Yau (Harvard U. and Qinghua U.) • Bankteshwar Tiwari (BHU, Varanasi) • Mukut Mani Tripathi (BHU, Varanasi)

Kulkarni@80: Symmetries. • Gurmeet K. Bakshi (Punjab University, Chandigarh) • Anupam K. Singh (IISER Pune) • Siddhartha Sarkar (IISER Bhopal) • Soham Pradhan (Haifa University, Israel) • Jagmohan Tanti (BBAU Lucknow) • Anthony Weaver (CUNY) • Devendra Tiwari (BP, Pune) • Vikas Jadhav (Wadia College, Pune)

Kulkarni@80: General Mathematics and Mathematical Philosophy. • Satya Deo (HRI) • Jugal Verma (IIT Bombay) • Rohit Parikh (CUNY) • M. A. Sofi (Kashmir Uni.) • Sudesh Kaur Khanduja (IISER Mohali).

This volume includes submitted and refereed articles written by speakers in the above conference. The authors include many of his friends, colleagues and former students who had interacted with him on these topics over the years. An expanded version of Prof Kulkarni's lecture during this program has also been included. An autobiographical account of Prof. Kulkarni in his own words has been added as well. This article is a modified version of a speech by Prof Kulkarni delivered at the Queen's College in 2013 in the conference "Conformal geometry and Riemann surfaces" in honor of his seventieth birthday during October 25–27, 2013 organized by his former colleagues Professor Sudeb Mitra and Professor Yunping Jiang.

We wholeheartedly thank all the speakers, authors and anonymous referees for their contributions. Without their support, it would not have been possible to publish this volume. Above all, we thank all the members of the organizing committee for the Kulkarni@80 conference and all the participants for their active engagements during the conference. The organizing committee for the conference is given below.

Organizing Committee of the Kulkarni@80 conference.

• S. A. Katre (Convener), S. P. Pune University and BP, Pune; • Krishnendu Gongopadhyay (Convener), IISER Mohali, Punjab; • V. M. Sholapurkar, S. P. College and BP, Pune; • V. V. Acharya, Fergusson College and BP, Pune; • Vikas Jadhav, Nowrosjee Wadia College, Pune.

Finally, with deep regret, we must mention that just after the conference, Professor Satya Deo, a speaker of the conference, passed away due to a sudden heart attack on June 19, 2022. Professor Satya Deo was very enthusiastic about this volume and even expressed confirmation to submit an article. We are very sorry that he is no more with us to see this volume in print.

Krishnendu Gongopadhyay
IISER Mohali, S.A.S. Nagar, India
February 29, 2024

Shashikant A. Katre
Bhaskaracharya Pratishthana, Pune, India
February 29, 2024

Contemporary Mathematics
Volume **811**, 2025
https://doi.org/10.1090/conm/811/16232

My pursuit of truth through mathematics

Ravindra S. Kulkarni

1. Greetings

I am very happy today. I did not know that so many people loved me enough to gather at Queens College to wish me a healthy, long, and productive life over and above the 71 years I have already lived. It includes my teacher Shlomo Sternberg, present here on skype, and my "almost"-teachers Hyman Bass, and Cliff Earle. Alex Lubotzky came from Israel, Ulrich Pinkall from Germany, and Shiga from Japan. If I have counted correctly there are 14 people among the speakers who are above 65, and 5 below 65, of which only 3 in their 30s to 50s. There are many more in the audience who are in their 50s and below. I interpret this as: we old people have done something right. And of course that something right, is that we have done mathematics.

The conference of this type is new for the Math department at Queens College, although it had many distinguished mathematicians like Arthur Sard, Leo Zippin, Banesh Hoffman, Edwin Moise, ... before, on its faculty.

I find this Conference especially gratifying since I already went back to India in 2001, enjoyed several leaves without pay, and finally retired from Queens College, in Feb 2008. However I keep coming back to Queens college and Graduate Center twice a year and enjoy my emeritus positions with all the office and library/computer advantages.

For a long time, I felt that people here thought that I was an Indian in America. I was even more surprised that the people in India thought that I was an American in India. But today I feel that the people from both the continents think of me as their own.

One sign of it is that I may tell you that for 70 years I never ran for any office like that of HoD. But in India, someone nominated me for the office of the President of the Ramanujan Mathematical Society. In April 2013, I was elected with 75 percent of the voters approving me!

I wish to use that position to invite many of you to India to develop more networks, and light many more lamps of higher mathematics in India.

To begin with, I must thank my many Queens College Colleagues: President Muyskens, Dean Liebovich, the HoD Prof. Wallace Goldberg, our untiring Math secretary Minnie, who has been like my elder sister taking care of me in many ways for the last 28 years. Then there is Valerie, and her whole family, her parents, husband, and her five children. Valerie gave me something which I consider as the highest honor I have received in my life. She and her husband Yuri gave their first son the first name Valerie's father's name, Lawrence, and the second name, my name, Ravi. So here is the most handsome, fairest, blue-eyed Lawrence Ravi I know in the world. He is thinking of a career in science.

For this Conference, I especially wish to thank Sudeb Mitra, Yunping Jiang, Zhe Wang, and Blendi Koroveshi, who put in days of efforts to arrange this Conference on Conformal Geometry and Riemann Surfaces. I have to spend at least a year to learn the high-tech they have used to make this Conference a success. I have worked in other areas, but these subjects have been the inspiration of my life.

We may not quite realize it in our day-to-day lives, but we are truly living in momentous times. Millennia have been squeezed into centuries, and centuries have been squeezed into decades. It is not yet 100 years since Ramanujan, a clerk working at an obscure place in Chennai, India, wrote to Hardy, a distinguished professor of mathematics at a world-renowned university in UK. He was invited by Hardy, who made arrangement for his travel and living expenses by way of a scholarship. One great, unexpected, result in Mathematics came from their fruitful collaboration. But far more than that it significantly added to the human consciousness about Mathematics and the sciences all over the world. Can you believe, all this happened less than 100 years, that is less than 40 thousand days, ago?

Unprecedented, great changes in our attitudes and ways of thinking are taking place every few years in our short lives. We all know, that Knowledge is Power, Knowledge is Money, Knowledge is Prestige, and all that, but far more than that, Knowledge remains the most ancient, and supreme human value.

In this transformation US has played, is playing, and is inspiring the whole world to play a great historic role.

Let me add a nuance to the romanticization of Ramanujan's story. I think, even Ramanujan, just a 100 years ago, probably, could not have been offered a position at a university in UK without breaking its written or unwritten statutes. But US broke down all such barriers. Harvard University supported me for 5 years as a student, and one year as a Research Fellow. Following that the Johns Hopkins University, Columbia University, Rutgers University, Indiana University, and finally the Queens College of the City University of New York offered me long-term positions, the last two, full professorships. In between, IHES, France, University of Bonn, Germany, Institute for Advanced Study in Princeton, Institut Mittag Leffler in Sweden, U of Helsinki in Finland, Princeton U, U of Michigan, U of Colorado, offered me long peaceful stays, where I could work. Following a recommendation of my adviser Shlomo Sternberg, Prof. Marcel Berger arranged a position for me in Paris, France, the land of Poincare. Shortly after that Prof. Klingenberg opened the doors of the land of Riemann, Hurwitz, and Weyl for me. These are my heroes in Mathematics. Then I came in contact with John Milnor, Michael Atiyah, Armand Borel, and S. S. Chern, and then my friends Dennis Sullivan, Bill Thurston, Misha Gromov, and Shing -Tung Yau. I really do not know how to count my blessings!

On this occasion, I do remember my loving parents, my first teachers. These are intensely personal feelings, and not of much interest to others. Let me only say that I do not owe my interest in Mathematics to my parents, but whatever else that you may have found good in me, I owe it entirely to my parents. I should share with you one thing: my father, who was a Gandhian in thinking, taught me before I finished my school, the Upanishads, and Bhagavad-Geeta, the Old and New

Testaments, and the Koran. This reading had a great impact on my thinking. In fact, when I try to analyze my subconscious mind, I feel that why I was attracted to Shlomo Sternberg as my thesis adviser was not just differential geometry, but also Shlomo's scriptural scholarship, which many of us greatly admire!

I would like to use this opportunity to say something which you may also like to hear

i) Thoughts on Mathematics at the age of 70+.
ii) Thoughts about the Mathematics profession.
iii) My Teachers.
iv) A philosophical note.

2. Thoughts on Mathematics

I have summarized my thoughts on Mathematics in Figure 1: A Mathematician's View of the Tree of Knowledge.

You know the Tree of Knowledge is a part of the Old Testament. It also occurs in the 15th chapter of the Bhagavad-Geeta. For our purposes, let us simply consider it as "Arts and Sciences". In Galileo's and Newton's times, what we now call natural sciences was a part of a single subject: "Natural Philosophy". In the picture, which is admittedly a mathematician's perspective, I have put the natural sciences below Mathematics, and emerging upwards from Philosophy. Downwards from Philosophy are the social sciences, and the Arts – this is just a mathematician's view, not a judgment on their significance.

You know, Mathematics is not a popular subject. But Music is. And Mathematics and Music are so closely related. Why should Mathematics be not as popular as Music? I think, that we should admit that our communication of mathematics is not up to the mark. We should consciously try to change this popular reaction to Mathematics. In our communication of Mathematics to intelligent laymen, or to our colleagues and students at a college or a university, before heading into a particular area, we should try to explain the overall logical structure of Mathematics and its larger role in science and society. The "logical structure" includes not only the deductive part we always present in our class-rooms, but also the inductive part, following the steps from the original experience to theory-building, which will potentially lead to new discoveries.

In particular, we should tell our students and interested people outside the profession, how we ourselves think in Mathematics. We think in Mathematics in terms of Pictures and Numbers. They symbolize the two categories, Space and Number, of mathematical thought. At the High School these develop into Plane Euclidean Geometry, Combinatorics, and School Algebra. At the College level these develop further into 2-dimensional synthetic geometries, Calculus, and so on. At the master's level and beyond these develop further into several disciplines which we separately recognize today.

Until the 18th century, existence of mathematical entities, their "ontology", was mainly based on sense-perception, the same as in Physics, and other natural sciences. But at the end of the 19th century, Dedekind and Cantor ushered us into Set Theory. Now we follow 2-valued Logic, and express all mathematics in terms of sets, structures on sets, and maps among sets with structures. (We need "categories" in order to avoid Russell's paradox about the set of all sets. But I won't go into this aspect.) This has clarified our expression enormously. There is

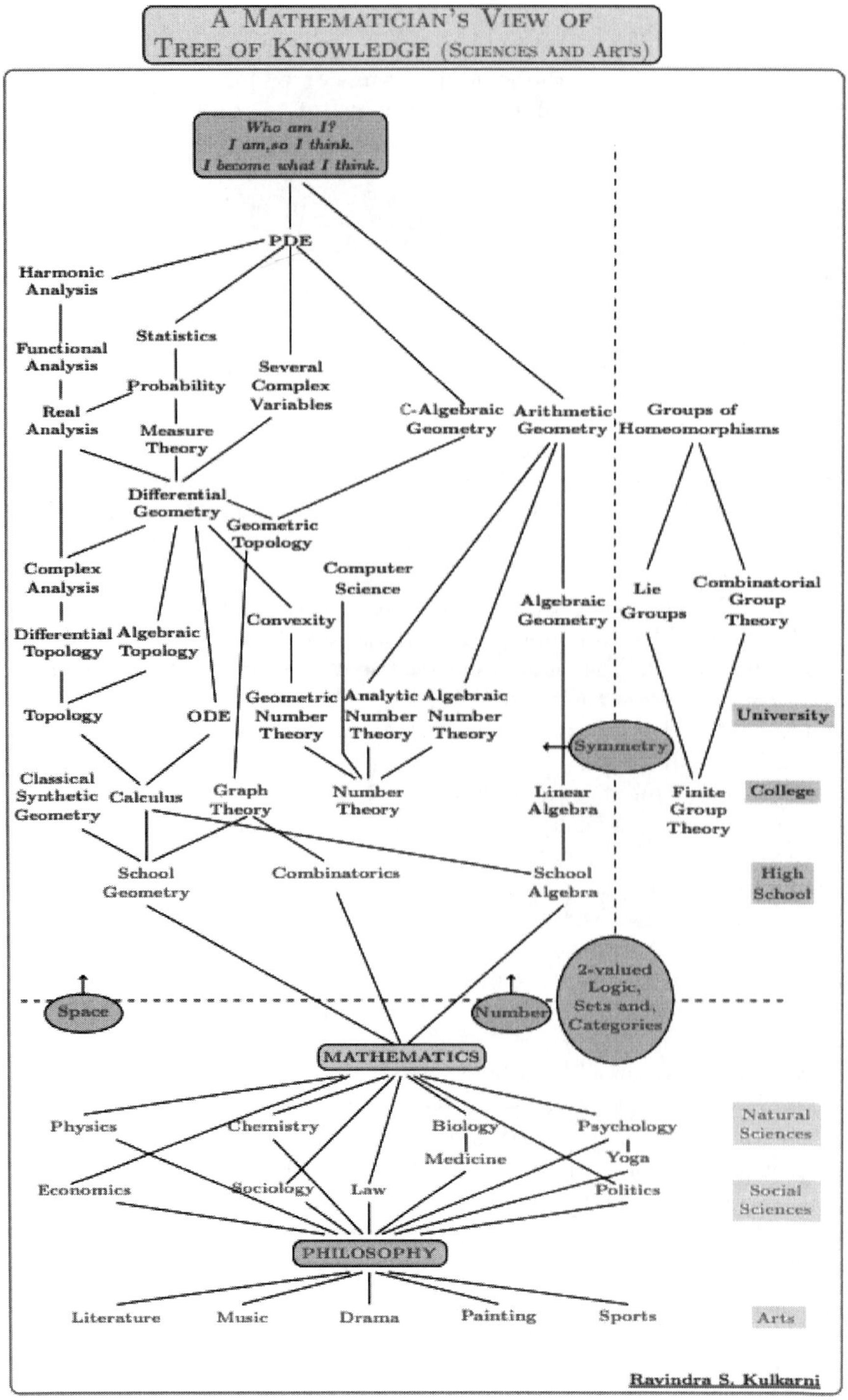
A MATHEMATICIAN'S VIEW OF
TREE OF KNOWLEDGE (SCIENCES AND ARTS)
Who am I?
I am,so I think.
I become what I think.
PDE
Harmonic Analysis
Statistics
Functional Analysis
Several Complex Variables
Probability
Real Analysis
Measure Theory
C-Algebraic Geometry
Arithmetic Geometry
Groups of Homeomorphisms
Differential Geometry
Geometric Topology
Complex Analysis
Computer Science
Lie Groups
Combinatorial Group Theory
Algebraic Geometry
Convexity
Differential Topology
Algebraic Topology
Topology
ODE
Geometric Number Theory
Analytic Number Theory
Algebraic Number Theory
University
Symmetry
Classical Synthetic Geometry
Calculus
Graph Theory
Number Theory
Linear Algebra
Finite Group Theory
College
School Geometry
Combinatorics
School Algebra
High School
2-valued Logic, Sets and, Categories
Space
Number
MATHEMATICS
Physics
Chemistry
Biology
Psychology
Natural Sciences
Medicine
Yoga
Economics
Sociology
Law
Politics
Social Sciences
PHILOSOPHY
Literature
Music
Drama
Painting
Sports
Arts
Ravindra S. Kulkarni

FIGURE 1

no mystery about "space-time". To begin with, locally, it is just 4-tuples of real numbers. "Velocity", "Mass", "Time", "Force" are not new, a priori, concepts in mathematics. "Mass" and "Time" are modelled on real numbers. (By the way, the "Real Line and Real Numbers", based on the Dedekind's notion of completeness of ordered sets, is one of the mysteries of mathematics, which, in my opinion, we cleverly hush up!) The "Velocity", "Force" are some sections of some fiber bundles and so on.

Hilbert described Set Theory as the paradise created by Cantor. I think, it is at least for the following three reasons.

First, it gave us the "definition" of infinity, and we came up with a startling conclusion that there are infinity of infinities – a view with which most traditional philosophers are still catching up.

Secondly, it changed the "ontology" of Mathematics. It helped us to get beyond the cruder pre-Cantor ontology based on sense-perception. For a mathematical entity to exist, we do not need to see it with our physical eyes, or touch it, or taste it, or smell it. It exists, so long as it can be constructed out of "known" sets using elementary axioms such as taking products, quotients, and maps among them. There is no issue that we shall never be able to see the real projective plane with our physical eyes. It exists as a mathematical entity as a quotient space of the 2-sphere in 3-space.

Thirdly, a great contribution of Set Theory is that it also gave us a "definition" of "Symmetry". The idea of "Symmetry" goes much beyond Mathematics. But with Set Theory in their repertoire, the mathematicians could make it precise in terms of groups (which is a set with an extra structure) and the actions of groups on sets. Verily, the classification of simple, connected, Lie groups, and the classification of finite simple groups, are the two great contributions of our fathers' generation and our generation. Hermann Weyl even advocated "Symmetry" as the third category of mathematical thought, besides "Space" and "Number".

So in the Tree-picture, I have put "Symmetry" vertically on the right, "looking" at all Mathematics at its left. At the "origin", the intersection of the horizontal and the vertical lines, I have put the "2-valued Logic, Set Theory, and Category Theory". That is how the mathematical community has agreed to express all its experience of all Mathematics.

This expression is extremely powerful, and is useful for all natural or social sciences. It is also concise, and has helped us to formulate large parts of mathematics in a deductive way, fulfilling Euclid's dream. In fact, set theory is really an extension of what Euclid called "General Notions". I have explained above that we do not need to introduce "velocity", "force", "mass", "time" as new undefined concepts. For theory-building, they can all be formulated once we accept "Real Line and Real Numbers" as "known" sets.

Let no one tell us "Math is not a science. It is merely a "Language" of science". Mathematics is a science. Only that we allow our "data" to be mental. Our data need not have a material model. For example, we find that some of the primes like 2, 5, 13, 17, . . . are sums of two squares, whereas the primes 3, 7, 11, . . . are not. This is our data. This data does not become a serious "Theorem" until Euler proved that the primes 2 and those congruent to 1, mod 4, are exactly the primes which are sums of two squares. Those congruent to 3, mod 4, are not. So the statement now applies to all primes, which are infinitely many.

Thus Mathematics is not only a Science, but it is a Science which gives ways to formulate theories for all physical and social sciences. So *the theory-building in all sciences is fundamentally mathematical in nature.*

So the sets, structures on sets, and maps among the sets with structures have provided an extremely powerful expression for the experience of mathematics. But is that experience equal to its expression? On this point, I may differ from some of you here. I can only give an analogy. I find the experience of mathematics is like the experience of seeing the Himalayan peaks, or the Atlantic ocean. In my opinion, Mathematics is deeper than its set-theoretic expression. It has to deal with the "Who Am I"-rectangle at the top of the Tree-picture. I shall return back to this point at the end.

But in practical terms, while teaching or communicating Mathematics, or while writing papers, let us be more generous to our potential audience. Let us make the subject more human. Let us communicate it with some semi-historical perspective. Not so much the biographies of eminent mathematicians, although that too, to explain the social background in which even the eminent mathematicians work. But more emphasis on how the mathematical ideas developed over time.

Let us explain our subject with pictures which explain equations, and equations which explain pictures, juxtaposing the "space" and "number" categories of mathematical thought. Let our students thoroughly know that mathematics is not just numbers. In fact if you think about it, most mathematical entities are often expressible in terms of numbers, but they themselves are not numbers. For example, a vector in an $n-$dimensional vector space is expressible as an $n-$tuple of numbers, after choosing a basis, but intrinsically it is not that! An operator and a quadratic form on an $n-$dimensional vector space may both be expressed by an $n \times n$-matrix, after choosing a basis, but they are different entities, and they are different from the representing matrix. You can hardly call, a connection on a fiber bundle, a number.

Our big sense of pride lies in using Set Theory and 2-valued Logic, and present large parts of mathematics in a deductive way. But human mind comprehends inductive reasoning more easily. From particular examples, often provable in a simpler way, to a general theory. Also even one schematic picture is worth a thousand words, which have to be written carefully in conformity with the dictates of Set Theory.

3. Thoughts on the mathematics profession

Let us be happy that Mathematics is a noble profession. All societies acknowledge its importance, and allow us to make a respectable living. Let us not be too much pre-occupied about writing papers and getting them published, at least after we have gained tenures! Let us also try to spread the message of Mathematics.

Since Galileo's and Newton's times the progress in mathematics has been extremely rapid. Just note that not even 150,000 days have passed since Galileo's times. So sooner or later, perhaps already in our life-times a plateau must be reached for the pace of new discoveries in Mathematics. But even when the plateau will be reached, we shall have an enormous task of digesting what is discovered, and communicating the whole edifice to the larger community. That will be our recurring, all-time, commitment.

In this regard, India offered me better opportunities. The authorities were happy that I was interested in teaching. So along with some of my friends, I recommended to the National Board of Higher Mathematics (NBHM) in India to start an Advanced Training In Mathematics (ATM)-Program, in 2005. It has taken root, and is developing very well. Its management has been recently given to the National Center of Mathematics (NCM), which is a joint venture of the Tata Institute of Fundamental Research (TIFR), and the Indian Institute of Technology in Mumbai, (IIT(Bombay)). Over 200 research mathematicians participate in it. It is aimed at post-master's doctoral students, and encourages them to gain an in-depth knowledge of 2 or 3 different fields before zeroing in on a thesis topic. NCM will soon have its own building situated in the IIT(Bombay) campus.

The Presidency of the Ramanujan Mathematical Society offers me a good opportunity to extend this philosophy to undergraduate and high school education. Professors Kumaresan and Ambat Vijaykumar of RMS have already been working in these areas for some time. Following the example of AMS, we intend to have many more regional meetings and many regional innovative Undergraduate Faculty workshops. At the undergraduate level, we are interpreting Mathematics broadly, wiping out the distinction between Pure and Applicable Mathematics, to include Engineering, Statistics, and Computer Science.

I invite all of you to come to India, enjoy the different culture, natural scenery, new foods, and also exchange notes and contribute to the development of Mathematics in many ways. Love of Mathematics unites us all!

Writing a paper and getting it published has its plus and minus points. It need not be the only way we can enjoy and contribute to Mathematics. In fact this singular focus has made us self-centered, and unduly competitive. I know some of my colleagues are so competitive, that they compete not only with their colleagues, but they compete with their own students. Many of them would not like their students to consult with their colleagues.

In India, in many universities, there are rules, which, with my experience in the freer atmosphere in the West, I find very strange. In India, after a student gets his masters, if he wants to go for a Ph.D., he needs to get attached to an adviser. Worse, he needs his adviser's certification for financial support every six months. The adviser often suggests some narrow research problem without providing its place in the broader mathematical landscape. He may discourage the student consulting with other mathematicians. Effectively, the adviser's limitations continue to remain the student's limitations for a long time. This hampers the progress of mathematics and the whole fun of doing mathematics is lost. That was partly the reason why the authorities in India agreed to start the ATM-activities mentioned above.

But these are complex issues. Some of them can be handled by devising alternate structures and reward systems. But to make changes in the established structures is usually very hard.

At the root, it really depends on the senior mathematicians recognizing that they should encourage a "co-operative" model over a "competitive" model of doing research. As much as all scientific research is finally a social activity, a "largely co-operative" and "moderately competitive" – just to ensure enough incentives – are the models which are, in my opinion, justifiable.

Considering the responsibility of the senior people in our profession, brings me to the consideration of the "priority issues" which have plagued our mathematical community, all the time. Just compare: Mahavir was a contemporary of Buddha. Both developed largely similar meditational techniques, still existing and practised today. But they never indulged in any priority battles! In fact, Buddha reputedly said: "if you "follow" Buddha, then you are not a buddhist! Just take whatever that suits you best." By contrast, in the scientific community, Newton and Leibniz had priority battles over the discovery of Calculus. And following that, the English mathematicians did not adopt Leibniz's superior notation for a century. The Bernoulli brothers had a priority battle. I even do not recall what the battle was about. In my opinion, Gauss acted meanly towards Lobachevsky and Bolyai and denying them the credit of discovery of Hyperbolic Geometry. He wanted all the credits to himself, for the accounts, he would not publish in his life-time. If we ask Buddha – or for that matter the larger world scientific community over the centuries – about priority issues, Buddha would smile, shrug his shoulders, and say that these issues are far from the real issues of life. They are not worth a discussion.

As a community, we need to develop a more mature view of the priority issues. As the group activity in Mathematics, is slowly increasing, I think, we shall develop such a view in the 21st century. Let the senior people among us consciously encourage group activity. Let us start talking to each other, crossing the boundaries of our specializations. Let us obliterate the difference between "Pure" and "Applicable" Mathematics. Let us appreciate different ways of thinking, looking at the same thing in different ways. That way, we shall increase the enjoyment of Mathematics.

Finally, I would like to express my concern over something very important, although some of you may consider that this is not the occasion to raise it. But since you have been listening to me, let me use the occasion.

I am concerned over the role big money has started to play in our social life. This raises a more fundamental philosophical issue: what should be our attitude towards acquisition of money? We, in the scientific community, are all very intelligent people. How we think, and how we express, is likely to have some impact. So we should express freely on these matters.

My comments are motivated by the following life-experiences.

(i) When I came to US, we could get a slice of pizza for 15 cents. Now it costs 2.25 dollars, if you are lucky! About 15-fold inflation. Our salaries have increased much less. We are all wondering, how has it come to pass?

(ii) More seriously, President Eisenhower talked about the power of the Military-Industrial Complex in the 50s, and our Grothendieck protested against getting research-funding from the same establishment in the 60s. I remember one of my teachers, Prof. George Mackey, refused to accept NSF-grants.

(iii) In our own times, Perelman refused to get the Fields Medal and the million-dollars prize. Some people called him a "nut". But I think, he was trying to say something more fundamental. He should elaborate it some time. I request his friends to get an answer from him to this question. Perhaps he wanted to suggest that the acquisition of knowledge is a supreme human value. But mere acquisition of money can never be a human value.

(iv) My concern is mainly coming from the rampant corruption in India in all walks of life, and the nexus between the politicians and the big corporations. The cost of education, and the cost of health-care are sky-rocketing, and a lot of corruption in these areas also. For example, the medical professionals are resorting to "cut-practice". In many colleges run by politicians, the "untenured" faculty have to sign a statement that they are getting X amount of salary, and actually they get much less.

(v) At the same time, I find it disturbing that many college students, waste their precious mornings, in reading the financial pages of newspapers to check how their investments are faring, or getting excited by the stories on the internet how some kid hit a fortune by some lucky investment.

(vi) Just see how our language has changed in the last 50 years: these days we don't even "earn" money, we "make" money!

(vii) A quote by my good friend, Sylvain Cappell: A mathematician can make a lot of money – after he leaves mathematics.

(viii) A quote by my good friend Todd Drum: In America, you can pay God to get anything. Praying will get you nowhere.

(ix) In my childhood, "speculation" was a bad word. Now it has become a very respectable word. The economists tell us that in a capitalist society, private industry needs capital, and industry creates jobs. So a citizen who has some money to spare, is encouraged to speculate. The argument given is: we have to do this, in order to keep up with the inflation, and the inflation is indeed very real. But is creating jobs the primary motivation of the people who are asking us to speculate? They seem to be just after making more and more money. There is no transparency in much of this financial talk. I get a feeling that an "insider" information is necessary if you wish to make big money.

I am recalling these experiences keeping in view only a limited aim: what should be the attitude towards acquisition of money in our scientific community?

Most of us have come from a lower middle-class background. We often thought in our childhood, "if only I had money, I would do this or that for the world!". What percentage of us have kept that idealism?

I think, it was about the age of 5, that I learnt that Knowledge (with a capital K) is important. That Knowledge (with a capital K) is symbolized by a Goddess, we call her Saraswati, who looks like an elementary school teacher. She plays music. There are books around her. That Knowledge is supposed to make you free, and give peace and happiness. It is hard to acquire. It is like walking on a razor's edge. Mathematics is a part of that Knowledge. The scriptural literature describes Mathematics as the highest communicable part of that Knowledge. To acquire it, what we really need is the intense desire, sincerity, character, and a good teacher. If we have intense desire, the good teacher may even come in our dream, and we get the instruction that is right for the stage we are in. But otherwise the Knowledge that makes one free, is absolutely free, like sun-shine and moon-light!

But that is definitely not the knowledge of financial markets. It does not make you free. It does not make you happy. It is like the chains of gold or platinum, but chains nevertheless. It binds you, and gives you tension, headaches, backaches, and sleepless nights. And it is not free. It always comes with a cost, which is picked up by the teams of brokers who sell the financial packages!

I am not saying anything new. I know, many of us are privately worrying about these deep issues of what should be the values we should promote in a society. I know, many of my colleagues in Queens College, are doing exemplary social service. Their children and students are inspired by these acts. I only wish, that we from the scientific community at large, should express our opinions more freely on these issues. We are very intelligent, but timid, people. We do not have the courage of a Gandhi. But collectively, we can make a difference.

I would like to mention here Mahatma Gandhi's thoughts on such issues, which unfortunately are not well-known even in India among the intellectual circles. However, thanks to Google, they are now at least readily available. Just Google: "Mahatma Gandhi, on Trustee-ship".

With a strong basis in the scriptural literature, and the "wisdom"-traditions of the world, Gandhi advocates the idea of the attitude of "Trustee-ship". Suppose you are super-intelligent. Then Gandhi recommends that a super-intelligent individual should consciously develop an attitude that "I don't have a personal claim on my intelligence. I am a Trustee. As a principle, I should keep my needs few, and after they are fulfilled, I should make my intelligence freely available for the benefit, first of my family, then of my community, and then of the society at large."

I think, if a majority of us consciously develop such attitude, that would solve the "priority" concerns in the scientific community – just as in the last forty years, we solved the problem of smoking in public spaces.

Gandhi was thinking not in the context of super-intelligence, but in the context of super-wealth. Just read it from the original. I recommend it.

4. My teachers

- Prof.s D. W. Kerkar, M. R. Railkar, and Mrs Apte (my undergraduate teachers)
- Prof.s Shlomo Sternberg, Raoul Bott, (my thesis advisers)
- Prof.s Brauer, Mackey, Zariski, Tate, Mumford, (I served as a TA in their undergraduate and graduate courses, and learnt a lot from them)
- Prof.s M. Berger (France), and Klingenberg (Germany)
- Prof.s Borel, Atiyah, Milnor, and Chern (IAS, Princeton)
- Prof. Lipman Bers (who brought me down from n-dimensions, to just 2-and 3-dimensions!)
- Prof. Max Zorn (who gave me a view of the French and German mathematical traditions)
- Prof.s Dennis Sullivan, William Thurston, Misha Gromov, and S. -T. Yau (In the US, I shake their hands. In india, I would touch their feet.)
- Prof.s Peter Shalen, Ulrich Pinkall, Hyman Bass, Ephraim, Constantine, Krishnendu Gongopdhyay, Vikas Jadhav, Rahul Kitture (my collaborators, from each of whom I have learnt a lot.)

5. A philosophical note

Let me come back to the top rectangle in the Tree of Knowledge, starting with "Who Am I?".

You know Descartes' famous statement "I think, therefore I am". This statement occurs in his "Discourse on the Method", the appendix of which contains his fundamental contribution to mathematics, which we call Cartesisn Geometry, which says that "Space" and "Number" are two categories of mathematical thought, but they are really two sides of the same coin.

Buddha's first discourse, after he got enlightenment, begins with apparently the opposite: "We are (or we become) what we think".

I think, the word "therefore" in Descartes' statement need not be construed as "implication". I think, Descartes is simply saying that the "existence" is recognized by its thinking capacity. If we go behind the usual dualistic formulation of "recognizing" something by "somebody else", then in my opinion, both Buddha and Descartes are saying the same thing: "Existence" is capable of thinking. So first of all, "Existence", that is "the experience of I", exists! No doubt about it. It comes first, then comes "thinking".

The old Buddhist/Vedantic image of "thoughts" is that they are like waves on the ocean of existence. More precisely, "thoughts" are finite in number, but the mystery of existence is at least as deep as the mystery of the Real Line and Real Numbers. A more precise mathematical image of the relationship of "existence" and "thoughts" is that the existence is like the unit interval $[0, 1]$ of real numbers, and thoughts are like the rational numbers in $[0, 1]$ whose denominators are bounded, say, by 10, 000, depending on the individual. They appear to be filling the whole interval, but they cannot fill the whole interval, for the simple reason that they are finite in number, and the whole interval is not.

Through the process of meditation, you can zoom out the interval $[\frac{1}{10{,}000}, \frac{1}{9{,}999}]$, and changing the scale, recognize, "Ah! this was my first thought this morning, then came that thought, ...". So what is the space in between these two rational numbers? It exists. That is all. It is also capable of evolving into more thoughts.

In other words, experience generates thoughts, and thoughts are expressed by a language. The correspondences "experience to thoughts" and "thoughts to expressions" are one-to-many. The expression may not be unique. Still, experience is deeper than any of its expressions. In particular, the experience of mathematics is deeper than its set-theoretic expression, although the set-theoretic expression is better than its earlier expressions!

In Buddha's writing, "being" and "becoming" are often not clearly distinguished. So Buddha's "I am what I think" should be really rendered as "I become what I think". So I have made two sentences out of Buddha's one sentence: "I am, so I think", and "I become, what I think." For us, this is quite easy to understand. Dennis thought about Topology, and became a topologist. Hyman thought of Algebra, and became an algebraist. Sagun thought of Analysis, and became an analyst. Shlomo and I thought of Differential Geometry, and we became differential geometers. It is as simple as that.

Coming out of the limited experience-field of Mathematics, we can appreciate Buddha's statement (in Buddha's words):

We become what we think,
With our thoughts we create our worlds,
Think of constructive thoughts, and our life becomes happy,
Think of crooked thoughts, and our life becomes messy,
This is the simple truth of life!

So today, I am very, very happy that all of you have come here. Coming from the Eastern background, I have no problem with the notion of God, as a grand and beautiful hypothesis, which simplifies so much in life. I have seen God in my grand-parents, my parents, especially my mother, my brethren, nephews, nieces, and their children, my wife who was unfortunately plucked away in the blossom of her youth, my teachers, and friends, many of them movers and shakers in the field of mathematics in the last 50 years, which includes all of you.

My salutations to all of you.

Contemporary Mathematics
Volume **811**, 2025
https://doi.org/10.1090/conm/811/16233

Philosophy of mathematics – four issues

Ravindra S. Kulkarni

Abstract. In this short paper, I have organised my thoughts on four issues in the Philosophy of Mathematics, which have bothered me throughout my professional career. The issues are: (1) Is Mathematics a science? (2) Is Mathematics objective? (3) Definition of Mathematics. (4) Mathematics and Mind. The paper contains my opinions on these issues which seem right to me in my 82nd year!

1. Four issues

(i). All our universities have departments of Mathematics, Physics, Chemistry, and Biology. Among these, Physics, Chemistry, and Biology are universally called "sciences". Mathematics and these departments are usually included in a Division of Science. The people working in Physics, Chemistry, and Biology are called scientists. The people working in Mathematics are called mathematicians. Although in common parlance, mathematicians are often considered as scientists, the physicists, chemists, and biologists are reluctant to call us scientists.

The theoreticians in Physics, Chemistry, and lately, Biology know a lot of abstract mathematics, but there is no Nobel Prize in Mathematics. Moreover, a Nobel Prize in Physics, Chemistry, and Biology is not given just for building a nice mathematical theory. They need "verification" of the theory in "nature".

So Einstein had to wait for 16 years before he was given the Nobel Prize. Moreover, his Nobel Prize was not for his theory of Special Relativity which he published in 1905, and the theory of General Relativity which he published in 1914. These theories are basically mathematical, and his iconic fame among scientists, and among the general public rests on these theories. The underlying mathematics was worked out by Riemann in 1864 in a lecture for an assistant professorial job!

Newton's main book [**7**] is called "The Mathematical Principles of Natural Philosophy". In his times, "Natural Philosophy" meant "Philosophy of Nature", that is "Philosophy of Matter", to be carefully distinguished from "Philosophy of Mind". With advances in the last four centuries, more precisely, it now mainly means Physics (including Astronomy) and Chemistry, and the "material" parts of Biology, such as Physiology with applications in Medicine. But note that Newton distinguishes "Mathematical Principles" from "Natural Philosophy".

The understanding that these mathematical principles are not to be confused with their specific applications in other rational human pursuits was recognized quite early in the history of mathematics. For instance, Mahavir (850 AD), [**5**], ch 1, describes the uses of Mathematics in mundane or sophisticated human affairs such as love-making, cooking, music, grammar, prosody, astronomy, business, engineering, statecraft, measurements of areas and their perimeters, ..., Roger Bacon (1214-1292 AD), [**4**], in alchemy, optics, ... A few decades before Newton, Galileo (1564-1642), in [**2**], almost said the same things about the use of Mathematics in the study of motion, with applications to the motions of planets.

So one issue in the philosophy of Mathematics is: Besides its applications to Physics, Chemistry, ..., is mathematics itself a science? This was the concern of a Nobel Laureate physicist, Eugene Wigner [**10**] in his famous paper: "Unreasonable Effectiveness of Mathematics in the Natural Sciences".

(ii). Mathematics is usually presented in a deductive manner. It gives the impression that everything is contained in the axioms one starts with. The mind of someone who constructs an axiom system is not a part of the axiom system! Mathematics is considered like pearls at the bottom of a sea. A diver merely discovers pearls. He does not invent them, nor in any way is he involved in their creation. In other words, the issue is: Does mathematics have an objective reality? Or more picturesquely, is Mathematics a Discovery or Invention?

(iii). What is a *definition* of Mathematics? We agree that Mathematics is not Physics, Chemistry, or Biology. Then what is Mathematics? *Define* it. Note that this use of the word "definition" is different from the way we use this word in teaching Mathematics. In a mathematics examination, we can ask definitions of a "group", a "ring", or a "topological space". But we do not ask a question "define Mathematics" – although, such a question may be posed in an interview of a person who has applied for a professorial or a Vice Chancellor's position. What is meant here is an examination of the way the word "mathematics" is used presently or in historical contexts and attempt to capture its essence in a succinct but non-technical way.

(iv). Mathematics and Mind. Galileo [**2**], Roger Bacon [**4**], or Newton [**7**], used the word Mathematics in the context of the "Philosophy of Nature". One may ask: can "Mathematics" be used in the context of the "Philosophy of Mind"? We have already seen in the last 50 years new areas emerging such as "Mathematical Economics", "Mathematical Biology", "Mathematical Medicine", ... after the 100-year-old "Mathematical Philosophy" by Bertrand Russell. In these areas, the "mind of an observer" plays a larger and larger role.

We take these issues in subsequent sections.

2. Is Mathematics a Science?

In any rational discourse, it is important to have definitions of the keywords or concepts. So the question "Is Mathematics a Science" cannot be answered until a careful definition of the word "Science" is given. Whatever it is, "science" certainly includes "Physics, Chemistry, Biology", as they are practiced in the universities in the 21st century. In my mind, Physics includes Astronomy also. Some laboratory work is expected for the students working in these fields. A typical physics laboratory includes pulleys, magnets, pendulums, lenses, prisms, springs, microscopes, ... A typical chemistry laboratory includes chemicals, gas for heating, refrigerators for cooling or preserving, ... Astronomy departments have observatories, ...

Labs or observatories are needed for amassing "data" in these respective fields. This "data" consists of *material* objects.

Mathematics departments have no labs and no observatories. If one considers labs/observatories dealing with *material* objects an essential part for being called a science, then mathematics is not a science. But why should one consider *material* objects only to constitute a "data"? Psychologists call Psychology a science. Economists call Economics a science. Moreover, any data of *material* objects would consist of finitely many objects. But a "data" can be mental also. It could be thoughts, or categories of thoughts. For instance, one can collect a data of prime numbers which can be written as sums of squares of two natural numbers. Euler deduced a wonderful defining pattern about these primes: these primes are either 2, or congruent to 1 mod 4. If one allows "science" to include such pattern-recognitions, then mathematics is a science!

Moreover, an essential part of any theory-building in the sciences for explaining any data is *counting*. For counting, the human mind created the abstract natural number system 1, 2, 3, ... For instance, "2" is a common abstraction of all pairs: two boys, two girls, two cups, ... This abstraction is part of what is called "mathematics" – "Gan'itam" – from ancient times. This is a mental process. As we know, this natural number system was extended to whole numbers (natural numbers including zero), integers (whole numbers and their negatives), rational numbers (inverting non-zero integers), real numbers ("completions" of rational numbers w.r.t. order or metric), complex numbers (adjoining all roots of polynomials, ...). Still, more generally, in the 20th century, mathematicians created commutative rings with identity. – This line of thinking may be considered as a "number"-thought category of Mathematics. In this sense, mathematics deserves to be called a "science".

In the last two decades of the 19th century, by further abstraction of the two operations of addition and multiplication, the theories of groups, (non-commutative) rings, and fields were created. Also until that time, "Number" was considered as providing a basic structural element for Mathematics. Remember Weierstrass's "arithmetization"-program for analysis. However around the same time "sets and structures on sets" was proposed as a basic structural element for Mathematics. This is how we view Mathematics today. A set with a structure captures the philosophical idea of a "Space". So Mathematics may be considered as a "theory-building methodology" not only for Physics, Chemistry, and Biology, but for all rational thought, in which the thought categories used are "number" and "space".

3. What is (or what should be) a theory?

(1). To arrive at a decision about the nature of Mathematics, it should not be considered in isolation. The questions like "Define Mathematics" cannot be answered "within mathematics" itself. We regard any human discourse (in particular Mathematics) as a "theory" in the sense discussed below. There are five salient ingredients of what deserves to be called a "theory".

(i) "Data" is taken from some domain of human experience.

(ii) Data generates "thoughts".

(iii) Thoughts are expressed in a (written) "language". The word "language" is used in the sense of Computer Science. It involves a finite "alphabet" called "letters". Finite sequences of letters are called "words", and finite sequences of words, are called "sentences" or "statements".

(iv) Statements are connected by an "implication relation", which accounts for a philosophical idea of "cause and effect". The implication relation constitutes the "epistemology of a theory".

(v) There must be some basic notions (words) that are not explained in terms of other notions (words), and some basic statements that are not implied by any other statement. Moreover, every statement is implied by a basic statement in finitely many steps. The basic notions and statements, that are not implied by any other statement constitute the "ontology of a theory".

Definition: A theory is a directed tri-partite graph: Data $\rightarrow$ Thoughts $\rightarrow$ Language, having its epistemology and ontology.

The condition v) is imposed to avoid what is known as the "defect of infinite regression".

For instance, the usual "binary logic" used in a large part of mathematics constitutes its epistemology, and the axioms of ZFC-set theory constitute its ontology.

The ontological notions or statements may be considered as "algebraic statements" $x, y, z \ldots$ Russell-Whitehead's "Principia Mathematica" is such a theory. In this sense, Russell used to say jocularly that "mathematicians do not know what they are talking about".

The ontological notions or statements may be given a *meaning* by a human mind.

(2). An important example: In Newton's "Principia" [**6**], Newton starts with some definitions and axioms. He flatly says that he is not going to define "time", "mass", and "space" as these things are well-known! In the language introduced above, "time", "mass", and "space" are ontological notions in Newton's theory.

His axiom about "time" and "space" is that they are "independent". This is an ontological statement. Newton does not define "independence".

In his Relativity theory, Einstein related time to light-signals. He knew that the speed of light is about 300,000 km per second. This number is very large but *finite*. Einstein intuitively grasped that the finiteness of the speed of light would have profound consequences for Newton's theory. What Einstein called a "space-time", much more precisely, mathematicians called a "Lorentzian manifold". In particular, Minkowski later showed the Lorentzian Geometry as the underlying geometry of Einstein's theory, which implies that "time" and "space" are not "independent" in the presence of "mass".

Thomas Kuhn, an American philosopher of Science, in his book, cf. [**3**], "Structure of Scientific Revolutions", coins the term "paradigm shift" for the passage from Newton's theory to Einstein's theory. This just amounts to a "change of ontology" between these two theories.

(3). One more important example: This example is from the Indian philosophy. This Philosophy is expressed in *darshans*, which literally mean viewpoints on many concerns of human life. Initially, there were six main darshans. A medieval text *Sarva-Darshan-Sangraha*(ca 1374-1380 AD) lists sixteen darshans. After the introduction of Judaism, Christianity, and Islam in India, their doctrines are also presented as darshans in a 19th-century text, *Satyarth-Prakash* (1882). All these darshans can be reformulated as *theories*. This would be a rational, logical formulation, and as such, it will be more accessible and can be better understood by the scientific community.

Among these darshans, the *Nyaaya Darshan* is a precursor of the modern scientific methodology. The *Saankhya Darshan* and *Vaisheshik Darshan*, are precursors of what we now call Physics and Chemistry. Browsing through these darshans suggested the formulation of the Philosophy of Mathematics, as a "theory", as presented above.

Among these darshans, *Vedanta Darshan* as formulated by Shankaraachaarya, cf. [**7**], in his commentary on Baadaraayan Vyaasa's *"Brahmasutras"*, in particular, its five-page introduction *"Adhyaas-Bhaashya"* is admittedly a very high point of the Indian philosophy.

Regarding *Vedanta Darshan* as a theory, we find that "time" and "space" ("dik-kaala") are also ontological concepts in this theory. Shankaraachaarya's basic statement in "Adhyaas-Bhashya" is that the "I am"-experience of an individual is independent of "time" and "space". This statement may be taken as an "axiom" of the theory. From this axiom, Shankaraachaarya's profound "Advaita"("Monism"), as an explanation of the whole human experience, can be deduced!

This relates to the "Consciousness"-issue mentioned by Wigner [**10**] at the end of his famous paper, and the musings of Schrodinger [**9**], and others, which are richly available on Google.

4. Final remarks on the four issues

(i). Is mathematics a science ? This was already resolved in section 2. Mathematics is a "science", if we allow "science" as rooted in theories based both on material and mental data. If one means by "science" as rooted in theories based only on the material (inanimate) data then Mathematics is not a science.

In any case, Mathematics provides a very important methodology for theory-building in any science. "Number", and "Space" are two thought categories that are essential parts of this methodology. They can be invoked in science, as well as in any rational discourse.

We may say that "time" and "space" are thought categories of physics. The mathematical thought category of "number" projects neatly onto the physical thought category of "time". The mathematical thought category of "space" is much more general than the physical thought category of "space".

In modern mathematics, a "space" is formulated in the language of "sets with structures", developed at the end of the 19th century. The notions of "sets with structures" gave a new ontology to Mathematics, replacing the older ontology consisting of only "numbers".

The effectiveness of Mathematics, in Physics, Chemistry, Biology, and in fact in any rational discourse, may be explained as follows: a mathematical viewpoint allows one to construct, using "number" and "space" thought categories, a mathematical model for any phenomenon or notion under discussion. For instance, consider the physical notions of "mass" and "space". Mathematics assigns to the "mass of any substance", its weight, which is a real number. In Newton's gravitation law "product of masses" means a product of the associated numbers. To the physical space of human experience, Descartes, making some choices, associated triples of real numbers. Newton-Leinbniz differential and integral calculus is a purely mathematical construction in terms of real numbers. Using calculus, Newton deduced Keplar's laws of motions of planets from a purely mathematical theory of differential equations, used in the formulation of his law governing gravitational force. Similar explanations can be given for the physical laws governing electricity and magnetism, which in turn explained the atomic theory of the 118 constituents of matter discovered in chemistry, and the three "quarks" which are the constituents of protons and neutrons, discovered in quantum mechanics. The seven SI-units are numbers associated with the seven ontological notions in mathematics and physics.

(ii). The "objectivity" of mathematics. This is a trickier issue, precisely because of the "unreasonable effectiveness" of mathematics in the theories of Physics and Chemistry, dealing with inanimate matter. However once one interprets mathematics as a "theory", the issue becomes tractable. For, as developed in section 3, no theory is absolute. For, a theory crucially depends on its ontology, and the meaning to its ontology is given by the human mind. So once one interprets mathematics as a "meaningful theory", it will inevitably have some subjective element.

In this connection, it is instructive to recall the title of Dedekind's famous paper on real numbers, "Was sind oder was sollen die Zahen". In this paper, he asserted that real numbers are "free creations of the human mind".

For example, in my opinion, a Riemannian metric is definitely an "invention", not a "discovery".

(iii). The definition of mathematics. We noted that the "number" and "space" are the thought categories of mathematics. Let us go a step further: mathematics is defined by the thought categories of "number" and "space". So

Definition: Mathematics is a theory in which its ontological notions and statements are assigned a meaning by a human mind using the thought categories of number and space.

In the 21st century, these thought categories are expressed in the language of sets and structures on sets. Mathematics that is taught in schools, colleges, and universities today follows binary logic and the ZFC-set theory. This is a theory with binary logic as its epistemology and the ZFC-set theory (with meaning attached by a human mind) as its ontology.

If we take fuzzy set theory and fuzzy logic, then it would be a different theory, as both its epistemology and ontology have changed. Still, it may be called mathematics, since its thought categories are number and space.

(iv). Mathematics and Mind. The issue is: Can mathematics be used in the context of the "Philosophy of Mind"?

We already see the disciplines like Mathematical Economics, Mathematical Biology, Mathematical Medicine ... arising in the 20th and 21st centuries, in which "mind" plays a role in various degrees. So the answer to the above question is "yes". But the success of Mathematics in these disciplines is not as decisive as in the disciplines coming under "Natural Philosophy". The citizens of the world in the 21st century need to learn more and more mathematics as a part of their education, but such educated citizens should learn to use mathematics for arriving at decisions on "values"— a subject which seems to be outside the ken of Mathematics!

References

[1] Richard Dedekind, *Was sind und was sollen die Zahlen?* (German), Friedr. Vieweg & Sohn, Braunschweig, 1960. 8te unveränderte Aufl. MR106846

[2] Galileo, Galilei: "*Dialogue Concerning the Two Chief World Systems*" (1632)

[3] Kuhn, Thomas: *"Structure of Scientific Revolutions"* (1962), University of Chicago Press.

[4] Lindberg D. C.: *"Roger Bacon's Philosophy of Nature: A Critical Edition"* (1983), `Mathhistory.st-andrews.ac.uk/biographies/Bacon`

[5] Mahavir, *"Ganitsarasangraha"* (850 AD)

[6] Newton, Isaac, *"Philosphiae Naturalis Principia Mathematica"* (1687)

[7] Shankaraachaarya, *"Brahma-Sutra-Bhashyam"* (ca 7th century AD), The first five pages of this book is known as "Adhyaa-Bhaashya".

[8] Schr¨odinger Erwin, *"What is Life"* (1944) with *"Mind and Matter"* and *"Autobiographical Sketches"*, (2013), Cambridge University Press.

[9] Whitehead A. N. and Russell B., *"Principia Mathematica"*, (1927), Cambridge University Press

[10] Wigner Eugene: *"The unreasonable effectiveness of mathematics in the natural sciences"*:, Communications on Pure and Applied Mathematics. (1960) 13 (1): 1–14.

Contemporary Mathematics
Volume **811**, 2025
https://doi.org/10.1090/conm/811/16234

Counting cusp excursions of reciprocal geodesics

Ara Basmajian and Robert Suzzi Valli

Dedicated to Ravi S. Kulkarni on his 80th birthday

ABSTRACT. For a fixed cusp neighborhood (determined by depth D) of the modular surface, we investigate the class of reciprocal geodesics that enter this neighborhood (called a cusp excursion) a fixed number of times. Since reciprocal geodesics traverse their image an even number of times such a geodesic only has an even number of cusp excursions, say $2n$. Denoting this class of geodesics as $\mathcal{E}_{2n}^D$, we derive several growth rate results for $|\{\gamma \in \mathcal{E}_{2n}^D : |\gamma| = 4t\}|$, where $|\gamma|$ is the word length of γ in $\mathbb{Z}_2 * \mathbb{Z}_3$ with respect to the generators of the factors.

1. Introduction

For a fixed cusp neighborhood determined by its depth D, the paper [**2**] was interested in counting the number of so called D-low lying reciprocal geodesics on the modular surface. That is, in counting the reciprocal geodesics that stay in the compact subsurface determined by chopping off the D-cusp neighborhood. In this paper, we are interested in counting the number of reciprocal geodesics on the modular surface that leave this compact subsurface, called a cusp excursion, a fixed number of times. Since reciprocal geodesics traverse their image an even number of times such a geodesic only has an even number of cusp excursions, say $2n$. Denoting this class of geodesics as $\mathcal{E}_{2n}^D$, we derive growth rate results (Theorem 1.1) for $|\{\gamma \in \mathcal{E}_2^D : |\gamma| = 4t\}|$, and $|\{\gamma \in \mathcal{E}_{2n}^1 : |\gamma| = 4t\}|$. Before we state our theorem we first define our terms and set up language.

Consider the group $G = \mathbb{Z}_2 * \mathbb{Z}_3$ with its natural symmetric generators a, b, and b^{-1} where $a \in \mathbb{Z}_2$ has order two, and $b \in \mathbb{Z}_3$ has order 3. The representation of G given by, $a \mapsto A$ and $b \mapsto B$ where,

$$A = \begin{pmatrix} 0 & -1 \\ 1 & 0 \end{pmatrix}, \text{ and } B = \begin{pmatrix} 1 & -1 \\ 1 & 0 \end{pmatrix}$$

is a discrete faithful representation with image the modular group $PSL(2, \mathbb{Z})$. The quotient $S = \mathbb{H}/PSL(2, \mathbb{Z})$, called the *modular surface*, is a generalized pair of pants of genus zero and signature $(2, 3, \infty)$. For γ a closed geodesic on S, its *word length*,

---

2020 *Mathematics Subject Classification.* Primary 20F69, 32G15, 57K20; Secondary 20H10, 53C22.

Key words and phrases. Reciprocal geodesic, modular group, modular surface, cusp excursion, word length.

The first author was supported by a PSC-CUNY Grant and a grant from the Simons foundation (359956, A.B.).

denoted $|\gamma|$, is the minimal word length (with respect to the natural generators of the factors in $\mathbb{Z}_2 * \mathbb{Z}_3$) among the elements of the conjugacy class of a lift in $PSL(2,\mathbb{Z})$.

Now, it is well-known that if an orbifold surface has a cusp then it has a cusp neighborhood, say $\mathcal{C}$, of area one with boundary a horocycle segment of length one. For D a positive integer, we define the *D-thick part* of S, denoted S_D, to be the smallest compact subsurface of S with horocycle boundary which contains all the closed geodesics which when written in normal form, $ab^{\epsilon_1}\dots ab^{\epsilon_t}$, $(\epsilon_i = \pm 1, \epsilon_1 \neq \epsilon_t)$ have at most D consecutive $\epsilon_i's$ of the same sign. The number of consecutive $\epsilon_i's$ of the same sign (called a run) correspond to how deep the geodesic goes into the cusp. (See for example Lemma 7.1 in [**2**]). Hence it is natural to associate the geometric depth into the cusp with the size of a run; that is, the highest power of the parabolic. The D-thick parts form a compact exhaustion of S.

We say that a closed geodesic (including reciprocal geodesics) γ has *k (cusp) excursions of depth greater than D* if during one periodic orbit γ enters $S - S_D$ exactly k-times. See Figure 1 for an example of a cusp excursion where $k = 1$.

A *reciprocal geodesic* on the modular surface is a geodesic that begins and ends at the order two cone point, traversing its image twice if it is primitive, and more generally traversing the image $2m$-times if it is the mth power of a reciprocal geodesic. Its lift is the conjugacy class in $PSL(2,\mathbb{Z})$ of a hyperbolic element with axis passing through an order two fixed point. In the paper [**2**], reciprocal geodesics that stay in S_D (so called D-low lying reciprocal geodesics) were studied. In this paper, we are interested in the reciprocal geodesics that leave the D-thick part a prescribed (necessarily even) number of times (cusp excursions). To that end, for each integer $n \geq 0$, define

$$\mathcal{E}_{2n}^D = \{\gamma : \gamma \text{ a reciprocal geodesic that enters } S - S_D \text{ exactly } 2n\text{-times}\}$$

Equivalently, $\gamma \in \mathcal{E}_{2n}^D$ if when represented in normal form its associated run sequence has exactly $2n$ parts each of size at least $D + 1$. With this in mind, the reciprocal geodesics are partitioned into the disjoint sets

$$\bigcup_{n=0}^{\infty} \mathcal{E}_{2n}^D$$

The set $\mathcal{E}_0^D$ is exactly the D-low lying reciprocals whose asymptotic growth rate is $d_D\alpha_D^t$, where d_D is given by the coefficient in expression (2.1), and α_D is the unique positive root of $z^D - z^{D-1} - \dots - 1$. In stark contrast, the set of all reciprocal geodesics has asymptotic growth 2^{t-1} (see [**2**] for details).

So the first non-trivial excursions occur in $\mathcal{E}_2^D$. In this note we study the growth of reciprocal geodesics with a fixed number of excursions. Specifically, we are interested in how the word lengths of the geodesics in $\{\gamma \in \mathcal{E}_{2n}^D : |\gamma| = 4t\}$ grow as a function of t. In this paper, we address the growth of reciprocal geodesics with any number of excursions greater than depth 1, and the growth of reciprocal geodesics with two excursions of depth greater than D.

THEOREM 1.1. *For any integers $n \geq 0$,*

$$\frac{\left|\{\gamma \in \mathcal{E}_{2n}^1 : |\gamma| = 4t\}\right|}{t^{2n}} \xrightarrow{t\to\infty} \frac{1}{2n!} \tag{1.1}$$

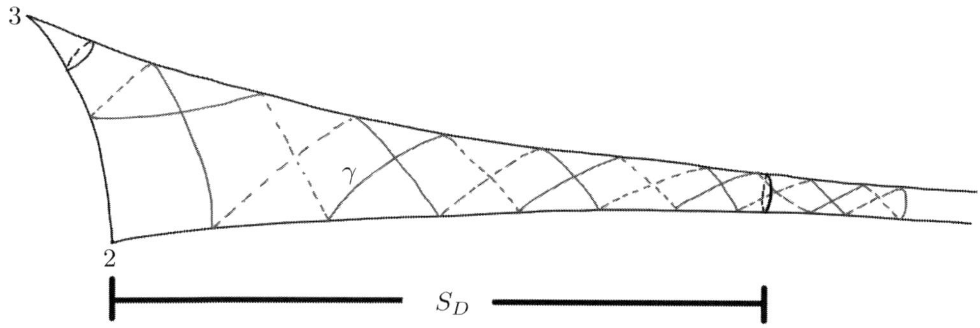

FIGURE 1. A cusp excursion of depth greater than D on S

For any integer $D \geq 2$,

$$\frac{\left|\{\gamma \in \mathcal{E}_2^D : |\gamma| = 4t\}\right|}{t\alpha_D^t} \xrightarrow{t\to\infty} \frac{d_D^2}{\alpha_D^D(\alpha_D - 1)} \tag{1.2}$$

where $d_D = \frac{\alpha_D - 1}{2+(D+1)(\alpha_D-2)}$, *and* α_D *is the unique positive root of* $z^D - z^{D-1} - \cdots - 1$.

In the sequel, Theorem 1.1 and its proof are done in two parts, Theorems 3.1 and 3.3. For the set $\{\gamma \in \mathcal{E}_{2n}^1 : |\gamma| = 4t\}$, we in fact obtain a precise formula, see Theorem 3.1. Moreover in the course of the proof of Theorem 3.3, we derive coarse bounds (in closed form) for the set $\{\gamma \in \mathcal{E}_2^D : |\gamma| = 4t\}$. As a result, in both cases, it is straightforward to write down the growth rates of $|\{\gamma \in \mathcal{E}_{2n}^1 : |\gamma| \leq 4t\}|$ and $|\{\gamma \in \mathcal{E}_2^D : |\gamma| \leq 4t\}|$, as $t \to \infty$. Finally, we note that the asymptotic constant in expression (1.2) is only dependent on the depth D and goes to zero as $D \to \infty$. The more general setting of counting any number of excursions of depth bigger than D will be addressed elsewhere.

2. Some background and preliminaries

There is an extensive history surrounding the modular group as well as an extensive history around the question of geodesic growth in a hyperbolic surface. For the study of growth from a geometric point of view we refer to the papers, [**3–6, 10, 12, 27, 33**]. From an algebraic point of view for the study of surface groups and free groups we refer to the papers, [**7, 9, 19, 25, 29, 32, 37**]. Papers involving normal forms, enumeration schemes for curves, Farey arithmetic, and generating elements in a non-abelian free group can be found in [**1, 13–18, 34**]. Papers specifically about reciprocal geodesics appear in [**2, 3, 5, 11, 25, 33**]. There is an extensive literature on cusp excursions by a random geodesic on a hyperbolic surface [**20–24, 26, 28, 30, 31, 35, 36, 38**]. In particular, the papers [**20–24, 36**] investigate the relation between depth, return time, and other invariants in various contexts including connections to number theory.

A closed geodesic in S corresponds to a conjugacy class of hyperbolic elements in G. Using the fact that ab and ab^{-1} are parabolic elements, a conjugacy class of hyperbolic elements in $PSL(2,\mathbb{Z})$ has a representative of the form $ab^{\epsilon_1}\dots ab^{\epsilon_t}$, where $\epsilon_i = \pm 1$ and $\epsilon_1 \neq \epsilon_t$. Being a product of the parabolic elements ab and ab^{-1}, if k consecutive ϵ_i have the same sign, then γ wraps around the cusp k-times, and

TABLE 1. Various classes of reciprocal geodesics and their growth

Reciprocal Geodesics	**Bijection To**	**Cardinality**
Length $4t$	C_t: compositions of t	$= 2^{t-1}$
Length $4t$ in $S_D = \mathcal{E}_0^D$	$C_{t,D}$: Compositions of t parts bounded by D	$\sim \left(\frac{\alpha_D - 1}{2+(D+1)(\alpha_D-2)}\right) \alpha_D^t$
Length $4t$ in $\mathcal{E}_{2n}^1$	$C_t^{2n,1}$: $2n$ excursions of depth > 1	$=\binom{t}{2n}$
Length $4t$ in $\mathcal{E}_2^D$	$C_t^{2,D}$: 2 excursions of depth $> D$	$\sim \left(\frac{d_D^2}{\alpha_D^D(\alpha_D-1)}\right) t\alpha_D^t$

hence the hyperbolic geometry dictates that the geodesic γ goes deeper into the cusp.

Now, to a t-tuple $(\epsilon_1, \ldots, \epsilon_t)$ we associate a *run* sequence (*composition* of t) $(r_1, \ldots, r_k)$, where r_1 is the number of consecutive ϵ_i of the same sign starting with ϵ_1, r_2 is the length of the next such sequence, and so on. For example, $(1,1,1,-1,1,-1,-1,1)$ corresponds to the composition of 8, $(3,1,1,2,1)$. Thus a composition of t, denoted C_t, is a sequence of positive integers that sum to t. The individual summands of a composition are called its *parts*. When the parts are at most m, we denote that subset of C_t by $C_{t,m}$. Lastly, we denote a composition with exactly n parts that are bigger than D, and all other parts at most D by $C_t^{n,D}$.

Geometrically the length of the run corresponds to how deep the geodesic goes into the cusp. (See lemma 7.1 in [**2**]). Namely, for $D > 0$, we say a closed geodesic (reciprocal geodesics are included among the closed geodesics) is said to have a *cusp excursion* (excursion for short) of depth bigger than D if there is a run longer than D in its normal form. For reciprocal geodesics note that the requirement that its orbit be traversed an even number of times is necessary to guarantee that the geodesic corresponds to an element of the modular group. Namely, a lift of such a geodesic corresponds to the axis of a hyperbolic element which passes through an order two fixed point. Since there is only one conjugacy class of order two fixed points in $PSL(2, \mathbb{Z})$, this hyperbolic element can be conjugated so that its axis passes through i, the fixed point of A. As a consequence, a word representing a reciprocal element (as an abstract group element) is conjugate to a word of the form $a(\gamma a \gamma^{-1})$ and thus its translation length is realized on the surface as a geodesic that retraces its steps. In terms of the generators of $G = \mathbb{Z}_2 * \mathbb{Z}_3$ it has the *normal form*, $ab^{\epsilon_1} \ldots ab^{\epsilon_t} ab^{-\epsilon_t} \ldots ab^{-\epsilon_1}$, where $\epsilon_i = \pm 1$. Note that the word length of a reciprocal geodesic is always a multiple of 4. The reciprocal word has a cusp excursion of depth bigger than D if there are at least (D+1) consecutive ϵ_i of the same sign. Table 1 has the association between various classes of reciprocal geodesics, the types of compositions they correspond to, and their growth rate.

It is well-known that the primitive elements in the modular group G are comprised of

- one conjugacy class of order two corresponding to the element a,
- one conjugacy class of order three corresponding to the element b and its inverse,

- one conjugacy class of parabolic elements corresponding to the element ab and its inverse, and
- infinitely many conjugacy classes of maximal cyclic hyperbolic subgroups.

We remind the reader that a hyperbolic element in $PSL(2,\mathbb{Z})$ that represents a reciprocal geodesic on S has axis passing through an order two fixed point. As a result, such a hyperbolic element is conjugate to itself in $PSL(2,\mathbb{Z})$.

Recall that $C_{t,D}$ denotes the set of compositions of t with parts bounded by D. Counting this set involves using the recursion relation: $|C_{t,D}| = \sum_{i=1}^{D} |C_{t-i,D}|$ for which details can be found in [**8**]. We have

$$|C_{t,D}| = rnd\left(\frac{\alpha_D - 1}{2+(D+1)(\alpha_D - 2)}\alpha_D^t\right) \tag{2.1}$$

where $rnd(x) = \lfloor x + \frac{1}{2} \rfloor$ and α_D is the unique positive root of the polynomial $z^D - z^{D-1} - \cdots - 1$ (see [**8**] and [**2**] for details). We remark that the coefficient $\frac{\alpha_D-1}{2+(D+1)(\alpha_D-2)}$ in equation (2.1) only depends on D and thus we denote it by d_D. We note that $2(1-2^{-D}) \leq \alpha_D < 2$ and the α_D are increasing as D increases. For the details see [**8**].

3. Growth rates of reciprocal geodesics

Let n be some non-negative integer, and $D \geq 1$ a positive integer. The focus of this work is on reciprocal geodesics with $2n$ excursions of depth bigger than D. Namely, we investigate the growth rates of such reciprocal geodesics and how they depend on the parameters n and D. Since reciprocal geodesics always have an even number of excursions, we fix a positive integer n and consider reciprocal geodesics with exactly $2n$ excursions of depth greater than D, and word length $4t$.

We begin by setting up a bijection between reciprocal geodesics of length $4t$ on the modular surface, and compositions of t. Let $\mathcal{N}_{4t}$ be the set of normal forms for a reciprocal word of length $4t$ in $G = \mathbb{Z}_2 * \mathbb{Z}_3$, that is, elements of the form $ab^{\epsilon_1}\dots ab^{\epsilon_t}ab^{-\epsilon_t}\dots ab^{-\epsilon_1}, \epsilon_i = \pm 1$. Let N_{4t} be the set of conjugacy classes of elements in $\mathcal{N}_{4t}$, i.e. the set of reciprocal geodesics with length $4t$. For each conjugacy class in N_{4t} there are exactly two distinct representatives from $\mathcal{N}_{4t}$, see either Sarnak [**33**] or Lemma 3.6 of [**2**]. Thus, the map $\xi : \mathcal{N}_{4t} \to N_{4t}$, which sends a normal form to its conjugacy class, is 2-1. Next, let $X_t = \{(\epsilon_1, \dots, \epsilon_t) : \epsilon_i = \pm 1\}$, i.e. the set of binary t-tuples. Then the map $\Phi : \mathcal{N}_{4t} \to X_t$, given by $\Phi(ab^{\epsilon_1}\dots ab^{\epsilon_t}ab^{-\epsilon_t}\dots ab^{-\epsilon_1}) = (\epsilon_1, \dots, \epsilon_t)$, is 1-1. Now if we projectivize X_t, that is $PX_t = X_t/\{\pm 1\}$ and an element of PX_t is the class, $\{(\epsilon_1, \dots, \epsilon_t), (-\epsilon_1, \dots, -\epsilon_t)\}$, we see that the map $\tau : X_t \to PX_t$ which sends a t-tuple to its corresponding class, is 2-1. Lastly, the set of projectivized binary t-tuples is in one to correspondence with the compositions of t. These correspondences are summarized in figure 3. Thus it follows that $|N_{4t}| = \frac{1}{2}|\mathcal{N}_{4t}| = \frac{1}{2}|X_t| = \frac{1}{2}(2|PX_t|) = |PX_t| = |C_t|$, and we have shown that reciprocal geodesics are in 1-1 correspondence with compositions, C_t. In fact, more is true. This 1-1 correspondence persists if we pass to the subclasses of reciprocal geodesics which are of interest in this article. Namely, the D-low lying reciprocal geodesics are in bijection with compositions with parts bounded by D, and the reciprocal geodesics with $2n$ excursions of depth bigger than D (that is, $\mathcal{E}_{2n}^D$) are in 1-1 correspondence with compositions with exactly n parts that are bigger than D (that is, $C_t^{n,D}$).

$$
\begin{array}{ccc}
\mathcal{N}_{4t} & \xrightarrow{1-1} & X_t & & \\
\big\downarrow{\scriptstyle 2-1} & & \big\downarrow{\scriptstyle 2-1} & & \\
N_{4t} & & PX_t & \xrightarrow{1-1} & C_t
\end{array}
$$

FIGURE 2. Correspondence between N_{4t} and C_t

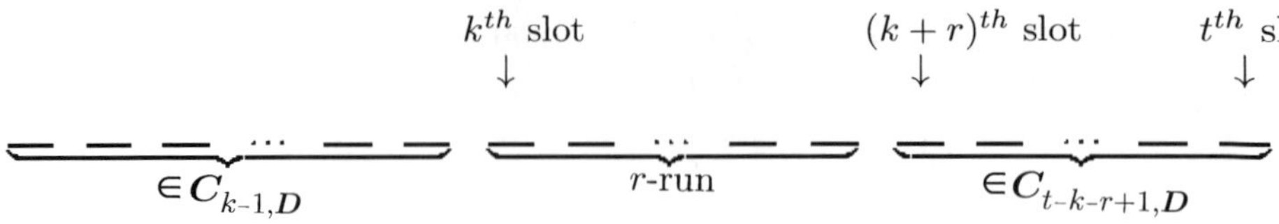

FIGURE 3. $2 \leq k \leq t - r$

3.1. The case of $2n$ excursions of depth bigger than 1. Denote the subset of N_{4t} consisting of reciprocal words with exactly $2n$ excursions of depth greater than one by $N_{4t}^{(2n)}$. The corresponding subset of PX_t consists of the classes of t-tuples in which there are exactly n runs of $(+1)$ or (-1) of length at least two starting from the left. Denote this subset by $PX_t^{(n)}$. (For example, the class $\{\pm(1,1,-1,1,-1,-1,-1)\} \in PX_7^{(2)}$, since there are seven entries with two runs of length at least two. The first is a run of two +1's and the second is a run of three -1's.) Restricting the mappings from the diagram to these subsets, we have that $|N_{4t}^{(2n)}| = |PX_t^{(n)}|$.

THEOREM 3.1. *For any integer $n \geq 0$,*

$$\left|\{\gamma \in \mathcal{E}_{2n}^1 : |\gamma| = 4t\}\right| = \binom{t}{2n} \quad \textit{and} \quad \frac{\left|\{\gamma \in \mathcal{E}_{2n}^1 : |\gamma| = 4t\}\right|}{t^{2n}} \xrightarrow{t\to\infty} \frac{1}{2n!}$$

PROOF. Fix integers $t > 0, n \geq 0$. We have reduced the problem to computing $|PX_t^{(n)}|$. Let $x \in PX_t^{(n)}$. Then x has length t with exactly n runs of +1 or -1 of length at least two. Each of the n runs has a starting position and ending position which amounts to choosing $2n$ of the t positions. Thus, any such x arises by selecting $2n$ slots among the t. That is, $|PX_t^{(n)}| = \binom{t}{2n}$. It follows that $|PX_t^{(n)}| \sim \dfrac{t^{2n}}{(2n)!}$, as $t \to \infty$ $(t \in \mathbb{N})$. □

3.2. The case of two excursions of depth bigger than D. We next focus on reciprocal geodesics with two excursions.

We, as usual, identify a reciprocal geodesic of length $4t$ with its normal form in $PX_t^{1,D}$, the t-tuples with one run of length greater than D and all other runs at most D. This projective binary sequence can in turn be identified with compositions of t with one part, which we denote by r, greater than D, and all other parts at most D. We denote this set by $C_t^{1,D}$. In order to count this set we decompose it in the following way. Let $r \geq D+1$ be the length of the run in $C_t^{1,D}$, and let k be the position in the t-tuple $PX_t^{1,D}$ where the r run starts. Note that $1 \leq k \leq t-r+1$.

For k and r fixed and in the permitted ranges listed above, we have

$$|C_t^{1,D}| = |C_{k-1,D}||C_{t-k-r+1,D}| \text{ for } 1 \leq k \leq t-r+1 \tag{3.1}$$

This follows by counting the number of possible compositions to the left and to the right of the r run and then observing that there is only one possible r run starting in the kth slot. See Figure 3 for a schematic. Here we use the convention that $|C_{0,D}| = 1$.

We remark that up until this point the discussion works for $D \geq 1$. For the rest of this section we assume $D \geq 2$. Summing over the admissible values of k and r, we have that

$$\big|\{\gamma \in \mathcal{E}_2^D : |\gamma| = 4t\}\big| = \sum_{r=D+1}^{t} \sum_{k=1}^{t-r+1} |C_{k-1,D}||C_{t-k-r+1,D}| \tag{3.2}$$

Plugging the appropriate values of expression (2.1) into (3.2) we obtain the lower estimate for $\big|\{\gamma \in \mathcal{E}_2^D : |\gamma| = 4t\}\big|$ as

$$\sum_{r=D+1}^{t} \sum_{k=1}^{t-r+1} \left(d_D\alpha_D^{k-1} - \frac{1}{2}\right)\left(d_D\alpha_D^{t-k-r+1} - \frac{1}{2}\right) \tag{3.3}$$

and the upper estimate as,

$$\sum_{r=D+1}^{t} \sum_{k=1}^{t-r+1} \left(d_D\alpha_D^{k-1} + \frac{1}{2}\right)\left(d_D\alpha_D^{t-k-r+1} + \frac{1}{2}\right) \tag{3.4}$$

The upper and lower estimates are of the form,

$$\sum_r\sum_k d^2\alpha^{t-r} + a\sum_r\sum_k d\alpha^{k-1} + a\sum_r\sum_k d\alpha^{t-k-r+1} + \sum_r\sum_k \frac{1}{4} \tag{3.5}$$

where $a = \pm\frac{1}{2}$ and, for ease of notation, we have set $d = d_D$ and $\alpha = \alpha_D$. We are interested in what happens to the terms of expression (3.5) as $t \to \infty$. To this end we first prove a lemma.

LEMMA 3.2. *The sums in the following are over the admissible values of r and k.*

(1) $d^2\sum_r\sum_k \alpha^{t-r} \sim \frac{d^2}{\alpha^D(\alpha-1)}t\alpha^t$
(2) $d\sum_r\sum_k \alpha^{k-1} = O(\alpha^t)$
(3) $d\sum_r\sum_k \alpha^{t-k-r+1} = O(\alpha^t)$

PROOF. To ease notation, we use the convention $d = d_D$ and $\alpha = \alpha_D$. To prove item (1), note that $d^2\sum_k \alpha^{t-r} = d^2\alpha^{t-r}(t-r+1)$. Summing this over the admissible r's we have,

$$d^2 \sum_{r=D+1}^{t} \alpha^{t-r}(t-r+1) =$$
$$d^2\alpha^t\left[t\sum_{r=D+1}^{t}\alpha^{-r} - \sum_{r=D+1}^{t}(r-1)\alpha^{-r}\right] =$$
$$\left(d^2\sum_{r=D+1}^{t}\alpha^{-r}\right)t\alpha^t - \left(d^2\sum_{r=D+1}^{t}(r-1)\left(\frac{1}{\alpha}\right)^r\right)\alpha^t =$$
$$d^2\left(\frac{\alpha^{-D}-\alpha^{-t}}{\alpha-1}\right)t\alpha^t + O(\alpha^t).$$

To prove item (2), we first note that a short computation yields

$$\alpha^{-1}\sum_{k=1}^{t-r+1}\alpha^k = \frac{\alpha^{t-r+1}-1}{\alpha-1}.$$

Now, summing over the admissible r's we obtain,

$$\sum_{r=D+1}^{t}\frac{\alpha^{t-r+1}-1}{\alpha-1} =$$

$$\left(\frac{1}{\alpha-1}\right)\left[\alpha^{t+1}\sum_{r=D+1}^{t}\alpha^{-r}-(t-D)\right] = O(\alpha^t).$$

The proof of item (3) follows in a similar fashion to item (2). □

We are now ready to prove

THEOREM 3.3. *For* $D \geq 2$,

$$\frac{\left|\{\gamma\in\mathcal{E}_2^D : |\gamma| = 4t\}\right|}{t\alpha_D^t} \xrightarrow{t\to\infty} \frac{d_D^2}{\alpha_D^D(\alpha_D-1)}.$$

PROOF. Consider the terms of expression (3.5). Lemma 3.2 and the fact that the fourth term grows like t^2, ensures that the first term, whose asymptotic is $t\alpha^t$, dominates the expression as $t\to\infty$. Since the first term does not depend on the constant a, we may conclude that the lower and upper estimates of $\left|\{\gamma\in\mathcal{E}_2^D : |\gamma| = 4t\}\right|$ have the same asymptotic. □

Acknowledgments

The first author would like to thank the University of Michigan and the University of Luxembourg for their support while parts of this paper were being completed. In particular, thanks go to Richard D. Canary (University of Michigan) and Hugo Parlier (University of Luxembourg) for their hospitality during the respective visits.

References

[1] A. Basmajian and M. Liu, *Low-lying geodesics on the modular surface and necklaces*, arXiv:2311.16041, 2023.

[2] Ara Basmajian and Robert Suzzi Valli, *Combinatorial growth in the modular group*, Groups Geom. Dyn. **16** (2022), no. 2, 683–703, DOI 10.4171/ggd/667. MR4502618

[3] Florin P. Boca, Vicenţiu Paşol, Alexandru A. Popa, and Alexandru Zaharescu, *Pair correlation of angles between reciprocal geodesics on the modular surface*, Algebra Number Theory **8** (2014), no. 4, 999–1035, DOI 10.2140/ant.2014.8.999. MR3248992

[4] Jean Bourgain and Alex Kontorovich, *Beyond expansion II: low-lying fundamental geodesics*, J. Eur. Math. Soc. (JEMS) **19** (2017), no. 5, 1331–1359, DOI 10.4171/JEMS/694. MR3635355

[5] Jean Bourgain and Alex Kontorovich, *Beyond expansion, III: Reciprocal geodesics*, Duke Math. J. **168** (2019), no. 18, 3413–3435, DOI 10.1215/00127094-2019-0056. MR4034890

[6] Peter Buser, *Geometry and spectra of compact Riemann surfaces*, Progress in Mathematics, vol. 106, Birkhäuser Boston, Inc., Boston, MA, 1992. MR1183224

[7] Danny Calegari and Joel Louwsma, *Immersed surfaces in the modular orbifold*, Proc. Amer. Math. Soc. **139** (2011), no. 7, 2295–2308, DOI 10.1090/S0002-9939-2011-10911-0. MR2784794

[8] Gregory P. B. Dresden and Zhaohui Du, *A simplified Binet formula for k-generalized Fibonacci numbers*, J. Integer Seq. **17** (2014), no. 4, Article 14.4.7, 9. MR3181762

[9] Viveka Erlandsson, *A remark on the word length in surface groups*, Trans. Amer. Math. Soc. **372** (2019), no. 1, 441–455, DOI 10.1090/tran/7561. MR3968775

[10] Viveka Erlandsson, Hugo Parlier, and Juan Souto, *Counting curves, and the stable length of currents*, J. Eur. Math. Soc. (JEMS) **22** (2020), no. 6, 1675–1702, DOI 10.4171/jems/953. MR4092896

[11] V. Erlandsson and J. Souto, *Counting and equidistribution of reciprocal geodesics and dihedral groups*, arXiv:2204.09956

[12] Viveka Erlandsson and Juan Souto, *Counting curves in hyperbolic surfaces*, Geom. Funct. Anal. **26** (2016), no. 3, 729–777, DOI 10.1007/s00039-016-0374-7. MR3540452

[13] Jane Gilman, *Informative words and discreteness*, Combinatorial group theory, discrete groups, and number theory, Contemp. Math., vol. 421, Amer. Math. Soc., Providence, RI, 2006, pp. 147–155, DOI 10.1090/conm/421/08034. MR2303833

[14] Jane Gilman, *Primitive curve lengths on pairs of pants*, Infinite group theory, World Sci. Publ., Hackensack, NJ, 2018, pp. 141–155. MR3586884

[15] Jane Gilman and Linda Keen, *Cutting sequences and palindromes*, Geometry of Riemann surfaces, London Math. Soc. Lecture Note Ser., vol. 368, Cambridge Univ. Press, Cambridge, 2010, pp. 194–216. MR2665010

[16] Jane Gilman and Linda Keen, *Enumerating palindromes and primitives in rank two free groups*, J. Algebra **332** (2011), 1–13, DOI 10.1016/j.jalgebra.2011.02.010. MR2774675

[17] Jane Gilman and Linda Keen, *Lifting free subgroups of $PSL(2,\mathbb{R})$ to free groups*, Quasiconformal mappings, Riemann surfaces, and Teichmüller spaces, Contemp. Math., vol. 575, Amer. Math. Soc., Providence, RI, 2012, pp. 109–122, DOI 10.1090/conm/575/11384. MR2933896

[18] Jane Gilman and Linda Keen, *Word sequences and intersection numbers*, Complex manifolds and hyperbolic geometry (Guanajuato, 2001), Contemp. Math., vol. 311, Amer. Math. Soc., Providence, RI, 2002, pp. 231–249, DOI 10.1090/conm/311/05455. MR1940172

[19] Victor Guba and Mark Sapir, *On the conjugacy growth functions of groups*, Illinois J. Math. **54** (2010), no. 1, 301–313. MR2776997

[20] Andrew Haas, *Excursion and return times of a geodesic to a subset of a hyperbolic Riemann surface*, Proc. Amer. Math. Soc. **141** (2013), no. 11, 3957–3967, DOI 10.1090/S0002-9939-2013-11767-3. MR3091786

[21] Andrew Haas, *Geodesic cusp excursions and metric Diophantine approximation*, Math. Res. Lett. **16** (2009), no. 1, 67–85, DOI 10.4310/MRL.2009.v16.n1.a8. MR2480562

[22] Andrew Haas, *Geodesic excursions into an embedded disc on a hyperbolic Riemann surface*, Conform. Geom. Dyn. **13** (2009), 1–5, DOI 10.1090/S1088-4173-09-00185-4. MR2476655

[23] Andrew Haas, *The distribution of geodesic excursions into the neighborhood of a cone singularity on a hyperbolic 2-orbifold*, Comment. Math. Helv. **83** (2008), no. 1, 1–20, DOI 10.4171/CMH/115. MR2365405

[24] Andrew Haas, *The distribution of geodesic excursions out the end of a hyperbolic orbifold and approximation with respect to a Fuchsian group*, Geom. Dedicata **116** (2005), 129–155, DOI 10.1007/s10711-005-9009-0. MR2195445

[25] Blanca Teresa Marmolejo, *Growth of Conjugacy Classes of Reciprocal Words in Triangle Groups*, ProQuest LLC, Ann Arbor, MI, 2020. Thesis (Ph.D.)–City University of New York. MR4187474

[26] María V. Melián and Domingo Pestana, *Geodesic excursions into cusps in finite-volume hyperbolic manifolds*, Michigan Math. J. **40** (1993), no. 1, 77–93, DOI 10.1307/mmj/1029004675. MR1214056

[27] Maryam Mirzakhani, *Growth of the number of simple closed geodesics on hyperbolic surfaces*, Ann. of Math. (2) **168** (2008), no. 1, 97–125, DOI 10.4007/annals.2008.168.97. MR2415399

[28] Ron Mor, *Excursions to the cusps for geometrically finite hyperbolic orbifolds and equidistribution of closed geodesics in regular covers*, Ergodic Theory Dynam. Systems **42** (2022), no. 12, 3745–3791, DOI 10.1017/etds.2021.101. MR4504680

[29] Peter S. Park, *Conjugacy growth of commutators*, J. Algebra **526** (2019), 423–458, DOI 10.1016/j.jalgebra.2019.02.024. MR3920909

[30] Mark Pollicott, *Limiting distributions for geodesics excursions on the modular surface*, Spectral analysis in geometry and number theory, Contemp. Math., vol. 484, Amer. Math. Soc., Providence, RI, 2009, pp. 177–185, DOI 10.1090/conm/484/09474. MR1500147

[31] Anja Randecker and Giulio Tiozzo, *Cusp excursion in hyperbolic manifolds and singularity of harmonic measure*, J. Mod. Dyn. **17** (2021), 183–211, DOI 10.3934/jmd.2021006. MR4251941

[32] Igor Rivin, *Growth in free groups (and other stories)—twelve years later*, Illinois J. Math. **54** (2010), no. 1, 327–370. MR2776999

[33] Peter Sarnak, *Reciprocal geodesics*, Analytic number theory, Clay Math. Proc., vol. 7, Amer. Math. Soc., Providence, RI, 2007, pp. 217–237. MR2362203

[34] Caroline Series, *The modular surface and continued fractions*, J. London Math. Soc. (2) **31** (1985), no. 1, 69–80, DOI 10.1112/jlms/s2-31.1.69. MR810563

[35] Bernd Stratmann, *A note on counting cuspidal excursions*, Ann. Acad. Sci. Fenn. Ser. A I Math. **20** (1995), no. 2, 359–372. MR1346819

[36] Dennis Sullivan, *Disjoint spheres, approximation by imaginary quadratic numbers, and the logarithm law for geodesics*, Acta Math. **149** (1982), no. 3-4, 215–237, DOI 10.1007/BF02392354. MR688349

[37] Charles Traina, *The conjugacy problem of the modular group and the class number of real quadratic number fields*, J. Number Theory **21** (1985), no. 2, 176–184, DOI 10.1016/0022-314X(85)90049-6. MR808285

[38] M. Trin, *Thurston's compactification via geodesic currents: The case of non-compact finite area surfaces*, `arXiv:2208.10763`, 2022.

PhD Program in Mathematics, The Graduate Center, CUNY, 365 Fifth ave., New York, New York 10016; and Department of Mathematics and Statistics, Hunter College, CUNY, 695 Park ave. New York, New York 10065

Email address: `abasmajian@gc.cuny.edu`

Department of Mathematics, Manhattan College, 4513 Manhattan College Pkwy., Riverdale, New York 10471

Email address: `robert.suzzivalli@manhattan.edu`

Contemporary Mathematics
Volume **811**, 2025
https://doi.org/10.1090/conm/811/16235

Elementary groups in PSL$(3, \mathbb{C})$

W. Barrera, A. Cano, J. P. Navarrete, and J. Seade

To Ravi Kulkarni, with great admiration

Abstract. In this paper, we give a classification of the subgroups of PSL$(3, \mathbb{C})$ that act on $\mathbb{P}^2_{\mathbb{C}}$ in such a way that their Kulkarni limit set has finitely many lines in general position lines. These are the elementary groups.

Introduction

Kleinian groups are discrete subgroups of PSL$(2, \mathbb{C})$ acting on $\mathbb{S}^2 \cong \mathbb{P}^1_{\mathbb{C}}$ in such a way that their limit set is not all of $\mathbb{S}^2$. These are classified into elementary and non-elementary groups. The elementary groups are those whose limit set consists of $0, 1$ or 2 points, and they are classified, see for instance (see [**20**]).

In this work we look at complex Kleinian groups, that is, discrete subgroups of PSL$(3, \mathbb{C})$ acting properly and discontinuously on some open subset of $\mathbb{P}^2_{\mathbb{C}}$. In [**11**] it is proved that the Kulkarni limit set Λ_{Kul} (see Definition 1.1) of every infinite complex Kleinian group contains at least one complex projective line. In fact we known from [**1**] that under some generic hypothesis on the group, the Kulkarni limit set is a union of complex projective lines in general position, *i.e.* no three of them intersect. By definition, the exception are the elementary groups. These have finitely many lines in general position in their limit set, and there are two types ef such groups: those of the first kind have a finite number of lines in their limit set, and the groups of the second kind have infinitely many lines in their limit set, but only finitely many in general position.

As an example of elementary groups of the first kind, take the cyclic group generated by a 3×3 diagonal matrix with non-zero eigenvalues of different norms; this has two projective lines as limit set in $\mathbb{P}^2_{\mathbb{C}}$. On the other hand (see [**9**]), take a $\mathbb{C}$-Fuchsian group in PSL$(3, \mathbb{C})$. Then its limit set in $\mathbb{P}^2_{\mathbb{C}}$ consists of a cone of projective lines with base a circle (see [**9**] ; it has infinitely many lines, but all of them pass through the vertex of the cone, so there are only two in general position. We remark that if we take an $\mathbb{R}$-Fuchsian group in PSL$(3, \mathbb{C})$, then it is non-elementary and its limit set actually has infinitely many lines in general position, see [**9**]

In this work we study and describe all the elementary groups in PSL$(3, \mathbb{C})$.

2020 *Mathematics Subject Classification.* Primary 37F30, 32Q45; Secondary 37F45, 22E40.

Key words and phrases. Kleinian groups, projective complex plane, discrete groups, limit set.

An interesting class of elementary groups are the discrete purely parabolic groups. One finds that the limit set of such a group consists of either one line, or a cone of lines over the circle, or the whole of $\mathbb{P}^2_{\mathbb{C}}$. We describe these in Section 2.

We know (see Theorem 1.5) that if $\Gamma \subset \mathrm{PSL}(3,\mathbb{C})$ is an infinite discrete subgroup, then the number of complex projective lines in its limit set $\Lambda_{\mathrm{Kul}}(\Gamma)$ is equal to $1, 2, 3$ or ∞, and the number of lines $\Lambda_{\mathrm{Kul}}(\Gamma)$ in general position can be $1, 2, 3, 4$ or ∞. Also, if the number of lines in $\Lambda_{\mathrm{Kul}}(\Gamma)$ is exactly 3, then these lines are in general position.

It follows from Theorem 1.5 and the results in this note, that if $\Lambda_{\mathrm{Kul}}(\Gamma)$ has more than one line, then it is a union of lines. The only case where $\Lambda_{\mathrm{Kul}}(\Gamma)$ can have isolated points is when this set actually consists of one line and one point.

Section 1 contains the basic definitions and preliminary results that we use in the sequel. Sections 4 and 7 are expository and we cite in each the appropriate bibliography. The results in these sections are taken from [**3**] and [**2**] and we include them here for completeness. In [**3**] the authors discuss the groups with limit set a line, and in [**2**] those with four lines in general position.

Section 5 focuses on the case of two lines. These can be of the first kind, with exactly 2 lines in Λ_{Kul}, or of the second kind, with limit set a cone of lines over a circle, but only 2 in general position.

In Section 6 we describe the groups with exactly three lines in their limit set, and Section 8 discusses the only case where the limit set is not a union of lines. Then Λ_{Kul} consists of a line and a point.

1. Preliminaries

1.1. Pseudo-projective transformations. We let $\mathbb{P}^2_{\mathbb{C}} := (\mathbb{C}^3 \setminus \{0\})/\mathbb{C}^*$, be the complex projective plane. This is a compact connected complex 2-dimensional manifold. Let $[\ ] : \mathbb{C}^3 \setminus \{0\} \to \mathbb{P}^2_{\mathbb{C}}$ be the quotient map. If $\beta = \{e_1, e_2, e_3\}$ is the standard basis of $\mathbb{C}^3$, we write $[e_j] = e_j$ and if $w = (w_1, w_2, w_3) \in \mathbb{C}^3 \setminus \{0\}$ then we write $[w] = [w_1 : w_2 : w_3]$. Also, $\ell \subset \mathbb{P}^2_{\mathbb{C}}$ is said to be a complex line if $[\ell]^{-1} \cup \{0\}$ is a complex linear subspace of dimension 2. Given $p, q \in \mathbb{P}^2_{\mathbb{C}}$ distinct points, there is a unique complex projective line passing through p and q, such complex projective line is called a *line*, for short, and it is denoted by $\overleftrightarrow{p,q}$.

Consider the action of $\mathbb{Z}_3$ (viewed as the cubic roots of the unity) on $\mathrm{SL}(3,\mathbb{C})$ given by the usual scalar multiplication, then the Lie group

$$\mathrm{PSL}(3,\mathbb{C}) = \mathrm{SL}(3,\mathbb{C})/\mathbb{Z}_3$$

is the group of automorphisms of $\mathbb{P}^2_{\mathbb{C}}$. Let $[[\ ]] : \mathrm{SL}(3,\mathbb{C}) \to \mathrm{PSL}(3,\mathbb{C})$ be the quotient map, $\gamma \in \mathrm{PSL}(3,\mathbb{C})$ and $\tilde{\gamma} \in \mathrm{GL}(3,\mathbb{C})$, we say that $\tilde{\gamma}$ is a lift of γ if there is a cubic root τ of $Det(\gamma)$ such that $[[\tau\tilde{\gamma}]] = \gamma$. Also, we use the notation (γ_{ij}) to denote elements in $\mathrm{SL}(3,\mathbb{C})$.

We denote by $M_{3\times 3}(\mathbb{C})$ the space of all 3×3 matrices with entries in $\mathbb{C}$ equipped with the standard topology. The quotient space

$$\mathrm{SP}(3,\mathbb{C}) := (M_{3\times 3}(\mathbb{C}) \setminus \{\mathbf{0}\})/\mathbb{C}^*$$

is called the space of *pseudo-projective maps of* $\mathbb{P}^2_{\mathbb{C}}$ and it is naturally identified with the projective space $\mathbb{P}^8_{\mathbb{C}}$. Since $\mathrm{GL}(3,\mathbb{C})$ is an open, dense, $\mathbb{C}^*$-invariant set of $M_{3\times 3}(\mathbb{C}) \setminus \{\mathbf{0}\}$, we obtain that the space of pseudo-projective maps of $\mathbb{P}^2_{\mathbb{C}}$ is a compactification of $\mathrm{PSL}(3,\mathbb{C})$. As in the case of projective maps, if $\mathbf{s} \in M_{3\times 3}(\mathbb{C}) \setminus \{\mathbf{0}\}$, then $[\mathbf{s}]$ denotes the equivalence class of the matrix $\mathbf{s}$ in the space of pseudo-projective maps of $\mathbb{P}^2_{\mathbb{C}}$. Also, we say that $\mathbf{s} \in M_{3\times 3}(\mathbb{C}) \setminus \{\mathbf{0}\}$ is a lift of the pseudoprojective map S, whenever $[\mathbf{s}] = S$.

Let S be an element in $(M_{3\times 3}(\mathbb{C}) \setminus \{\mathbf{0}\})/\mathbb{C}^*$ and $\mathbf{s}$ a lift to $M_{3\times 3}(\mathbb{C}) \setminus \{\mathbf{0}\}$ of S. The matrix $\mathbf{s}$ induces a non-zero linear transformation $s : \mathbb{C}^3 \to \mathbb{C}^3$, which is not necessarily invertible. Let $Ker(s) \subsetneq \mathbb{C}^3$ be its kernel and let $Ker(S)$ denote its projectivization to $\mathbb{P}^2_{\mathbb{C}}$, taking into account that $Ker(S) := \varnothing$ whenever $Ker(s) = \{(0,0,0)\}$. We refer to [**10**] for more details about this subject.

1.2. The limit set. There are two types of limit sets relevant for this work. These are the Kulkarni limit set Λ_{Kul} and the Myrberg limit set Λ_{Myr}. Let us define these. Let $\Gamma \subset \mathrm{PSL}(n+1,\mathbb{C})$ be a discrete subgroup.

DEFINITION 1.1. The *Kulkarni limit set* of Γ is:

$$\Lambda(\Gamma) = L_0(\Gamma) \cup L_1(\Gamma) \cup L_2(\Gamma),$$

where:

- $L_0(\Gamma)$ is the closure of the points in $\mathbb{P}^2_{\mathbb{C}}$ with infinite isotropy group.
- $L_1(\Gamma)$ is the closure of the set of accumulation points of the Γ-orbits of points in $\mathbb{P}^2_{\mathbb{C}} \setminus L_0(\Gamma)$.
- $L_2(\Gamma)$ is the closure of the union of accumulation points of $\{\gamma(K) : \gamma \in \Gamma\}$, where K runs over all the compact sets in $\mathbb{P}^2_{\mathbb{C}} \setminus (L_0(\Gamma) \cup L_1(\Gamma))$.

This is a closed Γ-invariant set. Its complement

$$\Omega_{\mathrm{Kul}}(\Gamma) = \mathbb{P}^n_{\mathbb{C}} \setminus \Lambda(\Gamma),$$

is the *Kulkarni discontinuity region* of Γ. We know from [**18**] that the Γ-action on $\Omega_{\mathrm{Kul}}(\Gamma)$ is properly discontinuous, and we further know from [**1**] that $\Omega_{\mathrm{Kul}}(\Gamma)$ contains the equicontinuity region $\mathrm{Eq}(\Gamma)$. However, as we will see later, $\Omega_{\mathrm{Kul}}(\Gamma)$ generically is the largest open subset of $\mathbb{P}^2_{\mathbb{C}}$ where the group acts properly and discontinuously, but this is not always so.

Recall that $\mathrm{Eq}(\Gamma)$ is the set of points in $\mathbb{P}^2_{\mathbb{C}}$ for which there is an open neighborhood where the family of transformations defined by Γ is a normal family.

Now, given $\Gamma \subset \mathrm{PSL}(n+1,\mathbb{C})$, a discrete subgroup, we let Γ' be the set of pseudo-projective maps of $\mathbb{P}^2_{\mathbb{C}}$ which are limits of sequences in Γ. That is:

$$\Gamma' := \{S \in \mathrm{SP}(3,\mathbb{C}) \mid S \text{ is an accumulation point of } \Gamma\}.$$

The following notion is due to Myrberg [**21**] (cf. [**1**, Definition 3.3]):

DEFINITION 1.2. The *Myrberg limit set* of Γ is:

$$\Lambda_{\mathrm{Myr}}(\Gamma) := \cup_{S\in\Gamma'} Ker(S),$$

where $Ker(S)$ is the kernel of the pseudoprojective transformation S defined above.

Notice that if a pseudoprojective map S is not in $\mathrm{PSL}(3,\mathbb{C})$, then its kernel is either a line or a point. The following useful notion is introduced in [**4**]:

DEFINITION 1.3. We say that a line $\ell \subset \mathbb{P}^2_{\mathbb{C}}$ is an effective line for Γ if there exists a pseudoprojective transformation $S \in \Gamma'$ with kernel ℓ.

The Myrberg limit set of Γ contains all the effective lines, and it is not hard to see that, by [**22**], it contains the Kulkarni limit set. It is immediate from [**21**] that Γ acts properly and discontinuously on the complement of $\Lambda_{\mathrm{Myr}}(\Gamma)$. Furthermore, one has the following theorem.

THEOREM 1.4. *The Myrberg limit set is the complement of the equicontinuity region:* $\mathrm{Eq}(\Gamma) = \mathbb{P}^2_{\mathbb{C}} \setminus \Lambda_{\mathrm{Myr}}(\Gamma)$. *Also, the set* $\Lambda_{\mathrm{Myr}}(\Gamma)$ *equals the union of all effective lines of* Γ *except when it is the disjoint union of one line and one point. Moreover,* $\Lambda_{\mathrm{Kul}}(\Gamma) \subset \Lambda_{\mathrm{Myr}}(\Gamma)$.

The first statement is a special case of [**10**, Lemma 4.1] (see also [**4**, Theorem 3.4]). The second statement is Corollary 4.5 in [**4**].

Throughout this work, when we say limit set we refer to the Kulkarni limit set, otherwise we will say it explicitly.

1.3. Classification of the elements in PSL(3, ℂ).

Recall that the elements of $\mathrm{PSL}(2,\mathbb{C})$ are classified as elliptic, parabolic or loxodromic: g is elliptic if regarded as a Möbius transformation in $S^2 \cong \mathbb{P}^1_{\mathbb{C}}$, up to conjugation it is a rotation; parabolic elements are translations up to conjugation and loxodromic elements are multiplication by a complex number of norm $\neq 1$. Equivalently, given $g \in \mathrm{PSL}(2,\mathbb{C})$ and a lift $\tilde{g}$ to $\mathrm{SL}(2,\mathbb{C})$, g is elliptic if $\tilde{g}$ is diagonalizable with unitary eigenvalues, it is parabolic if $\tilde{g}$ is not diagonalizable and it has unitary eigenvalues, and g is loxodromic otherwise. The classification given in these terms extends to all dimensions (see [**8**, **22**]).

The *elliptic* elements in $\mathrm{PSL}(3,\mathbb{C})$ are those elements γ that have a lift to $\mathrm{SL}(3,\mathbb{C})$ whose Jordan canonical form is

$$\begin{pmatrix} e^{i\theta_1} & 0 & 0 \\ 0 & e^{i\theta_2} & 0 \\ 0 & 0 & e^{i\theta_3} \end{pmatrix}.$$

The limit set for (the cyclic group generated by) γ elliptic is either empty or all of $\mathbb{P}^2_{\mathbb{C}}$, according to whether the order of γ is finite or infinite. The subgroups of $\mathrm{PSL}(3,\mathbb{C})$ containing an elliptic element of infinite order cannot be discrete.

The *parabolic* elements in $\mathrm{PSL}(3,\mathbb{C})$ are the elements γ with limit set $\Lambda_{\mathrm{Kul}}(\gamma)$ equal to one single complex line. If γ is parabolic then it has a lift to $\mathrm{SL}(3,\mathbb{C})$ with Jordan canonical form one of the following matrices:

$$\begin{pmatrix} 1 & 1 & 0 \\ 0 & 1 & 0 \\ 0 & 0 & 1 \end{pmatrix}, \begin{pmatrix} 1 & 1 & 0 \\ 0 & 1 & 1 \\ 0 & 0 & 1 \end{pmatrix} \begin{pmatrix} e^{2\pi i t} & 1 & 0 \\ 0 & e^{2\pi i t} & 0 \\ 0 & 0 & e^{-4\pi i t} \end{pmatrix}, \; e^{2\pi i t} \neq 1.$$

In the first case $\Lambda_{\mathrm{Kul}}(\gamma)$ is the complex line consisting of all the fixed points of γ, in the second case $\Lambda_{\mathrm{Kul}}(\gamma)$ is the unique γ-invariant complex line. In the last case $\Lambda_{\mathrm{Kul}}(\gamma)$ is the complex line determined by the two fixed points of γ.

There are four kinds of *loxodromic* elements in $\mathrm{PSL}(3,\mathbb{C})$:

- The *complex homotheties* are the elements that have a lift to $\mathrm{SL}(3,\mathbb{C})$ with Jordan canonical form:

$$\begin{pmatrix} \lambda & 0 & 0 \\ 0 & \lambda & 0 \\ 0 & 0 & \lambda^{-2} \end{pmatrix}, \quad |\lambda| \neq 1.$$

  The limit set $\Lambda_{\mathrm{Kul}}(\gamma)$ is the set of fixed points of γ, consisting of a complex line and a point.

- The *screws* are those elements $\gamma \in \mathrm{PSL}(3,\mathbb{C})$ that have a lift to $\mathrm{SL}(3,\mathbb{C})$ whose Jordan canonical form is
$$\begin{pmatrix} \lambda & 0 & 0 \\ 0 & \mu & 0 \\ 0 & 0 & (\lambda\mu)^{-1} \end{pmatrix}, \quad \lambda \neq \mu,\ |\lambda| = |\mu| \neq 1.$$
The limit set consists of a complex line l, on which γ acts as an elliptic transformation of $\mathrm{PSL}(2,\mathbb{C})$, and the fixed point of γ not lying in l.
- The *loxoparabolic* elements $\gamma \in \mathrm{PSL}(3,\mathbb{C})$ have a lift to $\mathrm{SL}(3,\mathbb{C})$ whose Jordan canonical form is
$$\begin{pmatrix} \lambda & 1 & 0 \\ 0 & \lambda & 0 \\ 0 & 0 & \lambda^{-2} \end{pmatrix}, \quad |\lambda| \neq 1.$$
The limit set $\Lambda_{\mathrm{Kul}}(\gamma)$ consists of two γ-invariant complex lines. The element γ acts on one of these complex lines as a parabolic element of $\mathrm{PSL}(2,\mathbb{C})$ and on the other as a loxodromic element of $\mathrm{PSL}(2,\mathbb{C})$.
- The *strongly loxodromic* elements $\gamma \in \mathrm{PSL}(3,\mathbb{C})$ have a lift to $\mathrm{SL}(3,\mathbb{C})$ whose Jordan canonical form is
$$\begin{pmatrix} \lambda_1 & 0 & 0 \\ 0 & \lambda_2 & 0 \\ 0 & 0 & \lambda_3 \end{pmatrix}, \quad |\lambda_1| < |\lambda_2| < |\lambda_3|.$$
This kind of transformation has three fixed points, one of them is attracting, another is repelling and the other point is a saddle. The limit set is the union of the complex line determined by the attracting and saddle points and the complex line determined by the saddle and repelling points.

1.4. Elementary groups in $\mathrm{PSL}(3,\mathbb{C})$**.** In the previous section we have examples of elements in $\mathrm{PSL}(3,\mathbb{C})$ where the limit set consists of one line, or one line and one point, or two lines. We have the following theorem from [**4**].

THEOREM 1.5. *Let* $\Gamma \subset \mathrm{PSL}(3,\mathbb{C})$ *be an infinite discrete subgroup. Then:*

(i) *The number of complex projective lines in* $\Lambda_{\mathrm{Kul}}(\Gamma)$ *is equal to* $1, 2, 3$ *or* ∞.
(ii) *The number of lines* $\Lambda_{\mathrm{Kul}}(\Gamma)$ *in general position can be* $1, 2, 3, 4$ *or* ∞.
(iii) *If the number of lines in* $\Lambda_{\mathrm{Kul}}(\Gamma)$ *is exactly 3, then these lines are in general position.*
(iv) *If there are infinitely many lines in* $\Lambda_{\mathrm{Kul}}(\Gamma)$*, then the effective lines form a perfect set in* $\Lambda_{\mathrm{Kul}}(\Gamma)$.
(v) *There can be at most one isolated point in* $\Lambda_{\mathrm{Kul}}(\Gamma)$*, and in that case this limit set is the disjoint union of that point and one line.*

Statements (i), (ii) and (iv) are theorems 1.1 and 1.2 in [**4**]. Statement (iii) is a corollary of [**4**, Proposition 5.4]. Statement (v) is not in the literature so we now give a short proof of it.

Recall that a discrete subgroup of $\mathrm{PSL}(2,\mathbb{C})$ is elementary if its limit set has finite cardinality. We remark that in the case of $\mathrm{PSL}(3,\mathbb{C})$ there are groups with infinitely many lines in their limit set Λ_{Kul}, but only two of them in general position. In view of the previous theorem, this leads to Definition 1.6:

DEFINITION 1.6. A discrete subgroup of $\mathrm{PSL}(3,\mathbb{C})$ is elementary of the first kind if its limit set Λ_{Kul} has finitely many lines. The group is elementary of the second kind if Λ_{Kul} has finitely many lines in general position.

So every elementary group of the first kind also is of the second kind, but not conversely. In the following sections we describe the classification of elementary groups.

1.5. The control group. We refer to [**8**] for a discussion about this section. Consider $\Gamma \subset \mathrm{PSL}(3,\mathbb{C})$ a (discrete or not) subgroup which acts on $\mathbb{P}^2_{\mathbb{C}}$ with a point p which is fixed by all of Γ. Choose an arbitrary line ℓ in $\mathbb{P}^2_{\mathbb{C}} \setminus \{p\}$, and notice we have a canonical projection:

$$\pi = \pi_{p,\ell} : \mathbb{P}^2_{\mathbb{C}} \setminus \{p\} \longrightarrow \ell,$$

given by $\pi(x) = \overleftrightarrow{x,p} \cap \ell$. It is clear that this map is holomorphic and it allows us to define a group homomorphism:

$$\Pi = \Pi_{p,\ell} : \Gamma \longrightarrow Bihol(\ell) \cong \mathrm{PSL}(2,\mathbb{C}),$$

by $\Pi(g)(x) = \pi(g(x))$. If we choose another line, say ℓ', one gets similarly a projection $\pi' = \pi_{p,\ell'} : \mathbb{P}^2_{\mathbb{C}} \setminus \{p\} \to \ell'$, and a group homomorphism $\Pi' = \Pi_{p,\ell'} : \Gamma \to \mathrm{PSL}(2,\mathbb{C})$. It is an exercise to see that Π and Π' are equivalent in the sense that there is a biholomorphism $h : \ell \to \ell'$ inducing an automorphism H of $\mathrm{PSL}(2,\mathbb{C})$ such that $H \circ \Pi = \Pi'$. As before, the line ℓ is called *the horizon.*

This leads to Definition 1.7:

DEFINITION 1.7. Let $\Gamma \subset \mathrm{PSL}(3,\mathbb{C})$ be a discrete group as above. We call $\Pi = \Pi_{p,\ell}$ the control morphism (or map) and its image $\Pi(\Gamma) \subset \mathrm{PSL}(2,\mathbb{C})$, is the *control group*. These are well-defined and independent of ℓ up to an automorphism of $\mathrm{PSL}(2,\mathbb{C})$.

The control map and the control group allow us to get information about the dynamics of Γ by looking at a subgroup of $\mathrm{PSL}(2,\mathbb{C})$, which is far easier to handle. The prize we pay is that the control group in $\mathrm{PSL}(2,\mathbb{C})$ may not be discrete.

2. Purely parabolic groups

We now follow [**5**] and look at the discrete subgroups in $\mathrm{PSL}(3,\mathbb{C})$ that, besides the identity, have only parabolic elements. These are called purely parabolic and there are five families of such groups; three of them split into various subfamilies according to their limit set (and their control group, see [**8**]). All of these are elementary.

The simplest purely parabolic groups are cyclic, generated by a parabolic element. As described above, there are three types of such elements in $\mathrm{PSL}(3,\mathbb{C})$, described by the Jordan normal form of their lifts to $\mathrm{SL}(3,\mathbb{C})$. Each of these belongs to a different type of the families we describe below. The first type generates torus groups (see definitions below), the second generates Abelian Kodaira groups and the ellipto-parabolic elements generate elliptic groups.

(i) Elliptic groups. These are the only purely parabolic groups that are not conjugate to subgroups of the Heisenberg group $\mathrm{Heis}(3,\mathbb{C})$ and they are subgroups of fundamental groups of elliptic surfaces. These have limit set a single line. Up to conjugation these groups are of the form:

$$\mathrm{Ell}(W,\mu)=\left\{\begin{bmatrix}\mu(w) & \mu(w)w & 0\\ 0 & \mu(w) & 0\\ 0 & 0 & \mu(w)^{-2}\end{bmatrix} : w\in W\right\},$$

where $W\subset\mathbb{C}$ is an additive discrete subgroup and $\mu : W\to\mathbb{S}^1$ is a group morphism.

(ii) Torus groups. These are subgroups of fundamental groups of complex tori. These are of the form:

$$\mathcal{T}(\mathfrak{L})=\left\{\begin{bmatrix}1 & 0 & a\\ 0 & 1 & b\\ 0 & 0 & 1\end{bmatrix} : (a,b)\in\mathfrak{L}\right\},$$

where $\mathfrak{L}$ is an additive discrete subgroup of $\mathbb{C}^2$. These groups also have a single line as limit set, so they are elementary of the first kind.

(iii) Dual torus groups,

$$\mathcal{T}^*(\mathfrak{L})=\left\{\begin{bmatrix}1 & a & b\\ 0 & 1 & 0\\ 0 & 0 & 1\end{bmatrix} : (a,b)\in\mathfrak{L}\right\}.$$

These split into three types: the first have Kulkarni limit set a complex projective line, so these are elementary of the first kind. The second type have limit set a cone of projective lines over a circle, so these are of the second kind. The third type have all $\mathbb{P}^2_{\mathbb{C}}$ as limit set and they are non-elementary.

(iv) Inoue groups and their extensions. Inoue groups are proper subgroups of fundamental groups of Inoue surfaces. To define these, let $\mathfrak{L}\subset\mathbb{C}^2$ be an additive discrete subgroup and consider a dual torus group.

$$\mathcal{I}=\mathcal{I}(u,v):=\left\langle\begin{bmatrix}1 & u & v\\ 0 & 1 & 0\\ 0 & 0 & 1\end{bmatrix},\ (u,v)\in\mathfrak{L}\right\rangle.$$

Inoue groups are obtained by taking a generator $\mathcal{I}$ as above and adding to it a generator of the form

$$\gamma_1=\gamma_1(x,y,z):=\begin{bmatrix}1 & x+z & y\\ 0 & 1 & z\\ 0 & 0 & 1\end{bmatrix}\quad x,y,z\in\mathbb{C}.$$

The limit set is a cone of lines over a circle, so these are elementary of the second kind.

Then one has the extended Inoue group, which are purely parabolic as well, with limit set all of $\mathbb{P}^2_{\mathbb{C}}$, so they are non-elementary and do not fall within the scope of this article. We refer to [**5**] for details.

(v) Finally one has the Kodaira groups Kod_0, which are Abelian, and their extensions. A Kodaira group is a discrete group in PSL$(3,\mathbb{C})$ such that each element h in the group can be written as:

$$\begin{bmatrix}1 & a & b\\ 0 & 1 & a\\ 0 & 0 & 1\end{bmatrix}.$$

One can show that these are extensions of dual torus groups. Their limit set is a line, so they are elementary of the first kind.

There are five types of extensions $\widetilde{\mathcal{K}_i}$, $i = 1, \cdots, 5$, which are purely parabolic and discrete. These are all non-Abelian and they split into five types according to their limit set and the control group. The first type $\widetilde{\mathcal{K}_1}$ have limit set a projective line, so they are elementary of the first kind; the second type $\widetilde{\mathcal{K}_2}$ have limit set a cone of projective lines over the circle, so they are of the second kind, while the remaining three types have limit set all of $\mathbb{P}^2_{\mathbb{C}}$ and they are non-elementary. We refer to [**5**] for details.

3. Solvable groups

An essential step towards understanding elementary groups in $\mathrm{PSL}(3,\mathbb{C})$ is studying the dynamics of solvable groups in $\mathrm{PSL}(3,\mathbb{C})$. Unlike the classical case of $\mathrm{PSL}(2,\mathbb{C})$, now there is great richness.

DEFINITION 3.1. A group H is called virtually solvable if it contains a finite index subgroups G and subgroups G_i, $i = 0, \ldots, k$, such that:

- $e = G_0 \subset G_1 \subset \ldots \subset G_k = G$, and
- G_{j-1} is normal in G_j, and G_j/G_{j-1} is an abelian group, for $j = 1, \ldots, k$.

For instance, for a subgroup of $\mathrm{PSL}(2,\mathbb{C})$, virtually solvable is equivalent to saying that the group is elementary, *i.e.* its limits set has finite cardinality, see[**11**] and [**20**], or in an equivalent way (via Tit's dichotomy) a Möbius subgroup is non-virtually solvable if it contains either a non-Ciclyc Schottky group or a purely elliptic free group whose rank is at least two, see [**7, 26**] for a general discussion or [**16, 20**] for standard arguments in the one-dimensional case.

One has the following theorem from [**27**]:

THEOREM 3.2. *For subgroups of* $\mathrm{PSL}(3,\mathbb{C})$, *solvable is equivalent to having a finite index subgroup which is conjugate to group of upper triangulable matrices.*

So the following are examples of solvable groups:

(i) All purely parabolic groups.
(ii) Hyperbolic Toral groups (we discuss these later, in Section 7).
(iii) Fundamental groups of Inoue surfaces, see [**2, 5, 11**].

So, after our previous Section 2, to have the complete picture of solvable groups we must look at solvable groups with loxodromic elements. This is done in [**27, 28**]. In those articles the author determined the corresponding limit set and the representations of solvable groups in $\mathrm{PSL}(3,\mathbb{C})$ containing loxodromic element. The main results can be summarized as follows. These summarize Theorems 1.1 to 1.4 in [**28**] and theorems 1 and 2 in [**27**]:

THEOREM 3.3. *The solvable groups in* $\mathrm{PSL}(3,\mathbb{C})$ *can be of three main types:*

(i) *Commutative.*
(ii) *Cyclic extensions of torus groups by a strongly loxodromic element.*
(iii) *Dual torus groups extensions.*

THEOREM 3.4. *If a solvable group* Γ *is commutative, then its limit set consists of either two lines, three lines in general position, or it is a line and a point, and* Γ *can be either triangular or a fake Hopf group:*

- *Diagonal groups:*
$$\{diag(\alpha^n, \alpha^m, 1) : n, m \in \mathbb{Z}\}$$
here $\alpha, \beta \in \mathbb{C}^*$, $|\alpha| \neq 1$ *or* $|\beta| \neq 1$.
- *Fake Hopf:*
$$\left\{ \begin{bmatrix} \mu(w) & w\mu(w) & 0 \\ 0 & \mu(w) & 0 \\ 0 & 0 & \mu(w)^{-2} \end{bmatrix} \right\}$$
where $W \subset \mathbb{C}$ *is a discrete group and* $\mu : W \to \mathbb{C}^*$ *is a group morphism, satisfying some technical conditions, see* [**28**].

THEOREM 3.5. *If the group is an extension of either a torus group or a dual torus group, then its limit set can be a single projective line, or two lines, or else it has infinitely many lines but either it is a cone of lines over a circle or it has four lines in general position. And one has:*

(i) *If the group* Γ *is a cyclic extension of a torus group, then it is a semi-direct product of a torus group and the cyclic group generated by a strongly loxodromic element.*
(ii) *If the group is an extension of a dual torus groups, then it can be:*
   - *An extension by a loxodromic element (of any type); or*
   - *An extension by two loxo-parabolic elements, and in this case the group is a semi-direct product of the dual torus group and the group generated by the two loxodromic elements.*

In the sequel we describe all groups with limit set as stated in this theorem.

4. Groups with limit set exactly one line

In the previous section we described several purely parabolic groups with limit set a single line. Here we follow [**3**] and describe all the discrete groups in PSL(3, ℂ) with limit set a line. In particular, every such group is virtually nilpotent.

Let Γ be complex Kleinian with limit set a line ℓ. Then Γ acts properly and discontinuously on the complement of ℓ. From the above described classification of the elements in PSL(3, ℂ) we known that the elements in Γ are all either elliptic, parabolic or loxoparabolic. If the group contains a loxoparabolic element, then one finds that Γ is conjugate to a group such that every element can be represented by a matrix of the form:
$$\begin{pmatrix} a & 0 & v \\ 0 & d & 0 \\ 0 & 0 & 1 \end{pmatrix},$$
where ℓ becomes the line $\overleftrightarrow{e_1, e_2}$. Moreover, Γ acts as an Euclidean group on the line $\overleftrightarrow{e_1, e_3}$. By well known facts on Euclidean groups (see [**20**]), a is a root of unity of order $1, 2, 3, 4$ or 6, and the rank of Γ considered as acting on $\overleftrightarrow{e_1, e_3}$ is equal to 1 or 2. One can show that a cannot be a root of unity of order $2, 3, 4$ nor 6, otherwise Γ would contain a complex homothety. Finally, the rank of Γ considered as acting on $\overleftrightarrow{e_1, e_3}$ is not equal to 1, otherwise, the Kulkarni limit set of Γ would be equal to two lines.

On the other hand, if the group does not contain any loxoparabolic elements, then there are two cases. If it acts on ℓ without parabolic elements then Γ can be considered as a discrete group of Euclidean isometries of $\mathbb{R}^4$. If some element acts on ℓ as parabolic element then the group can be identified with a group of triangular matrices. Then one can rule out the existence of irrational ellipto-parabolic elements and show that there exists a unipotent subgroup of finite index.

In this way we arrive to the following [**3**, Theorem 1.1]:

THEOREM 4.1. *Let* Γ *be a subgroup of* $\mathrm{PSL}(3,\mathbb{C})$ *such that its Kulkarni limit consists of precisely one complex projective line* ℓ. *Then:*

(i) *If* Γ *does not contain loxoparabolic elements nor an element which acts as a parabolic element on the line* ℓ, *then* Γ *is a group of isometries of* $\mathbb{C}^2$ *and it contains a free abelian normal subgroup of finite index and rank* ≤ 4.

(ii) *If* Γ *does not contain loxoparabolic elements but it does contain an element which acts as a parabolic element on* ℓ, *then* Γ *does not contain any irrational ellipto-parabolic elements and it is a finite extension of a unipotent subgroup that consists of unipotent parabolic maps. Hence it is a finite extension of a group of the form* $\mathbb{Z}, \mathbb{Z}^2, \mathbb{Z}^3, \mathbb{Z}^4$, Δ_k *or* G_k, *where*

$$\Delta_k = \left\langle A, B, C, D : C, D \text{ are central and } [A,B] = C^k \right\rangle, k \in \mathbb{N},$$

$$G_k = \left\langle A, B, C : C \text{ is central and } [A,B] = C^k \right\rangle, k \in \mathbb{N},$$

(iii) *If* Γ *does contain a loxoparabolic element, then* Γ *is isomorphic to the group* $\mathbb{Z} \oplus \mathbb{Z} \oplus \mathbb{Z}_{n_0}$, *where* $n_0 \in \mathbb{N}$ *is arbitrary. The* $\mathbb{Z}_{n_0}$ *summand is a group of complex reflections, while* $\mathbb{Z} \oplus \mathbb{Z}$ *is generated by a loxoparabolic element and another element which can be loxoparabolic or parabolic.*

5. Groups with limit set two lines

We must distinguish two cases:

(a) Groups that have exactly two lines in their limit set. These are elementary of the first kind; and

(b) Groups with more than two lines in the limit set, but only two in general position. In this case, we know from Theorem 1.5 that the total number of lines actually is infinite. These are elementary of the second kind.

For instance, cyclic groups generated by a strongly loxodromic element have exactly two lines in their limit set. On the other hand, for instance, the dual torus groups of the second type, described before, have limit set a cone of projective lines over a circle. In this case, there are infinitely many lines in the limit set, but only two in general position.

Let us describe these elementary groups. As in [4], for every open Γ-invariant set U we denote by $\lambda(U)$ the maximum number of complex projective lines contained in $\mathbb{P}^2_{\mathbb{C}} \setminus U$, and by $\mu(U)$ the maximum number of such lines in general position. If U is the Kulkarni discontinuity region $\Omega_{\mathrm{Kul}}(\Gamma)$, then its complement $\mathbb{P}^2_{\mathbb{C}} \setminus U$ is the limit set $\Lambda_{\mathrm{Kul}} = \Lambda_{\mathrm{Kul}}(\Gamma)$.

THEOREM 5.1. *Let* $\Gamma \subset \mathrm{PSL}(3,\mathbb{C})$ *be a group satisfying* $\mu(\Omega_{\mathrm{Kul}}(\Gamma)) = 2$, *then either the limit set is the union of two lines or it is a pencil of lines over a nowhere dense perfect set. Moreover, one of the following facts occurs:*

(i) *if* $\lambda(\Omega_{\mathrm{Kul}}(\Gamma)) = 2$, *then* Γ *is virtually solvable and contains a loxodromic element, see* [**27, 28**] *for a precise description.*

(ii) *if* $\lambda(\Omega_{\mathrm{Kul}}(\Gamma)) = \infty$, *then* Γ *either is virtually purely parabolic or contains a loxodromic element, and:*

(a) *If the group is virtually purely parabolic then it contains a finite index subgroup conjugate to one of the following groups: a dual torus group of type II, a Kleinian Inoue group, or an extended Kodaira group $\mathcal{K}_2$.*
(b) *If the group contains a loxodromic element, then either Γ is virtually conjugate to a solvable group (described in* [**28**]*), or it is a weakly controllable group whose control group is non-elementary.*

PROOF. Since $\mu(\Omega_{\text{Kul}}(\Gamma)) = 2$, we have that Γ has a global fixed point p, the point where those two lines intersect, and $\lambda(\Omega_{\text{Kul}}(\Gamma))$ is either 2 or ∞ by Theorem 1.1 [**4**]. In fact, will see in the next section that if $\lambda(\Omega_{\text{Kul}}(\Gamma)) = 3$, then also $\mu(\Omega_{\text{Kul}}(\Gamma)) = 3$. And we know from Theorem 1.5 that if $\lambda(\Omega_{\text{Kul}}(\Gamma)) > 3$, then this number actually is ∞.

Also, if $\lambda(\Omega_{\text{Kul}}(\Gamma)) = 2$, then, we claim, that Γ is virtually solvable. To see this notice that the group is weakly semi-controllable. Choose a line ℓ as the "horizon". Then Λ_{Kul} meets ℓ in two points which are invariant under the control group $\Pi(\Gamma)$. Hence $\Pi(\Gamma)$ is virtually diagonalizable and therefore Γ itself is virtually diagonalizable. Hence it is virtually solvable. Then, by Theorem 0.1 in [**5**], we deduce that Γ contains a loxodromic element. Then, by [**27, 28**], its limit set consists of exactly two lines.

In case $\lambda(\Omega_{\text{Kul}}(\Gamma)) = \infty$, we have that Γ is either virtually purely parabolic or it contains a loxodromic element. In the first case by part (b) item (2) of Theorem 0.1 in [**5**] we have that Γ contains a finite index subgroup conjugate to one of the following groups: a dual torus group of type II, a Kleinian Inoue group, or an extended Kodaira group $\mathcal{K}_2$, thus, in this case, the limit set is a cone of lines over a Euclidean circle, see section 2 of [**5**]. Now the proof splits into the following cases:

CASE 1. The group Γ is virtually solvable. In this case by Theorems 1.1, 1.2, 1.3, 1.4 in [**28**], the limit set of Γ is either two lines or a cone of lines over an Euclidean circle.

CASE 2. The group Γ is non-virtually solvable. In this case it is clear that the control group $\Pi_{p,\ell}\Gamma$ is non-elementary, here ℓ is any line not containing p, and

$$\Lambda_{\text{Kul}}(\Gamma) = \bigcup_{z \in \Lambda(\Pi(\Gamma))} \overleftrightarrow{p, z}.$$

That $\Pi_{p,\ell}\Gamma$ is non-elementary follows from the fact that, essentially by Borel's fixed point theory, a subgroup of PSL(2, ℂ) is elementary if and only if it is virtually solvable (see [**11**]).

□

6. Groups with limit set three lines

We now describe the subgroups of PSL(3, ℂ) with three lines in their limit set. As in the previous section, we denote by $\lambda = \lambda(\Omega_{\text{Kul}}(\Gamma))$ the maximum number of complex projective lines contained in Λ_{Kul}, and by $\mu = \mu(\Omega_{\text{Kul}}(\Gamma))$ the maximum number of such lines in general position. Then there are groups with $\mu = \lambda = 3$, and there are groups with $\lambda = \infty$ but $\mu = 3$. In the case $\lambda = \infty$ the groups in question are "suspensions" of non-elementary Kleinian groups in PSL(2, ℂ), see Chapter 5 in [**8**]:

THEOREM 6.1. *Let $\Gamma \subset$ PSL(3, ℂ) be a group for which $\mu = 3$. Then λ can be either 3 or ∞ and:*

(i) *If $\lambda = 3$, then these lines are in general position and Γ contains a finite index subgroup conjugate to:*

$$G_0 = \{diag[\alpha^n, \beta^m, 1] : n, m \in \mathbb{Z}\},$$

where $\alpha, \beta \in \mathbb{C}^*$ are non-unitary complex numbers.

(ii) *If $\lambda = \infty$, then the limit set is a cone of lines plus a non-concurrent line and Γ contains a finite index subgroup conjugate to:*

$$H_{\Sigma,\rho,\alpha} = \left\{ \begin{bmatrix} \alpha^n \rho([g]) & 0 \\ 0 & g \end{bmatrix} : n \in \mathbb{Z}, g \in \mathrm{SL}(2,\mathbb{C}),\ [g] \in \Sigma \right\},$$

where $\alpha \in \mathbb{C}^*$ is a non-unitary complex number, Σ is a non-elementary discrete group in $\mathrm{PSL}(2,\mathbb{C})$ with non-empty discontinuity region, and ρ is a function $\Sigma \to \mathbb{C}^*$. In this case

$$\Lambda_{\mathrm{Kul}}(H_{\Sigma,\rho,\alpha}) = \overleftrightarrow{[e_2],[e_3]} \cup \bigcup_{p \in \Lambda(\Sigma)} \overleftrightarrow{[e_1],p}.$$

Proof. We know from statement (iii) in Theorem 1.5 that if $\lambda = 3$ then $\mu = 3$. That is, if the group has exactly three lines in its limit set, then these lines are in general position.

Now we assume $\mu = 3$ and we choose three distinct lines in general position contained in $\Lambda_{\mathrm{Kul}}(\Gamma)$, say ℓ_1, ℓ_2, ℓ_3. We define

$$P = \{p_{ij} = \ell_i \cap \ell_j : 1 \leq i < j \leq 3\},$$

then any line contained in $\Lambda_{\mathrm{Kul}}(\Gamma)$ must intersect P.

We will say $p \in P$ is a vertex whenever there are infinite lines in $\Lambda_{\mathrm{Kul}}(\Gamma)$ passing through it. If the set of vertices is empty, then by Proposition 5.6 in [**4**] the set P is a Γ-invariant set. Then $H = Isot(P, \Gamma)$ is a finite index subgroup of Γ. Moreover, $\Lambda_{\mathrm{Kul}}(H) = \Lambda_{\mathrm{Kul}}(\Gamma)$ and H is conjugate to a group G_0 where every element has a diagonal lift. So by Theorem 1.1 in [**28**] we conclude $G_0 = \{diag[\alpha^n, \beta^m, 1] : n, m \in \mathbb{Z}\}$ for some $\alpha, \beta \in \mathbb{C}^*$ non-unitary complex numbers.

If the set of vertices is non-empty then by the proof of Proposition 6.5 in [**4**] we must have that this set is Γ-invariant. Observe that in case the number of vertices is at least two, we must have four lines in general position (compare with Proposition 6.5 in [**4**]) which is not possible, therefore we have exactly one vertex, say v, which is Γ-invariant. Let $\ell \in \{\ell_1, \ell_2, \ell_3\}$ be a line not containing v. Since $\mu(\Omega_{\mathrm{Kul}}(\Gamma)) = 3$, we conclude that ℓ is Γ-invariant. After conjugating with a projective transformation, if necessary, let us assume that $v = [e_1]$ and $\ell = \overleftrightarrow{[e_2],[e_3]}$, as in Section 1.5 take $\pi = \pi_{v,\ell}$ and $\Pi = \Pi_{v,\ell}$, now we claim:

Claim 1. If $\Pi(\Gamma)$ is non-virtually solvable, then $\Pi(\Gamma)$ is discrete. On the contrary let us assume that $\Pi(\Gamma)$ is non-discrete, thus by Theorem 1 and Proposition 12 in [**16**] the principal component of $\overline{\Pi(\Gamma)}$ is $\mathrm{PSL}(2,\mathbb{C})$ or conjugate to $\mathrm{SO}(3)$ or $\mathrm{PSL}(2,\mathbb{R})$. In the first two cases, the orbit of any line passing through v is dense in ℓ, and this is not possible. In the latter case, let $(g_n) \subset \Pi(G)$ be a sequence such that $\lim_{n\to\infty} g_n = Id$ and $h_n = [g_1, g_n] \neq Id$ for $n \geq 2$. If $G_m \in \Gamma$ satisfies $\Pi(G_m) = g_m$, then a straightforward computation shows $\lim_{n\to\infty}[G_1, G_n] = Id$, which is a contradiction.

Claim 2. The group Γ cannot be virtually solvable. On the contrary, let us assume that Γ is solvable, then by Theorem 0.1 in [**5**] the group Γ contains a loxodromic element. By Theorems 1.1, 1.2, 1.3 and 1.4 in [**28**], there is $W \subset \mathbb{C}$ a discrete additive subgroup with $rank(W) \leq 2$ and $\alpha, \beta \in \mathbb{C}^*$ such that $\alpha \neq 1$ and $\alpha\beta^2 \notin \mathbb{R}$ and Γ is conjugate to:

$$\left\{ \begin{bmatrix} 1 & 0 & 0 \\ 0 & 1 & w \\ 0 & 0 & 1 \end{bmatrix} : w \in W \right\} \rtimes \left\{ \begin{bmatrix} \alpha^{2n}\beta^n & 0 & 0 \\ 0 & \alpha^n\beta^{2n} & 0 \\ 0 & 0 & 1 \end{bmatrix} : n \in \mathbb{Z} \right\}.$$

Now a simple computation shows W is non-discrete, which is a contradiction.

By the previous claims, we conclude that $\Pi(\Gamma)$ is a discrete non-elementary with a non-empty ordinary region. Now the rest of the proof is trivial. □

7. Groups with limit set four lines in general position lines

In this section we provide an algebraic characterization of the subgroups of $\mathrm{PSL}(3,\mathbb{C})$ with exactly four lines lines in general position in their limit set Λ_{Kul}. Of course that by Theorem 1.5 this implies that Λ_{Kul} actually has infinitely many lines, but only four in general position. The basic reference is [**2**].

We recall first that an element $A \in \mathrm{SL}(2,\mathbb{Z})$ is a hyperbolic toral automorphism if none of its eigenvalues has norm 1 (see [**17**]).

Definition 7.1. A subgroup Γ of $\mathrm{PSL}(3,\mathbb{C})$ is a *hyperbolic toral group* if it is conjugate to a group of the form:

$$\Gamma_A = \left\{ \begin{pmatrix} A^k & b \\ 0 & 1 \end{pmatrix} \,\middle|\, b = \begin{pmatrix} b_1 \\ b_2 \end{pmatrix},\ b_1, b_2, k \in \mathbb{Z} \right\}$$

where $A \in \mathrm{SL}(2,\mathbb{Z})$ is a hyperbolic toral automorphism.

One has:

Theorem 7.2. *Let $\Gamma \subset \mathrm{PSL}(3,\mathbb{C})$ be a discrete group. The maximum number of complex lines in general position contained in its Kulkarni limit set is equal to four if, and only if, Γ contains a hyperbolic toral group whose index is at most 8. [In this case the limit set of the subgroup coincides with that of Γ, and the discontinuity region has four components each one diffeomorphic to $\mathbb{H} \times \mathbb{H}$, where $\mathbb{H}$ denotes the half open plane in $\mathbb{C}$.]*

To prove the theorem above, the first step is showing that a toral group Γ_A as above is the subgroup of $\mathrm{PSL}(3,\mathbb{Z})$ generated by the unipotent parabolic elements P_1, P_2 and the strongly loxodromic element L given by:

$$P_1 = \begin{pmatrix} 1 & 0 & 0 \\ 0 & 1 & 1 \\ 0 & 0 & 1 \end{pmatrix},\ P_2 = \begin{pmatrix} 1 & 0 & 1 \\ 0 & 1 & 0 \\ 0 & 0 & 1 \end{pmatrix},\ L = \begin{pmatrix} A & 0 \\ 0 & 1 \end{pmatrix}.$$

Therefore Γ_A is discrete. The dynamical properties of the hyperbolic toral group are used to show that $\Lambda_{\mathrm{Kul}}(L)$ is contained in Kulkarni limit set of Γ_A. Also, a straightforward computation shows that $\Lambda_{\mathrm{Kul}}(P_1)$ is contained in $\Lambda_{\mathrm{Kul}}(\Gamma_A)$. Thus,

$$\Lambda_{\mathrm{Kul}}(\Gamma_A) = \overline{\bigcup_{\gamma \in \Gamma_A} \Lambda_{\mathrm{Kul}}(\gamma)},$$

because both sets in this equation contain three lines in general position (see Theorem 1.2 in [**1**]).

Now, it is possible to compute the sets $\Lambda_{\text{Kul}}(\gamma)$ for every $\gamma \in \Gamma_A$. One finds that $\Lambda_{\text{Kul}}(\Gamma_A)$ can be described as a union of two pencils of lines over two distinct circles, with vertices two distinct points, which share the line determined by the two vertices. Therefore, the maximum number of lines in general position contained in $\Lambda_{\text{Kul}}(\Gamma_A)$ is equal to four. Moreover, this description shows that the Kulkarni discontinuity region of Γ_A is a disjoint union of four copies of $\mathbb{H} \times \mathbb{H}$, where $\mathbb{H}$ denotes the upper half plane in $\mathbb{C}$.

Conversely, if we assume that the maximum number of lines in general position contained in the Kulkarni limit set of a group Γ is equal to four, then one shows that there are two distinguished points (called vertices) contained in infinitely many lines in this limit set. The subgroup which stabilizes these two vertices is denoted by Γ_0 and it has index at most two in Γ. Furthermore, it can be proved that $\Omega_{\text{Kul}}(\Gamma_0)$ has four components, each one diffeomorphic to a copy of $\mathbb{H} \times \mathbb{H}$. Finally, the subgroup of Γ_0 stabilizing each of these components has finite index (at most four) in Γ_0, and it can be shown that it is a hyperbolic toral group.

A description of the quotient spaces $\Omega_{\text{Kul}}(\Gamma_A)/\Gamma_A$ is given by the following theorem (see [**6**].)

THEOREM 7.3. *If Γ_A is a hyperbolic toral group then*

(i) *The group Γ_A is isomorphic to a lattice of the group Sol (see Definition 7.4).*
(ii) *The quotient space $\Omega_{\text{Kul}}(\Gamma_A)/\Gamma_A$ is a disjoint union of four copies of*
$$(\text{Sol}/\Gamma_A) \times \mathbb{R}.$$

DEFINITION 7.4. The group Sol is $\mathbb{R}^3$ equipped with the operation
$$\left(\begin{pmatrix} x_1 \\ y_1 \end{pmatrix}, t_1\right)\left(\begin{pmatrix} x_2 \\ y_2 \end{pmatrix}, t_2\right) = \left(\begin{pmatrix} x_1 + e^{t_1}x_2 \\ y_1 + e^{-t_1}y_2 \end{pmatrix}, t_1 + t_2\right)$$

The proof of Theorem 7.3 is done by giving an explicit smooth foliation of $\mathbb{H} \times \mathbb{H}$ where the leaves are diffeomorphic to Sol, and by showing that $\mathbb{H} \times \mathbb{H}$ is diffeomorphic to Sol $\times \mathbb{R}$ by a Γ_A-equivariant diffeomorphism.

It can be shown that Γ_A is isomorphic to $\mathbb{Z}^2 \rtimes_A \mathbb{Z}$, and this group is determined by the conjugacy class in $\text{GL}(2,\mathbb{Z})$ of A (see [**23**] pages 96,97). It follows from Theorem 7.2 that there is a countable number of nonisomorphic Complex Kleinian groups such that the maximum number of lines in general position contained in its Kulkarni limit set is equal to four.

8. Groups with limit set having an isolated point

Let us suppose now that $\Gamma \subset \text{PSL}(3,\mathbb{C})$ is a group with at least an isolated point in its limit set. Then the results explained in this article show that Γ must have only one line in its limit set, for otherwise Λ_{Kul} is a union of lines.

Let ℓ be the unique line in $\Lambda_{\text{Kul}}(\Gamma)$ and let $p \in \Lambda_{\text{Kul}}(\Gamma)$ be an isolated point. Then ℓ is an invariant set and we look at the action of Γ on ℓ to deduce from it consequences on the global dynamics of Γ.

We note that this is somehow similar to the controllable groups, where we have a global fix point and a line L that we call the horizon, which is not invariant but we get an action on it. Then we know a lot about the action of the group on $\mathbb{P}^2_{\mathbb{C}}$ by looking at the induced action on L. In the case we now envisage, we have the invariant line ℓ and the action on it says a lot about the dynamics of the group on $\mathbb{P}^2_{\mathbb{C}}$. There are two possibilities as the action of Γ on ℓ can be either:

a. Virtually Solvable.
b. Non- Virtually solvable.

The idea of the proof consists in showing first that the non-virtually solvable case cannot happen, so the action must be virtually solvable. This uses the Tits Alternative.

In the solvable case, the dynamics implies that on the invariant line ℓ we must have either one or two fixed points. Then one shows that in the first case, when there is only one fixed point, one actually has an invariant flag and up to conjugation the elements are all of the form

$$\begin{bmatrix} \alpha & 0 & 0 \\ 0 & \beta & \delta \\ 0 & 0 & \gamma \end{bmatrix}.$$

In the second case we find that, up to conjugation, all the elements are diagonal, and the group consists of lodrodromic elements that are screws (see Section 1.3), so the limit set of each element consists of a point and a line. Then we show that the limit set of the whole group is a point and a line.

The above conisderations lead to the following:

THEOREM 8.1. *Let $\Gamma \subset \mathrm{PSL}(3,\mathbb{C})$ be a group with at least an isolated point in its limit set. Then the limit set consists of exactly one line and one point, and Γ is conjugate to either:*

(i)
$$G_0 = \{\mathrm{diag}[\alpha^n, e^{2\pi i\theta}, e^{-2\pi i\theta}] : n, m \in \mathbb{Z}\}$$
where $\theta \in \mathbb{R}$ and $\alpha \in \mathbb{C}^*$ is non-unitary complex number.

(ii) *or*
$$H_0 = \left\{ \begin{bmatrix} \mu(w)^{-2} & 0 & 0 \\ 0 & \mu(w) & w\mu(w) \\ 0 & 0 & \mu(w) \end{bmatrix} : w \in W \right\}$$
where $W \subset \mathbb{C}$ is a discrete additive group and $\mu : (\mathbb{C},+) \to (\mathbb{C}^*,\cdot)$ is a group morphism satisfying that for every sequence of distinct elements $(w_n) \subset W$, the limit $\lim_{n\to\infty} \mu(w_n)$ is either 0 or ∞.

PROOF. By Theorem 1.2 in [**2**] we have $\mu(\Omega_{\mathrm{Kul}}(\Gamma)) \leq 2$, and from Theorem 5.1 we conclude $\mu(\Omega_{\mathrm{Kul}}(\Gamma)) = 1$. On the other hand, by Theorem 0.1 in [**5**], the group Γ contains a Loxodromic element. Let ℓ be the unique line in $\Lambda_{\mathrm{Kul}}(\Gamma)$, $p \in \Lambda_{\mathrm{Kul}}(\Gamma)$ be an isolated point and $\gamma \in \Gamma$ a loxodromic element. We must consider the following cases:

CASE 1. The action of Γ restricted to ℓ is given by a virtually solvable group. In this case the group Γ is virtually solvable and the theorem follows from Claim 2 and Theorems 1,1, 1.2, 1.3 and 1.4 in [**28**], see also Section 1 above .

CASE 2. The action of Γ restricted to ℓ is given by a non- virtually solvable group. Under this assumption we claim:

CLAIM. There are $g_1, g_2 \in \Gamma$ such that the group generated by $\{\gamma_1|_\ell, \gamma_2|_\ell\}$ is a rank two free group which is either purely loxodromic or purely elliptic. If the action of Γ on ℓ is given by a discrete group H, then it is well know that H contains a rank two Shottky group, see Chapter X in [**20**], these will do the job. If the action on ℓ is given by a non discrete group G, thus by Theorem 1 and Proposition 12 in [**16**] the principal component of $\overline{G}$ is $\mathrm{PSL}(2,\mathbb{C})$, or conjugate to $\mathrm{SO}(3)$ or $\mathrm{PSL}(2,\mathbb{R})$. In any such case we deduce that G contains a non-virtualy solvable finitelly generated subgroup, as a consequence of Tit's alternative we deduce that G contains a rank two free subgroup which is either purely loxodromic or purely elliptic. Which proves the claim.

Now let g_1, g_2 be the elements given by the previous claim. After conjugating with a projective transformation we can assume that $\ell = \overleftrightarrow{[e_2],[e_3]}$ so $h = [g_1, g_2]$ is a loxodromic element and

$$h = \begin{bmatrix} 1 & 0 & 0 \\ c & d & e \\ f & g & h \end{bmatrix}$$

where $dh - eg = 1$ and $(d+h)^2 \in \mathbb{C} - \{4\}$. Thus h is either an elliptic element of infinite order, which cannot happen by the discreteness of Γ, or a strongly loxodromic element whose attracting and repelling points are contained in ℓ. We conclude that the attracting and repelling lines of h are distinct from ℓ. On the other hand, from the λ-Lemma (see Lemma 7.3.6 in [**8**]) we know that the attracting and repelling line of h are both contained in $\Lambda_{\mathrm{Kul}}(\Gamma)$, which is a contradiction.

□

References

[1] W. d. J. Barrera Vargas, A. C. Cordero, and J. P. N. Carrillo, *The limit set of discrete subgroups of* PSL(3, ℂ), Math. Proc. Cambridge Philos. Soc. **150** (2011), no. 1, 129–146, DOI 10.1017/S0305004110000423. MR2739076

[2] W. Barrera, A. Cano, and J. P. Navarrete, *Subgroups of PSL*(3, ℂ) *with four lines in general position in its limit set*, Conform. Geom. Dyn. **15** (2011), 160–176, DOI 10.1090/S1088-4173-2011-00231-1. MR2846306

[3] W. Barrera, A. Cano, and J. P. Navarrete, *One line complex Kleinian groups*, Pacific J. Math. **272** (2014), no. 2, 275–303, DOI 10.2140/pjm.2014.272.275. MR3284888

[4] W. Barrera, A. Cano, and J. Navarrete, *On the number of lines in the limit set for discrete subgroups of* PSL(3, ℂ), Pacific J. Math. **281** (2016), no. 1, 17–49, DOI 10.2140/pjm.2016.281.17. MR3459965

[5] W. Barrera, A. Cano, J. P. Navarrete, and J. Seade, *Discrete parabolic groups in* PSL(3, ℂ), Linear Algebra Appl. **653** (2022), 430–500, DOI 10.1016/j.laa.2022.08.027. MR4477082

[6] W. Barrera, R. Garcia-Lara, and J. Navarrete, *Three-dimensional Sol manifolds and complex Kleinian groups*, Pacific J. Math. **294** (2018), no. 1, 1–18, DOI 10.2140/pjm.2018.294.1. MR3743363

[7] E. Breuillard and T. Gelander, *A topological Tits alternative*, Ann. of Math. (2) **166** (2007), no. 2, 427–474, DOI 10.4007/annals.2007.166.427. MR2373146

[8] A. Cano, J. P. Navarrete, and J. Seade, *Complex Kleinian groups*, Progress in Mathematics, vol. 303, Birkhäuser/Springer Basel AG, Basel, 2013, DOI 10.1007/978-3-0348-0481-3. MR2985759

[9] A. Cano, J. R. Parker, and J. Seade, *Action of* ℝ*-Fuchsian groups on* $\mathbb{CP}^2$, Asian J. Math. **20** (2016), no. 3, 449–473, DOI 10.4310/AJM.2016.v20.n3.a3. MR3528828

[10] A. Cano and J. Seade, *On the equicontinuity region of discrete subgroups of* PU$(1, n)$, J. Geom. Anal. **20** (2010), no. 2, 291–305, DOI 10.1007/s12220-009-9107-6. MR2579511

[11] A. Cano and J. Seade, *On discrete groups of automorphisms of* $\mathbb{P}^2_{\mathbb{C}}$, Geom. Dedicata **168** (2014), 9–60, DOI 10.1007/s10711-012-9816-z. MR3158029

[12] H. Cartan and S. Eilenberg, *Homological algebra*, Princeton Landmarks in Mathematics, Princeton University Press, Princeton, NJ, 1999. With an appendix by David A. Buchsbaum; Reprint of the 1956 original. MR1731415

[13] C. Series, *A crash course on Kleinian groups*, Rend. Istit. Mat. Univ. Trieste **37** (2005), no. 1-2, 1–38 (2006). MR2227047

[14] L. J. Corwin and F. P. Greenleaf, *Representations of nilpotent Lie groups and their applications. Part I: Basic theory and examples*, Cambridge Studies in Advanced Mathematics, vol. 18, Cambridge University Press, Cambridge, 1990. MR1070979

[15] J. P. Fillmore and J. Scheuneman, *Fundamental groups of compact complete locally affine complex surfaces*, Pacific J. Math. **44** (1973), 487–496. MR328132

[16] L. Greenberg, *Discrete subgroups of the Lorentz group*, Math. Scand. **10** (1962), 85–107, DOI 10.7146/math.scand.a-10515. MR141731

[17] A. Katok and B. Hasselblatt, *Introduction to the modern theory of dynamical systems*, Encyclopedia of Mathematics and its Applications, vol. 54, Cambridge University Press, Cambridge, 1995. With a supplementary chapter by Katok and Leonardo Mendoza, DOI 10.1017/CBO9780511809187. MR1326374

[18] R. S. Kulkarni, *Groups with domains of discontinuity*, Math. Ann. **237** (1978), no. 3, 253–272, DOI 10.1007/BF01420180. MR508756

[19] A. Marden, *Outer circles*, Cambridge University Press, Cambridge, 2007. An introduction to hyperbolic 3-manifolds, DOI 10.1017/CBO9780511618918. MR2355387

[20] B. Maskit, *Kleinian groups*, Grundlehren der mathematischen Wissenschaften [Fundamental Principles of Mathematical Sciences], vol. 287, Springer-Verlag, Berlin, 1988. MR959135

[21] P. J. Myrberg, *Untersuchungen Über die Automorphen Funktionen Beliebig Vieler Variablen* (German), Acta Math. **46** (1925), no. 3-4, 215–336, DOI 10.1007/BF02564065. MR1555203

[22] J.-P. Navarrete, *The trace function and complex Kleinian groups in* $\mathbb{P}^2_{\mathbb{C}}$, Internat. J. Math. **19** (2008), no. 7, 865–890, DOI 10.1142/S0129167X08004868. MR2437075

[23] P. de la Harpe, *Topics in geometric group theory*, Chicago Lectures in Mathematics, University of Chicago Press, Chicago, IL, 2000. MR1786869

[24] J. G. Ratcliffe, *Foundations of hyperbolic manifolds*, Graduate Texts in Mathematics, vol. 149, Springer-Verlag, New York, 1994, DOI 10.1007/978-1-4757-4013-4. MR1299730

[25] T. Suwa, *Compact quotient spaces of* $\mathbf{C}^2$ *by affine transformation groups*, J. Differential Geometry **10** (1975), 239–252. MR369745

[26] J. Tits, *Free subgroups in linear groups*, J. Algebra **20** (1972), 250–270, DOI 10.1016/0021-8693(72)90058-0. MR286898

[27] M. Toledo-Acosta, *The dynamics of solvable subgroups of* PSL(3, $\mathbb{C}$), Bull. Braz. Math. Soc. (N.S.) **53** (2022), no. 1, 127–171, DOI 10.1007/s00574-021-00254-9. MR4379427

[28] M. Toledo-Acosta, *Representations of solvable subgroups of* PSL(3, $\mathbb{C}$), Bull. Braz. Math. Soc. (N.S.) **54** (2023), no. 4, Paper No. 55, 28, DOI 10.1007/s00574-023-00372-6. MR4654924

Facultad de Matemáticas, Universidad Autónoma de Yucatán, Anillo Periférico Norte Tablaje Cat 13615, Mérida, Yucatán, México
Email address: `bvargas@uady.mx`

Instituto de Matemáticas UNAM , Av. Universidad S/N, Col. Lomas de Chamilpa, C.P. 62210, Cuernavaca, Morelos, México
Email address: `angelcano@im.unam.mx`

Facultad de Matemáticas, Universidad Autónoma de Yucatán, Anillo Periférico Norte Tablaje Cat 13615, Mérida, Yucatán, México
Email address: `jp.navarrete@uady.mx`

Instituto de Matemáticas UNAM , Av. Universidad S/N, Col. Lomas de Chamilpa, C.P. 62210, Cuernavaca, Morelos, México
Email address: `jseade@im.unam.mx`

Contemporary Mathematics
Volume **811**, 2025
https://doi.org/10.1090/conm/811/16236

Quasilocal mass and angular momentum

Po-Ning Chen, Mu-Tao Wang, Ye-Kai Wang, and Shing-Tung Yau

It is my pleasure to give a talk in honor of my good friend Ravi Kulkarni. We have known each other since 1971, and I published a paper on understanding his famous thesis on curvature preserving diffeomorphism.

I visited India two times at Ravis invitation. They were great experiences. Through these trips, I learned much more about the deep culture of India. The first time, my wife went with me to Harish Chandra Institute, where Ravi was the director. My wife and I talked about this wonderful experience years later. The second time, I went alone to Mumbai. It was an eventful trip. I prepare a lecture concerning the history of mathematics, highlighting the part on the interaction between India and China. Ravi also came to visit China at my invitation several times.

Time passed by fast. We are celebrating his 80th birthday. I would have liked to be at his birthday party, but unfortunately, the pandemic made it impossible. I like to contribute this talk to express our friendship and say happy birthday to him!

This talk is about mathematical general relativity and is based on joint work with Po-Ning Chen, Jordan Keller, Mu-Tao Wang, and Ye-Kai Wang. The question is related to quasilocal mass, and angular momentum at null infinity, measuring the angular momentum carried away by the gravitational waves. The supertranslation ambiguity causes significant difficulty in adequately defining angular momentum at null infinity. We will discuss how to resolve this using the theory of quasilocal conserved quantities.

1. From Newtonian physics to General Relativity

Newton's equation for gravity is

$$\Delta u = \rho$$

where u is the gravitational potentials and ρ is the matter density.

Minkowski's geometrization of special relativity is given by the tensor:

$$\mathrm{d}s^2 = -\mathrm{d}t^2 + \mathrm{d}x^2 + \mathrm{d}y^2 + \mathrm{d}z^2$$

where photons travel along null geodesics and particles with non-zero mass travel along timelike geodesics. This led to the radical change of the concept of gravity to the dynamics of the geometry of four-dimensional spacetime.

Based on the work of Minkowski and the equivalence principle, Einstein was looking for a tensor $T_{\mu\nu}$, which should be considered as gravitational potential in spacetime.

In general relativity, spacetime is governed by Einstein's equation. The equation reads

$$R_{\mu\nu} - \frac{R}{2} g_{\mu\nu} = T_{\mu\nu}$$

where $g_{\mu\nu}$ is the gravitational potential for the spacetime, defined as a metric, $R_{\mu\nu}$ is the Ricci curvature tensor, obtained by differentiating the potential two times and contracted suitably and R is the trace of $R_{\mu\nu}$.

The left-hand side of the Einstein equation can be interpreted as the laplacian of the potential. On the other side of the equation, $T_{\mu\nu}$ is the stress-energy tensor measuring the distribution of matter in spacetime.

The Einstein equation relates gravity with matters similar to Newton's equation. In spacetime, particles with mass travel along timelike geodesics, while photons travel along null geodesics. Einstein's equation shows that the Ricci curvature of the spacetime metric accounts for matter distribution, but the gravitational force's content rests on the metric's full curvature tensor.

2. Mass and angular momentum in classical mechanics and special relativity

In classical mechanics, the angular momentum of a particle is defined as

$$J = m\mathbf{r} \times \mathbf{r}'.$$

If we translate the origin (or the coordinate system) from O to $\widetilde{O}$, the angular momentum J gets shifted by the linear momentum $\mathbf{p}$

$$J_{\widetilde{O}} = m(\mathbf{r} - \mathbf{a}) \times \mathbf{r}' = J_O - \mathbf{a} \times \mathbf{p}.$$

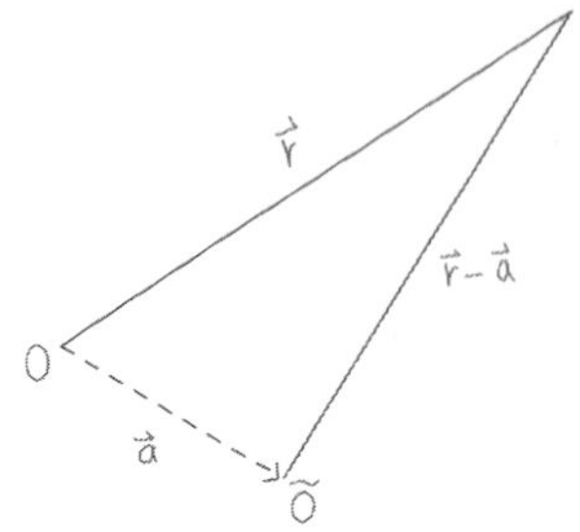

FIGURE 1. Angular momentum gets shifted by the linear momentum

In special relativity, the continuous symmetry of $\mathbb{R}^{3,1}$ consists of

$$\frac{\partial}{\partial t}, \quad \frac{\partial}{\partial x^i}, \quad x^i\frac{\partial}{\partial x^j} - x^j\frac{\partial}{\partial x^i}, i < j, \quad t\frac{\partial}{\partial x^i} + x^i\frac{\partial}{\partial t}$$

which generates the time translation, spatial translations, rotations, and Lorentz transformations respectively.

Noether's Theorem states that each continuous symmetry corresponds to a conserved quantity. For a particle γ in $\mathbb{R}^{3,1}$, we have

energy	$e = \langle \gamma', \frac{\partial}{\partial t} \rangle$
linear momentum	$p_i = \langle \gamma', \frac{\partial}{\partial x^i} \rangle$
angular momentum	$J_{ij} = \langle \gamma', x^i \frac{\partial}{\partial x^j} - x^j \frac{\partial}{\partial x^i} \rangle, i < j$
center of mass	$C_i = \langle \gamma', t \frac{\partial}{\partial x^i} + x^i \frac{\partial}{\partial t} \rangle$

If we translate the coordinate system $t \mapsto t + \alpha^0, x^i \mapsto x^i + \alpha^i$, the energy and linear momentum are the same but the angular momentum and center of mass get shifted

$$J'_{ij} = J_{ij} + \alpha_i p_j - \alpha_j p_i, i < j \tag{2.1}$$

$$C'_i = C_i + \alpha^0 p_i + \alpha^i e \tag{2.2}$$

3. Motivation for quasilocal mass and angular momentum

A fundamental difficulty with a local notion of mass in general relativity is that there is no density for gravitation, by Einstein's equivalence principle. Such a density would depend upon the first-order differentiation of the spacetime metric, but a choice of normal coordinates renders such a measurement trivial at a chosen point.

Instead, we associate conserved quantities to a spacetime region enclosed by a two-surface. There are many reasons to search for such a concept. Many important statements in general relativity make sense only with a good definition and understanding of quasilocal mass. For example, it allows us to discuss the gravitational binding energy of two bodies rotating around one another.

The importance of such a notion was evident to Penrose, who listed the search for a definition of quasilocal mass and a definition of quasilocal angular momentum as his number one and two problems in classical general relativity in 1982 [**28**].

In addition, a good definition of quasilocal mass should help to control the dynamics of the gravitational field. Such control would complement that of the energy method in hyperbolic equations, which encounters severe difficulties in the stability analysis of the Kerr metric.

In the following, we will focus on the Wang-Yau quasilocal [**37,38**] associated to a twodimensional spacetime surface, which is a topological two-sphere (the boundary of a threedimensional spacetime region). The quasilocal mass depends upon both the extrinsic and intrinsic geometry of the surface.

We suppose the surface is spacelike, i.e., the induced metric σ is Riemannian. In addition, we assume that the surface has a spacelike mean curvature vector field H, such that $|H| > 0$. Moreover, we consider the connection one-form of the surface's normal bundle by

$$\alpha_H(\cdot) := g\left(\nabla^N_{(\cdot)} \frac{J}{|H|}, \frac{H}{|H|}\right)$$

where (N, g) denotes the ambient spacetime and J is the future timelike vector given by reflection of H through the incoming light cone.

The Wang-Yau quasilocal mass associate a mass to a 2 -surface Σ with the physical data (σ, H, α_H). The main motivation is that surfaces in the Minkowski space contain no gravitational energy. In order to anchor such a background reference, one solves the optimal isometric embedding equation, which gives an embedding of Σ into the Minkowski spacetime with the image surface Σ_0 that has the same induced metric as Σ, i.e., σ.

One then compares the extrinsic geometries of Σ and Σ_0 and evaluates the quasilocal mass from $\sigma, |H|, |H_0|, \alpha_H$, and α_{H_0}.

4. Wang-Yau quasilocal mass and its properties

Given a physical surface Σ with physical data $(\sigma, |H|, \alpha_H)$, we consider an isometric embedding $X : \Sigma \to \mathbb{R}^{3,1}$ of σ into the Minkowski space. Let Σ_0 be the image $X(\Sigma)$ and T be a future timelike unit Killing field of $\mathbb{R}^{3,1}$ and define $\tau = -\langle X, T\rangle$.

On $\Sigma_0 = X(\Sigma)$ in $\mathbb{R}^{3,1}$, we extract the reference data (H_0, α_{H_0}) and introduce the quasilocal energy density and quasilocal momentum density

$$\rho = \frac{\sqrt{|H_0|^2 + \frac{(\Delta\tau)^2}{1+|\nabla\tau|^2}} - \sqrt{|H|^2 + \frac{(\Delta\tau)^2}{1+|\nabla\tau|^2}}}{\sqrt{1+|\nabla\tau|^2}}$$

$$j = \rho\nabla\tau - \nabla\left[\sinh^{-1}\left(\frac{\rho\,\mathrm{div}(\nabla\tau)}{|H_0|\,|H|}\right)\right] - \alpha_{H_0} + \alpha_H$$

The optimal embedding equation is

$$\begin{cases} \langle dX, dX\rangle & = \sigma \\ \nabla^a j_a & = 0 \end{cases}$$

For a solution of the embedding equation, the quasilocal mass is then

$$E(\Sigma, X, T) = \frac{1}{8\pi}\int_\Sigma \rho$$

A prototype form of the quasilocal mass (due to Brown-York [7, 8], Liu-Yau [23, 24] and Kijowski [**22**]) is

$$\frac{1}{8\pi}\int_\Sigma |H_0| - \frac{1}{8\pi}\int_\Sigma |H|$$

The positivity is proved by Shi-Tam [**34**]and Liu-Yau [**23, 24**]. However, for a surface in the Minkowski spacetime, the above expression may not be zero.

The optimal embedding equation gives the necessary correction so that the Wang-Yau quasilocal mass is positive in general, and zero for surfaces in the Minkowski spacetime.

The embedded surface Σ_0 is the "unique" surface in the Minkowski spacetime that best matches the physical surface Σ. If the original surface Σ happens to be a surface in the Minkowski spacetime, the above procedure identifies $\Sigma_0 = \Sigma$ up to a global isometry.

In general, the optimal isometric embedding equation is difficult to solve. However, in a perturbative configuration, when a family of surfaces limit to a surface in the Minkowski spacetime, the optimal isometric embedding equation is solvable, subject to the positivity of the limiting mass.

One way to test and apply the Wang-Yau definition is to consider its limit at spatial and null infinity, in hopes of recovering the ADM mass [**1**] and Bondi-Trautman mass [**5, 35**], respectively. In the following, we first describe how our quasilocal mass recovers the ADM mass at spatial infinity [**39**].

We consider an asymptotically flat initial data set (M, g, k); that is, there is a compact set $K \subset M$ such that $M \backslash K$ is diffeomorphic to a union of complements of balls in $\mathbb{R}^3$. Assuming $g - \delta = O(r^{-p})$ and $k = O(r^{-q})$ as $r \to \infty$, where $p > \frac{1}{2}$ and $q > \frac{3}{2}$, the ADM mass

$$\frac{1}{16\pi} \int_{S^2_\infty} \sum_{i,j} (g_{ij,j} - g_{jj,i}) \nu^i$$

is well-defined (Bartnik) and the positive mass theorem (Schoen-Yau [**32, 33**], Witten [**40**]) holds.

The limit of our quasilocal local mass calculated on the coordinate spheres S_r recovers the ADM definitions under the classical asymptotics $\left(p > \frac{1}{2} \text{ and } q > \frac{3}{2}\right)$.

For more general asymptotics (for example, $p \leq \frac{1}{2}$ or $q \leq \frac{3}{2}$) the ADM definitions are no longer valid. In these cases, our definition gives a well-defined conserved quantity that satisfies desirable rigidity and invariance properties.

The wilder asymptotic behavior of the physical data is captured by the optimal embedding equation and is countered by the reference data.

5. Description of null infinity and the Bondi-Sachs formalism

An idealized distant observer is situated at future null infinity $\mathscr{I}^+$, where light rays approach along null geodesics. Descriptions of $\mathscr{I}^+$of an isolated gravitating system include:

- Bondi-Sachs coordinates (Bondi et al. 1962 [**5**], Sachs 1962 [**31**])
- Penrose conformal compactification (Penrose 1965 [**29**])
- Christodoulou-Klainerman (1993) [**16**]

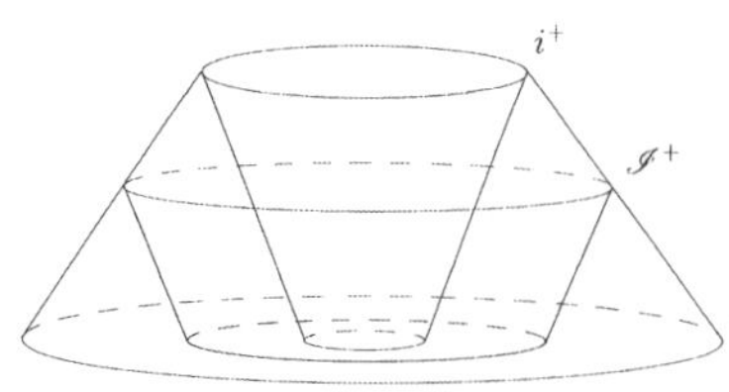

FIGURE 2. The future null infinity $\mathscr{I}^+$extends from spacelike infinity $i^0(u = -\infty)$ to timelike infinity $i^+(u = +\infty)$.

Bondi-Sachs and Penrose formalisms both imply the spacetimes have "peeling"-'an idealized property that may not hold in general, as demonstrated in the work of ChristodoulouKlainerman. In the following, we describe null infinity and distant observers using the BondiSachs formalism. Our result can be extended to "polyhomogeneity" (Chrusciel-MacCallumSingleton, 1995 [**15**] and the Christodoulou-Klainerman [**16**] setting.

Bondi and his collaborators postulate a coordinate system (Bondi-Sachs) in which the metric tensor is given by

$$-UV du^2 - 2U du dr + r^2 h_{AB} \left(dx^A + W^A du\right) \left(dx^B + W^B du\right)$$

The spacetime is asymptotically flat in the sense that $U, V \to 1, h_{AB} \to \sigma_{AB}$[1], $W^A \to 0$ as $r \to \infty$ and $\det h_{AB} = \det \sigma_{AB}$ (determinant condition) for r large.

Moreover, outgoing radiation condition is imposed so that all metric coefficients U, V, h_{AB}, W^A can be expanded into power series of $\frac{1}{r}$.

Bondi et al. found that the physics of gravitational field are encoded in the coefficients of power series expansion. Energy and linear momentum are well-understood in the Bondi-Sachs formalism. The expansion

$$V = 1 - \frac{2m(u,x)}{r} + O\left(r^{-2}\right)$$

gives the mass aspect $m(u, x)$ on $\mathscr{I}^+$.

In terms of the mass aspect function, the Bondi-Trautman energy-momentum $\left(E, P^k\right), k = 1, 2, 3$ associated with a $u =$ const. slice is

$$E(u) = \int_{S^2} 2m(u, \cdot), \quad P^k(u) = \int_{S^2} 2m(u, \cdot)\widetilde{X}^k$$

where $\widetilde{X}^1 = \sin\theta\cos\phi, \widetilde{X}^2 = \sin\theta\sin\phi, \widetilde{X}^3 = \cos\theta$ are the restriction of standard coordinate functions of $\mathbb{R}^3$ to S^2.

From the evolution of the mass aspect function, it follows that

$$\frac{d}{du}E(u) = -\frac{1}{4}\int_{S^2} |N|^2 \leq 0$$

The total energy radiated away is then

$$\delta E = E(\infty) - E(-\infty) = -\frac{1}{4}\int_{-\infty}^{\infty}\int_{S^2} |N|^2$$

The "mass loss formula" involves the coefficients of power series expansion $h_{AB} = \sigma_{AB} + \frac{1}{r}C_{AB}(u, x) + O\left(r^{-2}\right)$. The symmetric and traceless 2-tensors C_{AB} and $N_{AB} = \partial_u C_{AB}$ are called shear and news respectively.

Limit of the Wang-Yau quasilocal mass at null infinity recovers the Bondi-Trautman mass [**11**].

For surfaces near the null infinity, the optimal isometric embedding equation has a unique solution.

For this unique solution, the limit of our quasilocal mass recovers the Bondi-Trautman mass. Non-negativity of the Bondi-Trautman mass follows.

This process is robust and we carried it out on an asymptotically hyperbolic initial data sets where the null infinity is rougher and more erratic due to radiation. In particular, we don't need any peeling structure of null infinity.

6. Symmetry of $\mathscr{I}^+$: BMS group and supertranslation

The symmetry of $\mathscr{I}^+$is the Bondi-Metzner-Sachs (BMS) group. These are the coordinate transformations at null infinity which keep the Bondi-Sachs coordinate form. For any smooth function $f(x)$ on S^2, the change of coordinates $u = \bar{u} + f(x)$ is called *supertranslation*. Geometrically, supertranslation changes the foliation of $\mathscr{I}^+$.

[1] σ_{AB} is the standard metric of unit sphere in $\mathbb{R}^3$

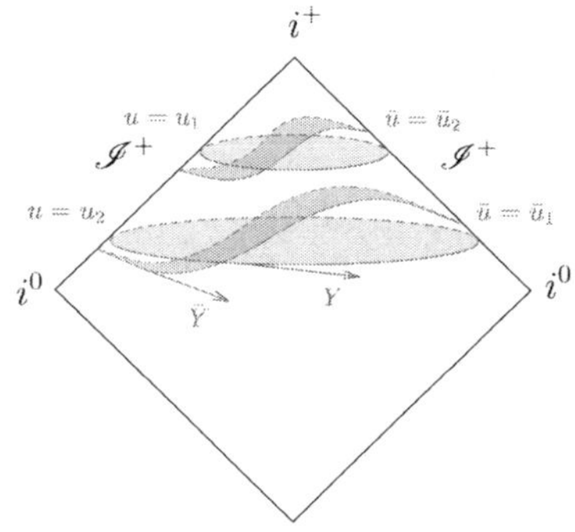

FIGURE 3. A supertranslation that maps $\bar{u}$ cuts to $u = \bar{u} + f(x)$ cuts. Y and $\bar{Y}$ are rotation BMS fields such that Y is tangent to u cuts and $\bar{Y}$ is tangent to $\bar{u}$ cuts.

We decompose f into

$$f = \alpha_0 + \sum_{i=1}^{3} \alpha_i \tilde{X}^i + S$$

where $\alpha_0 + \sum_{i=1}^{3} \alpha_i \tilde{X}^i$ corresponds to (4-dimensional) ordinary translation and S is referred to as *pure supertranslation.*

There is no preferred coordinate system (in particular no preferred u coordinate) when radiation is present. What if we use a different Bondi-Sachs coordinate system?

For a supertranslation $u = \bar{u} + f(x)$, we have the following transformation rules for the Bondi data:

$$\begin{aligned}
\bar{m}(\bar{u}, x) =& m(\bar{u} + f, x) + \frac{1}{2} \left(\nabla^B N_{BD}\right) (\bar{u} + f, x) \nabla^D f \\
& + \frac{1}{4} \left(\partial_u N_{BD}\right) (\bar{u} + f, x) \nabla^B f \nabla^D f \\
& + \frac{1}{4} N_{BD}(\bar{u} + f, x) \nabla^B \nabla^D f \\
\bar{C}_{AB}(\bar{u}, x) =& C_{AB}(\bar{u} + f(x), x) - 2\nabla_A \nabla_B f + \Delta f \sigma_{AB} \\
\bar{N}_{AB}(\bar{u}, x) =& N_{AB}(\bar{u} + f(x), x)
\end{aligned} \tag{6.1}$$

Consequently, the total flux of energy

$$\delta E = E(+\infty) - E(-\infty) = -\frac{1}{4} \int_{-\infty}^{+\infty} \int_{S^2} |N_{AB}|^2$$

is supertranslation invariant and the energy radiated away is without ambiguity.

7. Angular momentum at null infinity and the supertranslation

The definition of angular momentum turns out to be more subtle largely because of the convergence of the expansions above and the choice of the Bondi-Sachs coordinate.

The corresponding Killing fields are the rotation fields, which are of higher order near infinity in comparison to the translating Killing fields used in the definition of the energy and linear-momentum.

As a result, a very precise comparison with the Minkowski spacetime is needed in order to find an appropriate definition of angular momentum.

Different approaches (and descriptions of null infinity) lead to different definitions by AshtekarHansen [**2**], Barnich-Troessaert [**3**], Bramson [**6**], Chrusciel-Jezierski-Kijowski [**14**], DouganManson [**18**], Dray-Streubel [**19**], Hawking-Perry-Strominger [**20**], Ludvigsen-Vickers [**25**], Rizzi [**30**], Winicour-Tamburino [**36**], etc.

The Dray-Streubel angular momentum of a u cut is defined to be:

$$\widetilde{J}(u,Y)=\int_{S^2} Y^A\left(N_A-\frac{1}{4}C_{AB}\nabla_D C^{DB}\right)(u,\cdot)$$

for a rotation Killing field Y. N^A from the expansion of W^A is referred to as the angular momentum aspect.

According to Penrose (1982) [**28**], the very concept of angular momentum gets shifted by these supertranslations, and "it is hard to see in these circumstances how one can rigorously discuss such questions as the angular momentum carried away by gravitational radiation". In other word, if one choose another Bondi-Sachs coordinates $(\bar{u},\bar{x})$ related to (u,x) by a supertranslation f, Penrose worries that $\delta\widetilde{J}$ may not be the same as $\delta\widetilde{J}_f$.

We shall refer to the difference

$$\delta\widetilde{J}_f-\delta\widetilde{J}$$

as the supertranslation ambiguity of the angular momentum.

In the following theorem, we evaluate the supertranslation ambiguity for the Dray-Streubel angular momentum:

THEOREM 7.1 (Chen-Keller-Wang-Wang-Y, 2021 [**9**]). *Suppose the news tensor decays as*

$$N_{AB}(u,x)=O\left(|u|^{-1-\varepsilon}\right)$$

and two Bondi-Sachs coordinate systems are related by a supertranslation f. Then

$$\delta\widetilde{J}_f-\delta\widetilde{J}=-\int_{S^2}\left(2fY^A\nabla_A(m(+)-m(-))\right)$$

Here

$$m(\pm)=\lim_{u\to\pm\infty} m(u,x)$$

In particular, if $Y^A\nabla_A(m(+)-m(-))$ has any $\ell\geq 2$ spherical harmonics component, the total flux of the classical angular momentum $\delta\widetilde{J}$ can assume any value by supertranslations.[2]

There were also efforts to eliminate supertranslation ambiguity by choosing special null foliations, in particular "nice sections"[3] by Moreschi (1986) [**17**] and "preferred cuts"[4] by Rizzi (1997) [**30**]. However, both work only in special cases.[5]

[2]In fact, none of the definitions listed above were known to be supertranslation invariant.

[3]with zero Bondi-Sachs momentum

[4]with zero electric part of the shear

[5]$|\dot{\sigma}|^2<\sqrt{\frac{27}{4}}$ and $|\Xi|^2<16$, respectively.

8. Quasi-local angular momentum and a new definition of angular momentum at null infinity

The Chen-Wang-Yau quasilocal angular momentum was defined based on the theory of Wang-Yau quasilocal mass. The quasilocal formulation is essential in this work on angular momentum at null infinity. This was pointed out by Penrose in 1982 [**28**] as he considered the definition of quasilocal mass and quasilocal angular momentum as his number one and number two unsolved problems in general relativity.

For surfaces near null infinity, the optimal embedding equation admits a unique solution. Transplant a Killing field Y of $\mathbb{R}^{3,1}$ through the unique solution of the optimal embedding equation, we define the corresponding conserved quantity to be

$$\int_\Sigma -\rho \left\langle T_0, Y\right\rangle + j\left(Y^T\right)$$

In particular, the quasilocal angular momentum (Chen-Wang-Y 2015 [**12**]) is

$$\int_\Sigma j\left(Y^T\right)$$

At the quasilocal level, the angular momentum satisfies several important invariant properties, suggesting that their limit at infinity would give total angular momentum with the desired invariant properties.

We computed the limit of the quasilocal angular momentum in Bondi coordinate (KellerWang-Y 2018 [**21**]).

We consider the decomposition of C_{AB} into

$$C_{AB} = \nabla_A\nabla_B c - \frac{1}{2}\sigma_{AB}\Delta c + \frac{1}{2}\left(\epsilon_A^E\nabla_E\nabla_B \underline{c} + \epsilon_B^E\nabla_E\nabla_A \underline{c}\right)$$

The CWY angular momentum associated to a rotation Killing field is

$$J(u,Y) = \int_{S^2} Y^A\left(N_A - \frac{1}{4}C_{AB}\nabla_D C^{DB} - c\nabla_A m\right)(u,\cdot)$$

One should note that c has never occurred in any previous definition of angular momentum. It is a "non-local term".

9. Supertranslation invariance of the new angular momentum

We proved the supertranslation invariance of the new angular momentum.

THEOREM 9.1 (Chen-Keller-Wang-Wang-Y, 2021 [**9**]). *Suppose* $N_{AB}(u,\cdot) = O\left(|u|^{-1-\epsilon}\right)$ *as* $|u|\to\infty$. *Under a supertranslation* $f = \alpha_0 + \alpha_i\widetilde{X}^i + f_{l\geq 2}$, *the total flux of the CWY angular momentum*

$$\delta J(Y) = J(+\infty, Y) - J(-\infty, Y)$$

transforms according to

$$(\delta J)_f - \delta J = \alpha_i\epsilon^{ik}{}_j\delta P^j \text{ for } Y^A = \epsilon^{AB}\nabla_B\widetilde{X}^k$$

where δP^j *is the total flux of Bondi-Sachs linear momentum.*

In particular, δJ is invariant under pure supertranslation and the transformation law is the same as the special-relativistic angular momentum.

The limit of the quasilocal center of mass integral defined by Chen-Wang-Yau also gives a new definition of center of mass integral at null infinity.

We obtain a complete set of ten conserved quantities (E, P^k, J^k, C^k) at null infinity (all as functions of the retarded time u) that satisfy the following properties [**9, 10**]:

- (E, P^k, J^k, C^k) all vanish for any Bondi-Sachs coordinate system of the Minkowski spacetime where there is no gravity.
- In a Bondi-Sachs coordinate system of the Kerr spacetime, P^k and C^k vanish, and E and J^k recover the mass and angular momentum.
- On a general spacetime, the total fluxes of (E, P^k, J^k, C^k) are supertranslation invariant.
- (E, P^k, J^k, C^k) and their fluxes transform according to basic physical laws under ordinary translations.

References

[1] R. Arnowitt, S. Deser, and C. W. Misner, *Coordinate invariance and energy expressions in general relativity*, Phys. Rev. (2) **122** (1961), 997–1006. MR127946

[2] Abhay Ashtekar and R. O. Hansen, *A unified treatment of null and spatial infinity in general relativity. I. Universal structure, asymptotic symmetries, and conserved quantities at spatial infinity*, J. Math. Phys. **19** (1978), no. 7, 1542–1566, DOI 10.1063/1.523863. MR503432

[3] Glenn Barnich and Cédric Troessaert, *BMS charge algebra*, J. High Energy Phys. **12** (2011), 105, 22, DOI 10.1007/JHEP12(2011)105. MR2935590

[4] Peter G. Bergmann and Robb Thomson, *Spin and angular momentum in general relativity*, Phys. Rev. (2) **89** (1953), 400–407. MR54394

[5] H. Bondi, M. G. J. van der Burg, and A. W. K. Metzner, *Gravitational waves in general relativity. VII. Waves from axi-symmetric isolated systems*, Proc. Roy. Soc. London Ser. A **269** (1962), 21–52, DOI 10.1098/rspa.1962.0161. MR147276

[6] B. D. Bramson, *Relativistic Angular Momentum for Asymptotically Flat EinsteinMaxwell Manifolds*, Proc. R. Soc. London, Ser. A, 341, 463490 (1975).

[7] J. David Brown and James W. York Jr., *Quasilocal energy in general relativity*, Mathematical aspects of classical field theory (Seattle, WA, 1991), Contemp. Math., vol. 132, Amer. Math. Soc., Providence, RI, 1992, pp. 129–142, DOI 10.1090/conm/132/1188439. MR1188439

[8] J. David Brown and James W. York Jr., *Quasilocal energy and conserved charges derived from the gravitational action*, Phys. Rev. D (3) **47** (1993), no. 4, 1407–1419, DOI 10.1103/PhysRevD.47.1407. MR1211109

[9] Po-Ning Chen, Jordan Keller, Mu-Tao Wang, Ye-Kai Wang, and Shing-Tung Yau, *Evolution of angular momentum and center of mass at null infinity*, Comm. Math. Phys. **386** (2021), no. 1, 551–588, DOI 10.1007/s00220-021-04053-7. MR4287194

[10] Po-Ning Chen, Mu-Tao Wang, Ye-Kai Wang, and Shing-Tung Yau, *Supertranslation invariance of angular momentum*, Adv. Theor. Math. Phys. **25** (2021), no. 3, 777–789, DOI 10.4310/ATMP.2021.v25.n3.a4. MR4403306

[11] PoNing Chen, Mu-Tao Wang, and Shing-Tung Yau, *Evaluating quasilocal energy and solving optimal embedding equation at null infinity*, Comm. Math. Phys. **308** (2011), no. 3, 845–863, DOI 10.1007/s00220-011-1362-2. MR2855542

[12] Po-Ning Chen, Mu-Tao Wang, and Shing-Tung Yau, *Conserved quantities in general relativity: from the quasi-local level to spatial infinity*, Comm. Math. Phys. **338** (2015), no. 1, 31–80, DOI 10.1007/s00220-015-2381-1. MR3345371

[13] Po-Ning Chen, Mu-Tao Wang, and Shing-Tung Yau, *Quasilocal angular momentum and center of mass in general relativity*, Adv. Theor. Math. Phys. **20** (2016), no. 4, 671–682, DOI 10.4310/ATMP.2016.v20.n4.a1. MR3573024

[14] Piotr T. Chruściel, Jacek Jezierski, and Jerzy Kijowski, *Hamiltonian field theory in the radiating regime*, Lecture Notes in Physics. Monographs, vol. 70, Springer-Verlag, Berlin, 2002. MR1903925

[15] Piotr T. Chruściel, Malcolm A. H. MacCallum, and David B. Singleton, *Gravitational waves in general relativity. XIV. Bondi expansions and the "polyhomogeneity" of* $\mathcal{I}$, Philos. Trans. Roy. Soc. London Ser. A **350** (1995), no. 1692, 113–141, DOI 10.1098/rsta.1995.0004. MR1325206

[16] Demetrios Christodoulou and Sergiu Klainerman, *The global nonlinear stability of the Minkowski space*, Princeton Mathematical Series, vol. 41, Princeton University Press, Princeton, NJ, 1993. MR1316662

[17] Sergio Dain and Osvaldo M. Moreschi, *General existence proof for rest frame systems in asymptotically flat spacetime*, Classical Quantum Gravity **17** (2000), no. 18, 3663–3672, DOI 10.1088/0264-9381/17/18/305. MR1791097

[18] A. J. Dougan and L. J. Mason, *Quasilocal mass constructions with positive energy*, Phys. Rev. Lett. **67** (1991), no. 16, 2119–2122, DOI 10.1103/PhysRevLett.67.2119. MR1129012

[19] Tevian Dray and Michael Streubel, *Angular momentum at null infinity*, Classical Quantum Gravity **1** (1984), no. 1, 15–26, DOI 10.1088/0264-9381/1/1/005. MR730979

[20] Stephen W. Hawking, Malcolm J. Perry, and Andrew Strominger, *Superrotation charge and supertranslation hair on black holes*, J. High Energy Phys. **5** (2017), 161, front matter+32, DOI 10.1007/JHEP05(2017)161. MR3664137

[21] Jordan Keller, Ye-Kai Wang, and Shing-Tung Yau, *Evaluating quasi-local angular momentum and center-of-mass at null infinity*, Adv. Theor. Math. Phys. **24** (2020), no. 6, 1423–1473, DOI 10.4310/ATMP.2020.v24.n6.a3. MR4285615

[22] Jerzy Kijowski, *A simple derivation of canonical structure and quasi-local Hamiltonians in general relativity*, Gen. Relativity Gravitation **29** (1997), no. 3, 307–343, DOI 10.1023/A:1010268818255. MR1439857

[23] Chiu-Chu Melissa Liu and Shing-Tung Yau, *Positivity of quasilocal mass*, Phys. Rev. Lett. **90** (2003), no. 23, 231102, 4, DOI 10.1103/PhysRevLett.90.231102. MR2000281

[24] Chiu-Chu Melissa Liu and Shing-Tung Yau, *Positivity of quasi-local mass. II*, J. Amer. Math. Soc. **19** (2006), no. 1, 181–204, DOI 10.1090/S0894-0347-05-00497-2. MR2169046

[25] M. Ludvigsen and J. A. G. Vickers, *Momentum, angular momentum and their quasilocal null surface extensions*, J. Phys. A **16** (1983), no. 6, 1155–1168. MR706691

[26] T. Mädler and J. Winicour, *Bondi-Sachs formalism*, Scholarpedia, 11 (12): 33528, 2016. arXiv:1609.01731

[27] R. Penrose, *Quasilocal mass and angular momentum in general relativity*, Proc. Roy. Soc. London Ser. A **381** (1982), no. 1780, 53–63, DOI 10.1098/rspa.1982.0058. MR661716

[28] R. Penrose, *Some unsolved problems in classical general relativity*, Seminar on Differential Geometry, Ann. of Math. Stud., No. 102, Princeton Univ. Press, Princeton, NJ, 1982, pp. 631–668. MR645761

[29] Roger Penrose, *Republication of: Conformal treatment of infinity [MR0195547; MR0168334]*, Gen. Relativity Gravitation **43** (2011), no. 3, 901–922, DOI 10.1007/s10714-010-1110-5. MR2773545

[30] Anthony Rizzi, *Angular momentum in general relativity: a new definition*, Phys. Rev. Lett. **81** (1998), no. 6, 1150–1153, DOI 10.1103/PhysRevLett.81.1150. MR1638716

[31] R. K. Sachs, *Gravitational waves in general relativity. VIII. Waves in asymptotically flat space-time*, Proc. Roy. Soc. London Ser. A **270** (1962), 103–126, DOI 10.1098/rspa.1962.0206. MR149908

[32] Richard Schoen and Shing Tung Yau, *On the proof of the positive mass conjecture in general relativity*, Comm. Math. Phys. **65** (1979), no. 1, 45–76. MR526976

[33] Richard Schoen and Shing Tung Yau, *Proof of the positive mass theorem. II*, Comm. Math. Phys. **79** (1981), no. 2, 231–260. MR612249

[34] Yuguang Shi and Luen-Fai Tam, *Positive mass theorem and the boundary behaviors of compact manifolds with nonnegative scalar curvature*, J. Differential Geom. **62** (2002), no. 1, 79–125. MR1987378

[35] A. Trautman, *Boundary conditions at infinity for physical theories*, Bull. Acad. Polon. Sci. Sér. Sci. Math. Astr. Phys. **6** (1958), 403–406. Reprinted as arXiv:1604.03144. MR97265

[36] J. Winicour and L. Tamburino, *Lorentz-covariant gravitational energy-momentum linkages*, Phys. Rev. Lett. **15** (1965), 601–605, DOI 10.1103/PhysRevLett.15.601. MR194177

[37] Mu-Tao Wang and Shing-Tung Yau, *Quasilocal mass in general relativity*, Phys. Rev. Lett. **102** (2009), no. 2, no. 021101, 4, DOI 10.1103/PhysRevLett.102.021101. MR2475769

[38] Mu-Tao Wang and Shing-Tung Yau, *Isometric embeddings into the Minkowski space and new quasi-local mass*, Comm. Math. Phys. **288** (2009), no. 3, 919–942, DOI 10.1007/s00220-009-0745-0. MR2504860

[39] Mu-Tao Wang and Shing-Tung Yau, *Limit of quasilocal mass at spatial infinity*, Comm. Math. Phys. **296** (2010), no. 1, 271–283, DOI 10.1007/s00220-010-0990-2. MR2606635

[40] Edward Witten, *A new proof of the positive energy theorem*, Comm. Math. Phys. **80** (1981), no. 3, 381–402. MR626707

Contemporary Mathematics
Volume **811**, 2025
https://doi.org/10.1090/conm/811/16237

Geodesics on the modular surface and continued fraction expansions of their endpoints

S. G. Dani

Abstract. To each (forward) trajectory of geodesic on the modular surface there corresponds an "endpoint", which is either a real number or infinity; the endpoint is defined only upto equivalence under the action of the modular group acting as a group of fractional transformations. There is a close relationship between the Diophantine properties and the continued fraction expansions of the endpoint, on the one hand, and the geometrical/dynamical behaviour of the geodesic, on the othe hand. Apart from the classical simple continued fraction expansion, we consider more general expansions, and especially what is called the Hurwitz or nearest integer expansion. This article is aimed at giving an exposition on the theme of these interconnections.

1. Introduction

There is an intricate connection between the geometry of the modular surface and arithmetic of real numbers, which manifests in various ways, too vast for it to be recalled here. Our focus here will be on a specific connection arising as follows. To every geodesic trajectory (viz. the forward segment of a geodesic, starting from a point) in the Poincaré half-plane $\mathcal{H}$ there corresponds a point on the real line $\mathbb{R}$, or infinity, which is its endpoint when $\mathbb{R}\cup\{\infty\}$ is considered as the boundary of $\mathcal{H}$. This gives a natural correspondence between geodesic trajectories on the modular surface M and equivalence classes of real numbers under the action of the modular group Γ on $\mathbb{R}\cup\{\infty\}$. One of the main themes here is the relation between the behaviour of the geodesic trajectories and the arithmetical properties of the numbers in the corresponding equivalence classes. At a basic level this is classical, and folklore; see Proposition 2.1, where for ready reference we also indicate the proofs involved. The arithmetical, or Diophantine, properties themselves have been analysed classically, through the simple continued fraction expansions (henceforth CFEs for short) of the numbers, and the behaviourial aspects in question are known to have neat characterizations in terms of simple CFEs; see Proposition 3.1. After recalling the underlying theory in some detail, together with the perspective involved, we take up further analysis from there on, along two themes. Firstly, we recall a more general theory of CFEs that is emerging, and endeavour to characterize the properties involved in the broader framework, in place of the simple CFE involved in the

This is an expanded version of a talk presented (online) in the International Confernece on Geometry, Groups and Mathematical Philosophy, in Honour of Ravi S. Kulkarni's 80th Birthday, at Bhaskaracharya Pratishthana, Pune, during May 21-25, 2022, on 21 May 2022.

classical formulation. Analogous results are discussed in the extended framework (see Proposition 4.1 and Theorem 4.2). One particular CFE, called the Hurwitz CFE, stands out among these for which the characterizations hold. Our second theme concerns a closer look at the Hurwitz CFE and establishing a direct geometric link between the study of the geodesic trajectories as above and the Hurwitz CFE. This correspondence provides also applications to values of binary quadratic forms, which are recalled in the last section. Along the way we recall also some historical tidbits, from medieval Indian mathematics, related to the Hurwitz CFE.

2. The modular surface

We realize the modular surface as $M = \mathcal{H}/\Gamma$ where

$$\mathcal{H} = \{z = x + iy \in \mathbb{C} \mid y > 0\}$$

is the upper half-plane, and $\Gamma = \mathrm{PSL}(2, \mathbb{Z})$ is the "modular group" consisting of 2×2 matrices with integer entries, and determinant 1; the action of $\gamma = \begin{pmatrix} a & b \\ c & d \end{pmatrix}$, is given by $z \mapsto \dfrac{az + b}{cz + d}$ for all $z \in \mathcal{H}$. The quotient space M is then a surface, with two singular points corresponding to images of i and ω (the cube-root of unity in $\mathcal{H}$); the singular points will be mostly left out of consideration in the discussion here. The surface is noncompact, with a single "cusp"; see below for an elaboration of this term. The set

$$F := \left\{ z \in \mathcal{H} \mid |z| \geq 1 \text{ and } |\operatorname{Re} z| \leq \frac{1}{2} \right\}$$

is a fundamental domain for the action of Γ on $\mathcal{H}$, in the sense that $\mathcal{H} = \cup_{\gamma \in \Gamma} \gamma F$, and $\gamma F \cap F$ is of measure zero for all $\gamma \neq \pm I$ (see [**17**]); we note that F is a triangle in terms of hyperbolic geometry (see below for more on this), and the intersections $\gamma F \cap F$, $\gamma \neq \pm I$, are either empty or consists of one of the sides of the boundary of the triangle. Figure 1 illustrates the tessellation formed, with the fundamental domain, marked as **F**, some of the translates under elements of Γ including all adjacent ones, and the general pattern of the tessellation.

The quotient space M can therefore also be realized as the quotient of F, under the identifications $z' \approx z$ for $z' \neq z$ if and only if $|\operatorname{Re} z' - \operatorname{Re} z| = 1$ or $z' = -z^{-1}$; only the points on the boundary of F (in the usual sense) are involved in nontrivial identifications, and these consist of identifying points at the same height (imaginary part) on the vertical segments of the boundary, and those on the segment of the unit circle that are symmetrically placed with respect to the imaginary axis. It may be observed from this that a subset E of M is bounded (has compact closure) if and only if the heights of the representitatives of the elements of E in F are bounded above. Thus in particular there exists a sequence of compact subsets of M with connected complements, which covers M, which signifies that M has only one end.

We consider $\mathcal{H}$ equipped with the Poincaré metric $y^{-2}(dx^2 + dy^2)$. The metric on $\mathcal{H}$ induces canonically a metric on M. The geodesics on $\mathcal{H}$ are (Euclidean) semicircles with endpoints on the real axis, together with their orientation (positive direction), or vertical lines joining a point on the real axis to ∞, which also may be thought of as semicircles in the Riemann sphere $\mathbb{C} \cup \{\infty\}$; the subset $\mathbb{R} \cup \{\infty\}$ which may be realized as the *circle at infinity* will be denoted by C. We note that C is the boundary of $\mathcal{H}$ in the Riemann sphere, and the geodesics in $\mathcal{H}$ are in

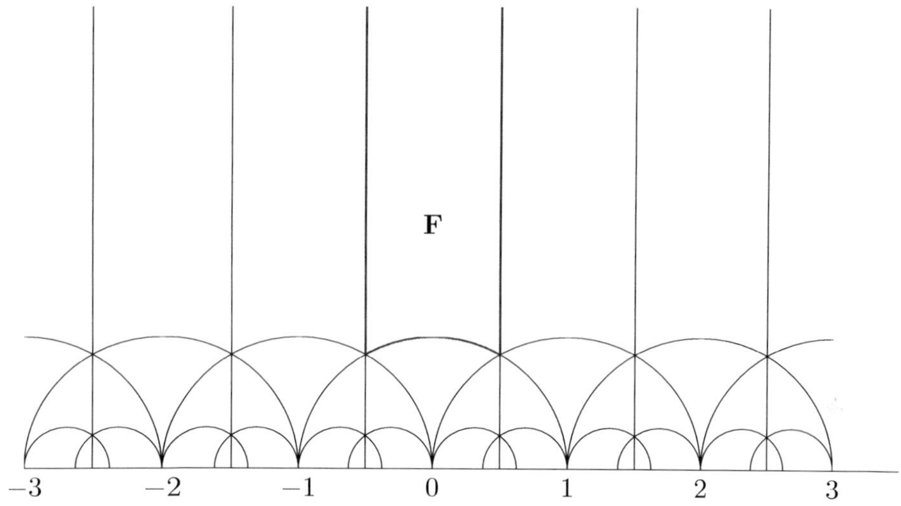

FIGURE 1. The tessellation by the modular group

one-one correspondence with semicircles joining a pair of points of C (see e.g. [**12**] for various details in this resepct).

The geodesics on M are the images of those in $\mathcal{H}$ under the canonical quotient map. We note that the action of Γ on $\mathcal{H}$ extends canonically to a Γ-action on C, and is given by $t \mapsto \dfrac{at+b}{ct+d}$ for all $t \in \mathbb{R}$ (to ∞ if $ct+d=0$), and $\infty \mapsto \dfrac{a}{c}$ (to ∞ if $c=0$).

Geodesic trajectories

A geodesic trajectory, or forward geodesic, say starting from a point $z \in \mathcal{H}$ and tangent to a unit direction ξ at z, is the segment of the semicircle passing through z and tangential to ξ, in the direction of the latter. It has an endpoint at the boundary C. We call it the endpoint of the trajectory. As each geodesic trajectory in M corresponds to an equivalence class of trajectories in $\mathcal{H}$ under the Γ-action, with each such trajectory is associated an *equivalence class of the endpoints* from C, under the Γ-action on C; we shall refer to each of the points in the equivalence class as an endpoint of the trajectory.

A complete geodesic in $\mathcal{H}$ corresponds to a (Euclidean) semicircle joining two points in C, and thus has two (distinct) endpoints corresponding to the backward and forward segments; a geodesic consisting of a vertical line has one endpoint on the real axis and another at ∞. Complete geodesics in M correspond to an equivalence classes of these under the Γ-action on $\mathcal{H}$, and hence they are in one-one correspondence with the equivalence classes of the Γ-action on $(C \times C) \setminus \Delta$, the space of pairs of distinct points of the endpoints (Δ being the diagonal subset), under the componentwise action.

A correspondence

Here are some well-known interrelations between certain properties of the endpoint, and the behaviour of the trajectories with that as the endpoint.

Recall that a geodesic $\{\varphi(t)\}_{t\geq 0}$ is said to be *periodic*, if there exists a $p > 0$ such that $\varphi(t+p) = \varphi(t)$ for all t; when this holds there exists a smallest positive number p for which it holds, and that is called the *period* of the geodesic. We shall say that a geodesic trajectory $\{\psi(t)\}_{t\geq 0}$ is *spiral* if it is asymptotic to a periodic geodesic; that is, if there exists a periodic geodesic $\{\varphi(t)\}_{t\geq 0}$ such that $d(\psi(t), \varphi(t)) \to 0$ as $t \to \infty$; in particular, for convenience, we count periodic trajectories also as being "spiral". Also, we say that a trajectory is *divergent* if it has no limit point in M.

PROPOSITION 2.1. *Let ψ be a geodesic trajectory in M and e be an endpoint of ψ.*

(i) *ψ is divergent if and only if e is rational;*

(ii) *ψ is a spiral trajectory if and only if $e = a + \sqrt{b}$ for some $a, b \in \mathbb{Q}$, and $\sqrt{b} \notin \mathbb{Q}$. (viz. a quadratic surd); it is periodic if and only if the endpoint of the backward trajectory is $a - \sqrt{b}$, the conjugate of e;*

(iii) *ψ is bounded (contained in a compact subset of M) if and only if e is a badly approximable number; (i.e. there exists a $\delta > 0$ such that $|e - \frac{p}{q}| > \frac{\delta}{q^2}$ for all $p, q \in \mathbb{Z}$, $q \neq 0$).*

These results may now be considered folklore. For want of convenient references we sketch here the arguments for them. The reader is referred to **[12]** for the relevant facts from basic hyperbolic geometry.

(i) Let $\alpha > 1$ and and $\Omega = \{z = x + iy \in \mathcal{H} \mid y \geq \alpha\}$. Then we see that the distinct members of the family $\{\gamma(\Omega)\}_{\gamma\in\Gamma}$ are pairwise disjoint. Consider a divergent trajectory ψ. It has to be eventually contained in the union of these sets, and by omitting an initial part we may assume that ψ is contained in the union. As the members are pairwise disjoint, it must be contained in one of them. Thus there exists $\gamma \in \Gamma$ such that $\psi \subset \gamma(\Omega)$. Then $\gamma^{-1}(\psi)$ is a trajectory contained in Ω. Hence the endpoint of $\gamma^{-1}(\psi)$ is ∞, the point at infinity, which implies that the endpoint of ψ is $\gamma(\infty)$, so it is a rational number. Conversely if ψ is a trajectory with a rational endpoint, then there exists $\gamma \in \Gamma$ such that $\gamma(\psi)$ has ∞ as its endpoint, which means that it is divergent.)

(ii) Let ψ be a geodesic in $\mathcal{H}$ whose image in M is periodic. Then there exists $\gamma \neq \pm I$ such that $\gamma(\psi) = \psi$. This implies that the endpoints of ψ are fixed points of γ. Since the entries of γ are integers this implies that the two endpoints are the roots of a quadratic polynomial, and since they are not rational (by (i)) they must be quadratic conjugates. Conversely given a pair of quadratic conjugates an element of Γ can be constructed with those as the fixed points. The geodesic joining the two points is therefore a periodic geodesic. This proves the part of the assertion about periodic trajectories. We note that two geodesic trajectories in M can be asymptotic only they have lifts in $\mathcal{H}$ which are asymptotic, and that two geodesic trajectories in $\mathcal{H}$ are asymptotic only if they have the same endpoint. This, together with the preceding conclusion, proves the assertion about the spiral trajectories.

(iii) For each rational $\frac{p}{q}$ with $p, q \neq 0$ and p, q coprime to each other, there exists a $\gamma \in \Gamma$ such that $\frac{p}{q} = \gamma(\infty)$. For each $\delta > 0$, let $\Omega_\delta = \{z = z + iy \mid y > \delta^{-2}\}$. Then it can be seen that $\gamma(\Omega_\delta)$ is the (Euclidean) disc contained in $\mathcal{H}$ tangenial to $\mathbb{R}$ at the point $\frac{p}{q}$, with diameter δ/q^2. It follows that for a geodesic trajectory $\{\varphi_t\}_{t\geq 0}$ in $\mathcal{H}$, with an endpoint other than ∞ and 0, the image in M will be bounded if and only if there exists $\delta > 0$ such that the trajectory avoids all these balls corresponding to the δ, at each of the rational numbers $\frac{p}{q}$. It is straightforward to verify that this would hold if an only if the endpoint is a badly approximable number. □

REMARK 2.2. The results on characterization of divergent or bounded trajectories on the modular surface also have analogues in the framework of flows on homogeneous spaces with finite invariant measure, involving suitable spaces as counterparts for the role played by $\mathbb{R}$ here. An interested reader is referred to [**3**] and [**4**] for some of the instances of this (and follow up from there).

3. Continued fraction expansions

Extending the idea of the usual fractions we can make sense of continued fractions. Let us consider an expression of the form

$$a_0 + \cfrac{1}{a_1 + \cfrac{1}{a_2 + \cfrac{1}{. + \cfrac{1}{. + \cfrac{1}{. \quad + \dots}}}}}$$

where $\{a_n\}$ is a sequence of real numbers; a similar discussion applies also to complex numbers, but here we shall limit to real numbers, and shall soon be specializing to integers. Consider the sequence of expressions obtained by truncating the above at a_n. We may then consider simplifying the expression successively, and hope to get a real number for it; one issue that could arise is that at some step in the simplification the process may involve division by zero (such as if $a_{n-1}a_n + 1 = 0$), in which case we would not get a real number as a value for the expression. If, on the other hand, for each n the truncated expression represents a real number, say t_n, and the t_ns converge to a real number t, then (in analogy with the case of infinite sums), we can associate the number t with the "continued fraction" as above and think of the latter as a continued fraction expansion (CFE for short) of t. We shall henceforth consider only sequences $\{a_n\}$ as above consisting of nonzero integers. In this case the truncated expressions, when they are definable (as clarified above), are rational numbers, say $\frac{p_n}{q_n}$, where $p_n, q_n \in \mathbb{Z}$ – these are called convergents of the CFE (as also of its limit). While we have described here the conceptual framework of CFE, finding systematic procedures to produce these involves further work in terms of identifying sequences $\{a_n\}$ which would correspond to a CFE of a number, and describing CFEs for a given number. We shall next take up some of the theory in this respect.

Ensuring definability and convergence

If we restrict to $a_n \geq 1$ for all $n \geq 1$ then the truncated expressions necessarily represent rational numbers and the desired convergence always holds (see e.g. [**10**]). Moreover each irrational number has a unique expansion as the limit of a sequence satisfying the condition, and it is called the *simple* CFE of the number. Starting with any irrational number t its simple CFE can be obtained by a simple algorithm, consisting of subtracting the largest integer not exceeding the number and inverting the remainder, and then repeating the process; thus we set $t_1 = (t-a_0)^{-1}$, where a_0 is the largest integer not exceeding t, and repeat the procedure to get the sequence $\{a_n\}$; it may be observed that for t irrational the sequence continues ad infinitum and that $a_n \geq 1$ for $n \geq 1$, automatically; see e.g. [**10**] for details.

For a rational number the same procedure as above leads to a *finite* expression of the above form. These CFE are classical and have played an immense role in various areas of mathematics in the recent centuries.

One of the objectives here is to discuss more general possibilities in this overall scheme of things involving CFE, and also to explore relations of the theme with geometry and dynamics of the geodesic trajectories. Before we proceed with this it would be good to recall some of the interesting properties of the simple CFE, that would motivate further explorations.

The following Proposition characterizes the conditions on real numbers that turned up in Proposition 2.1 in terms of the simple CFE; the results are classical and the proofs for them may be found, for example, in [**10**].

PROPOSITION 3.1. *Let $t \in \mathbb{R}$ and $\{a_n\}$ be its simple CFE. Then the following holds:*

(i) *t is rational if and only if the CFE is finite;*

(ii) *t is of the form $a + \sqrt{b}$, with a, b rationals and $\sqrt{b}$ irrational, if and only if the CFE is eventually periodic; i.e. there exist $m, p \in \mathbb{N}$ such that $a_{n+p} = a_n$ for all $n \geq m$. (**Lagrange's theorem**).*

(iii) *t is badly approximable if and only if the CFE sequence $\{a_n\}$ is bounded.*

We recall also the property of simple CFE that it gives the "best approximati-ion" for any number in terms rationals, in the sense as in the following Proposition (see [**15**], Theorem 7.13); for later reference we note that in the statement the rational numbers t are also included.

PROPOSITION 3.2. *Let $t \in \mathbb{R}$ and $\{\frac{p_n}{q_n}\}$ be the sequence of its convergents with respect to simple CFE. Then for any $n \in \mathbb{N}$ and $q \in \mathbb{N}$ such that $q < q_n$, $|t - \frac{p_n}{q_n}| < |t - \frac{p}{q}|$ for all $p \in \mathbb{Z}$.*

We may mention here that in fact the conclusion in the above proposition has a stronger version that $|q_n t - p_n| < |qt - p|$ for all $p \in \mathbb{Z}$, and moreover such approximations necessarily correspond to convergents; see [**15**]. However, from a perspective as involved in the discussion here the above weaker statement is more instructive.

4. More general CFEs and extension of the correspondence

Let $t \in \mathbb{R}$. We can construct more general CFEs for it as follows (see [**7**],[**5**], and [**8**] for general reference on the following discussion): let a_0 be an integer such that $|t - a_0| < 1$ (it has two possibilities, except when t is an integer, in which case a_0 is unique, and $t = a_0$); we terminate the sequence if t is an integer, and put $t_1 = (t - a_0)^{-1}$. Repeating the procedure we get a CFE for $\{a_n\}$ for t; while it is not quite obvious, a careful analysis shows that the truncated expressions represent rational numbers and that the requisite convergence holds; for a general idea of the proof of this statement we refer the reader to the proof of Theorem 2.1 in [**6**], which presents such a result for complex numbers, with respect to expansions in terms of general discrete subrings for which CFE are defined – we note that in the case of real numbers the argument involved can be simplified, using that, in the notation of [**6**], the set corresponding to Φ consists of integers.

Always choosing the closest integer on the left (viz. not exceeding the successive t_ns), yields the classical simple CFE.

Always choosing the closest integer, whether it is on the left or the right on the number line, is called the nearest integer CFE, and the corresponding CFE the *nearest integer* CFE. It is also known as Hurwitz CFE, after Adolph Hurwitz who first studied it in 1880's, in the framework of complex numbers [**11**].

As in [**6**] we shall refer to the continued fraction expansions obtained as above as *lazy* CFE. We shall say that a lazy CFE is algorithmic, or given by an algorithm, if the choice to be made at each stage is independent of the stage, and depends only on the number arrived at the previous stage, viz. t_k as in the above discussion; that is, there exists a function $f : \mathbb{R}\backslash\{0\} \to \mathbb{Z}$ such that $|t - f(t)| < 1$ for all $t \in \mathbb{R}$, and with $t_0 = t$ we have $t_k = (t_{k-1} - f(t_{k-1}))^{-1}$ and $a_k = f(t_k)$ for all $k \in \mathbb{N}$. We note that given any function $f : \mathbb{R}\backslash\{0\} \to \mathbb{Z}$ such that $|t - f(t)| < 1$ for all $t \in \mathbb{R}$, the preceding relations yield a CFE for t (which may be finite or infinite).

Interestingly (though this may seem just a matter of curiosity), for any t irrational, it is possible to have a CFE such that $a_n = \pm 1$ or ± 2 for all $n \geq 1$(!), and moreover, it can also be chosen to be algorithmic; see [**6**, § 3.5].

Extending the correspondence

The preceding observation shows in particular that the characterization as in Proposition 3.1, with respect to the simple CFE, does not hold with respect to a general CFE. We note however the following.

PROPOSITION 4.1. *Let $t \in \mathbb{R}$. Then t is rational if and only if every lazy CFE is finite.*

PROOF. If t has a CFE which is finite, then clearly it is a rational number; note that the convergents with respect to a lazy CFE are always defined (cf. [**6**]). Conversely let t be a rational number, say $t = \frac{p}{q}$, $p, q \in \mathbb{Z}$, $q \neq 0$. If t is an integer the CFE terminates with it, and we are through. Suppose t is not an integer and let a_0 be an integer such that $|\frac{p}{q} - a_0| < 1$ and hence $|p - qa_0| < |q|$. Then we have $t_1 = (t - a_0)^{-1} = \frac{q}{p - qa_0}$ which shows that t_1 is a rational number whose denominator has absolute value less than that of the denominator of t. Repeating the argument successively, we see that at each successive step, if the process does not terminate, the next t_k has a denominator of strictly smaller size. Thus for some $k < q$, t_k has to be an integer, and the process must terminate. □

The possible conditions on the CFE under which the characterizations of conditions (ii) and (iii) as in Proposition 3.1 hold have been explored in some of my papers, in a general framework of complex numbers and CFE with entries in certain discrete subrings, partly in collaboration with A. Nogueira and (more recently with) Ojas Sahasrabudhe; see [**7**], [**5**], and [**8**]. In the case of real numbers applying the ideas involved in these works analogous results can be proved, which in fact take a significantly simpler form; the details in this regard are presented in [**9**]. In particular the following is proved in respect of the correspondences (ii) and (iii) as above.

THEOREM 4.2. *Let $t \in \mathbb{R}$ and $\{a_n\}$ be an algorithmic lazy CFE of t corresponding to an algorithm f. Suppose that there exists $0 < c < \frac{2}{3}$ such that $|t - f(t)| \leq c$ for all $t \in \mathbb{R}$ and that $|a_n| \geq 2$ for all $n \geq 1$. Then the following holds:*

(i) *t is a quadratic irrational if and only if $\{a_n\}$ is eventually periodic; i.e., there exist $m, p \in \mathbb{N}$ such that $a_{n+p} = a_n$ for all $n \geq m$.*

(ii) *t is badly approximable if and only if $\{a_n\}$ is bounded.*

We note that the conditions as in the hypothesis of the theorem hold for the Hurwitz CFE of real numbers, with $c = \frac{1}{2}$; in fact this condition with $c = \frac{1}{2}$ characterizes the Hurwitz CFE (except for the amibiguities for half-integers, which do not concern us here). It may be noted also that for the condition as in the theorem to hold c has to be at least $\frac{1}{2}$, which shows that the Hurwitz CFE is an optimal CFE from the class of CFE as above.

5. Some historical tidbits

It may be noticed that the algorithm for finding the simple CFE (for rationals) is closely related to the Euclidean algorithm (division algorithm, or the process of mutual division) involved in finding the gcd (the greatest common divisor) of integers.

In India the process of mutual division was also used by astronomers for finding the gcd, needed more specifically for reducing fractions to reduced forms, as well as for finding approximations for fractions involving large denominators by those with smaller ones. The latter is perhaps unique to the Indian tradition, and may go back to Bhāskarāchārya (see [**16**]; the latter part needs confirmation, and is currently conjectural). The earliest known source with clear description of the process involved is the work *Karaṇapaddhati* of Putumana Somayājī [**19**], which is believed to have been composed sometime during 1532 to 1566 (see [**19**], page xxxv; we note however that there have been other, somewhat later, periods proposed earlier in this respect, recalled in [**19**]). The task involved is to find, given a fraction $\frac{P}{Q}$ with large values of P and Q, a fraction $\frac{p}{q}$ with smaller integers p and q, such that $\frac{p}{q}$ is "sufficiently" close (depending on the level of tolerence considered admissible for the context) to the given fraction $\frac{P}{Q}$. While the original is formulated differently, and involves no specific reference to "continued fractions", in our modern framework the ideas involved may be formulated as follows; for a detailed discussion on this aspect the reader is referred to [**16**].

Consider $t = \frac{P}{Q}$ where P and Q are large integers, which for simplicity we shall also assume to be positive;"largeness" is of course a relative term - the significance of the usage will become clear once the procedure is described; in the interim one may consider any P, Q greater than 1. Let $\{\frac{p_n}{q_n}\}$ be the sequence of convergents of t with respect to the simple CFE; note that as t is rational, it is a finite sequence and the last term, say $\frac{p_k}{q_k}$, is exactly $\frac{P}{Q}$. Recall that by Proposition 3.2 for any $n = 1, \dots, k-1$, $\frac{p_n}{q_n}$ is closer to $t = \frac{P}{Q}$ than any $\frac{p}{q}$ with $q < q_n$. Now if $\frac{P}{Q}$ is meant to be used as a multiplier in some computations, but one is willing to absorb a certain amount error with regard to the output and would like to replace $\frac{P}{Q}$ by a $\frac{p}{q}$ with p, q small compared to P, Q, then the preceding observation shows that $\frac{p_n}{q_n}$'s are better suited candidates, among all such fractions $\frac{p}{q}$; the choice of the particular $\frac{p_n}{q_n}$ may then be made depending on the error that one is willing to accommodate.

Thus the sequence of convergents with respect to the simple CFE play an important role in this practice.

Interestingly, for any t, the sequence of convergents of the CFE with respect to Hurwitz CFE form a subsequence of the sequence of convergents of the simple CFE [**18**] (this holds also for all $t \in \mathbb{R}$ – not only rationals). Viewed in another way, for a rational number the complete sequence of convergents can in general be computed in fewer steps with the Hurwitz CFE! In some form this phenomenon, or more specifically its significance to the computation of approximations as above, is seen to have been noted by the author of *Dṛkkaraṇa*, an astronomy text from 1608, composed in a Sanskrized version of Malayalam called Maḍipravālaṃ. The author correspondingly advocates the computation of the pairs of what corresponds to numerators and denominators of the convergents in terms of what corresponds to the Hurwitz algorithm. This method has been described in [**16**] by Venkateshwar Pai and M.S. Sriram. The authors illulstrate it with the example of the fraction $\frac{21741684881}{599082677500}$, which is involved in the computation of Moon's anomaly. In this case it is seen that while the simple continued fraction expansion is of length 24, the sequence corresponding the Hurwitz algorithm has length 17, involving about 30% less effort in arriving at a range of possible fractions, viz. convergents, which may be used as good approximations for the original fraction; the choice to be made would depend on the tolerance level in the context at hand. It would be interesting to have a quantitative analysis of how much efficiency can be achieved in general by using the Hurwitz algorithm, rather than algorithm as in simple CFE, in this respect. It may be worth recalling here that reduction in the length depends on the occurrence of 1's as partial quotients in the simple continued fraction expansion of the rational number, and may be obtained by adding, over the (maximal) blocks consisting of 1's, half the length of the block when the length is even and half of one more than the length when the length is odd (see [**18**]).

6. A geometrical perspective

In this section we shall analyse the issue of a geodesic trajectory being bounded or unbounded, from a direct geometric point of view. Each geodesic trajectory in $\mathcal{H}$ can be segmented as follows. Let

$$A = \{z = x + iy \in \mathcal{H} \mid |z| = 1, |x| \leq \frac{1}{2}\} \text{ and } B = \bigcup_{n \in \mathbb{Z}} (A + n).$$

We note that B is a curve separating $\mathcal{H}$ into two parts, an upper and a lower one. Let U denote the upper part, viz. $U = \{b + ti \mid b \in B, t \geq 0\}$. As U contains a fundamental domain for the Γ-action, every geodesic in M is represented by one which intersects U. Thus to consider boundedness of geodesic trajectories in M it suffices to consider the images of geodesic trajectories in $\mathcal{H}$ with initial point in U, which furthemore we may take to be on B, and more specifically in A, with the tangent direction pointing into the interior of U; we may leave out of consideration the geodesic joining -1 to 1, of which both the backward and forward trajectories are divergent. We may further assume the trajectory to be a semicircle, and not vertical; any trajectory in M has such a representative.

Consider such a trajectory $\{\varphi_t\}_{t\geq 0}$ in $\mathcal{H}$ starting from a point in A and traversing into U. We shall denote by Φ the image of the trajectory in M. We associate with $\{\varphi_t\}_{t\geq 0}$ a sequence $\{n_k\}$ of integers, finite or infinite, and a sequence $\{\psi_k\}$

of geodesic segments in $\mathcal{H}$, each starting from a point in A and ending at a point $A+n_k$, $n_k \neq 0$, with the property that the images of ψ_k in M are continuations of each other; see Figure 2 for illustration. As the trajectory is a semicircle centered on the real axis, it intersects $A+n_0$ for a $n_0 \in \mathbb{Z}\backslash\{0\}$; when this holds with two (consecutive) integers (when the point is a vertex of B), we shall choose n_0 to be the one with the larger absolute value. Let $t_0 \in \mathbb{R}$ be such that φ_{t_0} is the point of intersection as above, and $\psi_0 = \{\varphi_t \mid 0 \leq t \leq t_0\}$.

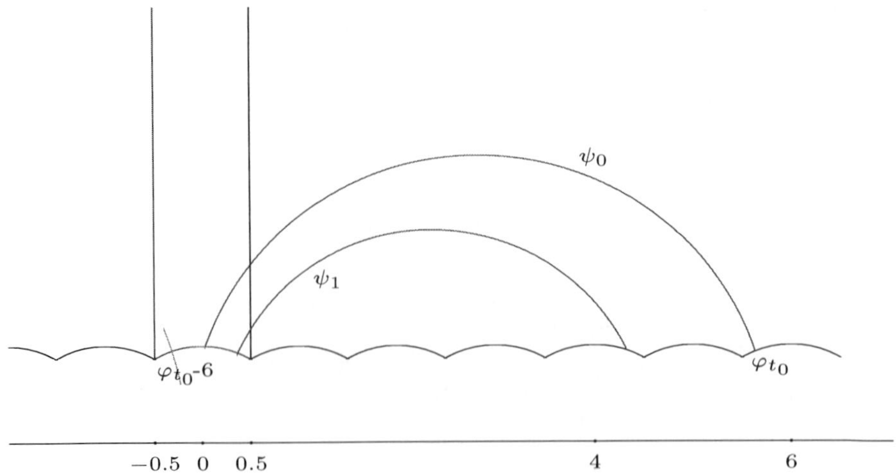

FIGURE 2. Segmentation of trajectories in U

Now consider the remaining part $\{\varphi_t\}_{t\geq t_0}$. We see that on application of $\gamma = \begin{pmatrix} 0 & -1 \\ 1 & 0 \end{pmatrix}\begin{pmatrix} 1 & -n_0 \\ 0 & 1 \end{pmatrix}$ to it (namely, applying to it the transformation $z \mapsto -(z-n_0)^{-1}$) we get a trajectory, say $\{\varphi_t^{(1)}\}_{t\geq 0}$, starting from a point in A and pointing into the interior of U. Its image in M is the continuation of the image of ψ_0. If $\{\varphi_t^{(1)}\}_{t\geq 0}$ is vertical then we stop the sequence with n_0. Otherwise repeating the process as above with $\{\varphi_t^{(1)}\}_{t\geq 0}$ we get $n_1 \in \mathbb{Z}\backslash\{0\}$ and a geodesic segment ψ_1 in $\mathcal{H}$ strarting at a point in A and ending at a point in $A+n_1$, whose image in M is a continuation of the image of ψ_0. In this way successively we get sequences $\{n_k\}$ of nonzero integers and $\{\psi_k\}$ of geodesic segments as proposed, with lengths $\{t_k\}$ respectively. If the sequence is finite then Φ is the union of the images of the segments together with the image of a trajectory that is entirely contained in F. We show that when the sequence is infinite the images of $\{\psi_k\}$ cover the whole of Φ. To this end it suffices to show that t_k can be small only on a set of k which does not contain any pair of consecutive integers. Let k be such that t_k is small. Then the endpoints of the segment ψ_k are close to $\pm\frac{1}{2}+\frac{\sqrt{3}}{2}i$. Suppose they are close to $\frac{1}{2}+\frac{\sqrt{3}}{2}i$; the other case works similarly. In this case the endpoint of the trajectory in $\mathcal{H}$ extending ψ_k may be seen to be greater than 1, and applying the steps as above we see that the next segment ψ_{k+1} starts close to $\frac{1}{2}+\frac{\sqrt{3}}{2}i$ and the endpoint of the extended trajectory in $\mathcal{H}$ containing it, is negative. It follows that t_{k+1} is at least $\log \frac{2\sqrt{5}}{1+\sqrt{5}}$, which is the distance of i from the geodesic $\{-\frac{1}{2}+e^t i \mid t \in \mathbb{R}\}$. This shows that the images of the segments ψ_k in M cover Φ.

The sequence $\{n_k\}$ may be thought of as a geometric version of CFE, associated with the trajectory (but not just the endpoint, though the latter is indeed involved). We note in this respect the following result which may be compared with Proposition 3.1 (iii), keeping in mind the equivalence in Proposition 2.1 (iii).

PROPOSITION 6.1. *Let the notation be as above. Then Φ is bounded in M if and only if $\{n_k\}$ is infinite and bounded.*

PROOF. It may be observed that the height (maximum size of the imaginary part) attained by the segment ψ_k is between $\frac{1}{2}(|n_k|-1)$ and $\frac{1}{2}(|n_k|+1)$. It may be recalled that a subset of M is bounded if and only if its preimage in $\mathcal{H}$ is contained in $\{z = x+iy \in \mathcal{H} \mid y \le h\}$ for some $h > 0$. The proposition is an immediate consequence of this, together with the fact that when the sequence is finite the remaining part of the trajectory is divergent (and hence not bounded). □

This geometric version of CFE for a trajectory would in general differ from the arithmetic CFE of its endpoint introduced earlier. There is however a modification of the above model, which relates to the arithmetic CFE. For simplicity we shall discuss here only the one which corresponds to the Hurwitz CFE, though a more general variation is possible. For this we recall the notion of reduced geodesics (H-reduced in the nomenclature in [**13**]). A geodesic σ in $\mathcal{H}$, with u, v as the endpoints of the backward and forward trajectories respectivly, v irrational, is said to be *reduced* if $|v| > 2$ and $r - 1 \le \dfrac{uv}{|v|} \le r$, where $r = \frac{1}{2}(3-\sqrt{5})$; a segment of a reduced geodesic will also be referred to as reduced. It was proved by A. Hurwitz that every geodesic in $\mathcal{H}$ has a representative (in its equivalence class under the Γ-action) which is reduced. A modified proof of this statement is given in [**13**], where the notion played an important role in the construction of symbolic dynamics for the geodesic flow (we shall however not concern ourselves with it here).

The reduced geodesics may not intersect the curve B, and may lie entirely in the complement of U; a geodesic with endpoints at r (as above) and $2+\epsilon$ for small $\epsilon > 0$ would be such an example. We therefore modify the geometric picture as above and introduce the following substitutes C and D in place of A and B respectively. Let $\mu = (23 - 3\sqrt{5})/22 \approx 0.74$; the point about this μ is that it is the real part of the intersection of the geodesic joining r and 2, with the unit circle. Let

$$C = \{z = x+iy \in \mathcal{H} \mid |z| = 1, |x| \le \mu\} \text{ and } D = \bigcup_{n \in \mathbb{Z}} (C+n).$$

The set D as above is no longer a simple continuous curve, and geodesic trajectories can have more than one point of intersection with D. However, a segmentation analogous to what was done before can be carried out with respect to it, following a suitable convention for the choice of the endpoint of the segment when there is ambiguity arising on account of the given trajectory intersecting more than one point of D; see below. In analogy with the earlier model consider a trajectory $\{\varphi_t\}_{t\ge 0}$, with an irrational endpoint, say e, starting from a point in C (a more natural choice now in place of A). Suppose that $\{\varphi_t\}_{t\ge 0}$ intersects D in two points, say $\varphi_\tau \in m + C$ and $\varphi_{\tau'} \in m' + C$; then m and m' differ by 1 and only one of

$|\varphi_\tau - m|$ and $|\varphi_{\tau'} - m'|$ is less than $\frac{1}{2}$; we choose t_0 to be τ or τ', and correspondingly n_0 to be m or m', so that $|\varphi_{t_0} - n_0| < \frac{1}{2}$; see Figure 3 for an illustration of the choice. If the point of intersection is unique then we chose t_0 and n_0 as before; in this case the point of intersection φ_{t_0} is contained in $n_0 + A$. We note that in view of the trajectory being reduced we have $|n_0| \geq 2$.

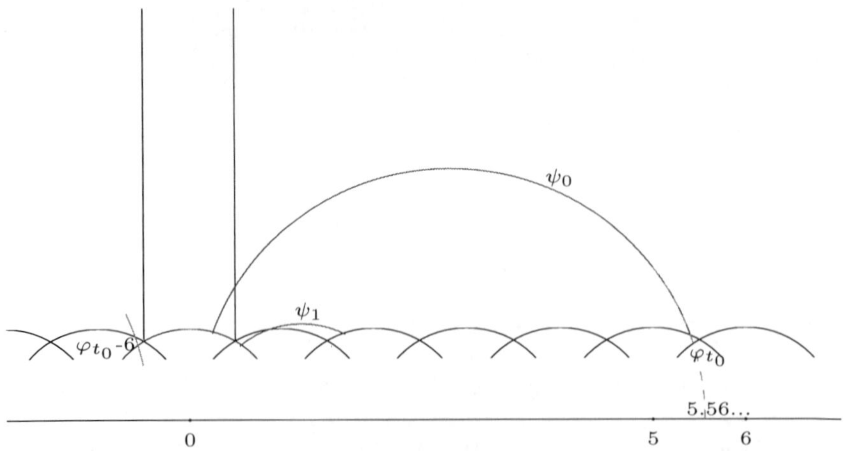

FIGURE 3. Segmentation of reduced trajectories

Now consider the remaining part $\{\varphi_t\}_{t \geq t_0}$. As in the earlier model we apply to it the transformation $z \mapsto -(z - n_0)^{-1}$ which produces a new trajectory starting from a point in C, whose image in M is the continuation of the image of the segment $\{\varphi_t\}_{0 \leq t \leq t_0}$. It turns out that the new trajectory is also reduced; a proof of this may be found in [**2**]. We now repeat the procedure as above successively and get a sequence $\{n_k\}_{k=0}^\infty$ of integers with $|n_k| \geq 2$, and a sequence of reduced geodesic segments $\{\psi_k\}_{k=0}^\infty$ in $\mathcal{H}$, with each ψ_k starting at a point in C and ending in $C + n_k$, whose images in M are continuations in succession. The condition $|n_k| \geq 2$ implies also that the length of each ψ_k is at least as much as the distance between C and $C + 2$, and as seen before this implies that the images of ψ_k together cover the entire trajectory in M.

It turns out that the sequence $\{n_k\}_{k=0}^\infty$ as above agrees with the "minus" Hurwitz CFE of the endpoint of the trajectory $\{\varphi_t\}_{t \geq 0}$ (see [**2**] for details); *minus* Hurwitz CFE of a number is its Hurwitz CFE with the signs of the odd terms changed to the opposite of the original sign. This correspondence of Hurwitz CFE with geometric version of CFE leads to interesting geometric and arithmetical consequences. In particular the correspondence provides a geometric illustration of the characterization of bounded trajectories, in terms of the Hurwitz CFE $\{n_k\}$ being bounded.

Aside from the issue of boundedness of the geodesics in M the model can also be used to analyse the quantitative aspect of how much time a geodesic spends deep into the cusp, at heights exceeding a given value, in terms of the Hurwitz CFE of the endpoint. The reader is referred to [**2**] for details in this respect; see also [**1**] for further results along the theme. In the next section we describe an application of such an analysis to values of indefinite binary quadratic forms at integral pairs.

7. Quantitative aspects and applications

We shall now briefly discuss some quantitative aspects. Specifically, given a geodesic trajectory of length T we may ask how frequently it reaches beyond a given height M, in terms of a CFE of the endpoint of the corresponding complete trajectory. This question turns out to be relevant, in particular, in studying, given a binary quadratic form $Q(x,y) = (ax+by)(cx+dy)$ which is not a rational form, the number of integer solutions to inequalities

$$|Q(x,y)| < \epsilon \text{ and } \|(x,y)\| \leq N,$$

for (small) $\epsilon > 0$ and (large enough) N, where $\|(x,y)\|$ denotes the Euclidean norm $(x^2+y^2)^{\frac{1}{2}}$. The connection arises as follows:

The unit tangent bundle of the modular surface, which is the space on which the geodesic flow is defined, can be realized as the space of lattices in $\mathbb{R}^2$ with discriminant 1; we shall denote the space by Ω. The geodesic flow is then the flow given by the pointwise action of the diagonal subgroup $\operatorname{diag}(e^t, e^{-t}), t \in \mathbb{R}$, under the natural action.

Also, for a lattice Γ from Ω, there exists a nonzero point $x \in \Gamma$ with $\|x\| < \delta$ for $\delta > 0$, if and only if the corresponding point on the modular surface is at a height $> \delta^{-2}$. (This is a stronger form in dimension 2, of what is called the Mahler criterion for lattices, which holds in all dimensions.)

Comparing the two models and using the above criterion we get the correspondence between the number of non-zero integer pairs on which a given quadratic form takes values less than a given $\epsilon > 0$, on the one hand, and the amount of time a geodesic associated with the quadratic form spends in the part of the cusp corresponding to the height exceeding $c\epsilon^{-2}$, for a constant $c > 0$ depending on the quadratic form. The reader is referred, for further details, to [**2**] where this approach was pursued, with respect to the Hurwitz CFE, proving in particular the following.

THEOREM 7.1. *Let $Q(x,y) = (ax+by)(cx+dy)$, where $a,b,c,d \in \mathbb{R}$, $b \neq 0$ and $ad-bc = 1$. Let $(a_0, a_1, \dots)$ be the Hurwitz CFE of a/b. For $\epsilon > 0$ and $r > 0$ let $N(\epsilon, r)$ denote the number of solutions of the inequalities*

$$|Q(x,y)| < \epsilon \quad \text{and} \quad \|(x,y)\| \leq r.$$

Then the following conditions are satisfied.

(i) *For any $\epsilon > 0$ there exists a constant $C > 0$ and $r_0 \geq 0$ such that $N(\epsilon, r) \leq C \log r$ for all $r \geq r_0$.*

(ii) *If $T := \limsup \frac{1}{n} \sum_{k=1}^{n-1} \log |a_k| < \infty$ and the set $\{j \geq 0 \mid |a_j| \geq \epsilon^{-1}\}$ has positive lower density $\Delta > 0$, then there exists $r_0 \geq 0$ such that $N(\epsilon, r) \geq \dfrac{\Delta}{T+3} \log r$ for all $r \geq r_0$.*

More specific information is provided in [**2**] on the constants involved in the bounds in the above statements, than we have chosen to recall here (in the interest of simplicity). The results complement, in various ways, the general theory of values of indefinite quadratic forms in $n \geq 3$ variables at integer n-tuples, for which there is a well-developed theory, emerging originally from the work of G.A. Margulis, on what was known as the Oppenheim conjecture; see [**14**] for an exposition, with a detailed historical background.

References

[1] Manoj Choudhuri, *On certain orbits of geodesic flow and (a, b)-continued fractions*, Proc. Indian Acad. Sci. Math. Sci. **131** (2021), no. 1, Paper No. 2, 19, DOI 10.1007/s12044-020-00593-5. MR4216402

[2] Manoj Choudhuri and S. G. Dani, *On values of binary quadratic forms at integer points*, Math. Res. Lett. **22** (2015), no. 4, 1023–1045, DOI 10.4310/MRL.2015.v22.n4.a4. MR3391875

[3] S. G. Dani, *Divergent trajectories of flows on homogeneous spaces and Diophantine approximation*, J. Reine Angew. Math. **359** (1985), 55–89, DOI 10.1515/crll.1985.359.55. MR794799

[4] S. G. Dani, *Bounded orbits of flows on homogeneous spaces*, Comment. Math. Helv. **61** (1986), no. 4, 636–660, DOI 10.1007/BF02621936. MR870710

[5] S. G. Dani, *Continued fraction expansions for complex numbers—a general approach*, Acta Arith. **171** (2015), no. 4, 355–369, DOI 10.4064/aa171-4-4. MR3430769

[6] S. G. Dani, *Lazy continued fraction expansions for complex numbers*, Math. Student **85** (2016), no. 1-2, 143–149. MR3559681

[7] S. G. Dani and Arnaldo Nogueira, *Continued fractions for complex numbers and values of binary quadratic forms*, Trans. Amer. Math. Soc. **366** (2014), no. 7, 3553–3583, DOI 10.1090/S0002-9947-2014-06003-0. MR3192607

[8] S.G. Dani and Ojas Sahasrabudhe, *Generalized continued fraction expansions of complex numbers, and applications to quadratic and badly approximable numbers*, (Preprint).

[9] S.G. Dani and Ojas Sahasrabudhe, *Lazy continued fraction expansions of real numbers*, (Preprint).

[10] G. H. Hardy and E. M. Wright, *An introduction to the theory of numbers*, 6th ed., Oxford University Press, Oxford, 2008. Revised by D. R. Heath-Brown and J. H. Silverman; With a foreword by Andrew Wiles. MR2445243

[11] A. Hurwitz, *Über die Entwicklung complexer Grössen in Kettenbrüche* (German), Acta Math. **11** (1887), no. 1-4, 187–200, DOI 10.1007/BF02418048. MR1554754

[12] Svetlana Katok, *Fuchsian groups*, Chicago Lectures in Mathematics, University of Chicago Press, Chicago, IL, 1992. MR1177168

[13] Svetlana Katok and Ilie Ugarcovici, *Arithmetic coding of geodesics on the modular surface via continued fractions*, European women in mathematics—Marseille 2003, CWI Tract, vol. 135, Centrum Wisk. Inform., Amsterdam, 2005, pp. 59–77. MR2223106

[14] G. A. Margulis, *Oppenheim conjecture*, Fields Medallists' lectures, World Sci. Ser. 20th Century Math., vol. 5, World Sci. Publ., River Edge, NJ, 1997, pp. 272–327, DOI 10.1142/9789812385215_0035. MR1622909

[15] Ivan Niven, Herbert S. Zuckerman, and Hugh L. Montgomery, *An introduction to the theory of numbers*, 5th ed., John Wiley & Sons, Inc., New York, 1991. MR1083765

[16] Pai R. Venketeswara and M. S. Sriram, *Nearest-integer continued fractions in Dṛkkaraṇa*, Gaṇita Bhāratī **41** (2019), no. 1-2, 69–89. MR4353082

[17] J.-P. Serre, *A course in arithmetic*, Graduate Texts in Mathematics, No. 7, Springer-Verlag, New York-Heidelberg, 1973. Translated from the French. MR344216

[18] David Simmons, *The Hurwitz continued fraction expansion as applied to real numbers*, Enseign. Math. **62** (2016), no. 3-4, 475–485, DOI 10.4171/LEM/62-3/4-5. MR3692894

[19] Putumana Somayājī, *Karaṇapaddhati*, with translation and Explanatory Notes by Venketeswara Pai, K. Ramasubramanian, M.S. Sriram and M.D. Srinivas, Hindustan Book Agency, 2017.

UM-DAE Centre for Excellence in Basic Sciences, Vidyanagari Campus of University of Mumbai, Kalina, Santacruz, Mumbai 400098, India

Email address: shrigodani@cbs.ac.in

Contemporary Mathematics
Volume **811**, 2025
https://doi.org/10.1090/conm/811/16238

Finite group actions on 1-complexes and homology

Allan L. Edmonds

In honor of Ravi Kulkarni on the occasion of his eightieth birthday

ABSTRACT. To what extent is a finite group action on a finite connected 1-complex determined up to equivariant homotopy by the induced action on integral homology? How many actions could induce the same representation on homology? What representations arise? What sorts of conditions on the homology representation could guarantee a global fixed point? We give good answers to these questions in the case of semifree actions, including, in particular, actions of cyclic groups of prime order.

1. Introduction

We consider a few of the most basic questions about actions of a finite group G on a finite, connected 1-complex X, focusing on what the representation on the first integral homology group $H_1(X)$ might be and on how many ways a given representation can be realized.

The study of such group actions is essentially equivalent to the study of homomorphisms $G \to \operatorname{Out} F_n$, the outer automorphism group of the free group on n letters, where $n = b_1(X)$, as we see from basic work of B. Zimmermann [**8**] and of M. Culler [**2**].

Moreover, the study of such homomorphisms $G \to \operatorname{Out} F_n$ is essentially equivalent to the study of group extensions

$$1 \to F_n \to E \to G \to 1$$

via the universal extension

$$1 \to F_n \to \operatorname{Aut} F_n \to \operatorname{Out} F_n \to 1$$

Here we focus on a purely topological approach.

1.1. Informal summary of the main results. We first present the appropriate equivariant analog of collapsing an edge with distinct end points, providing a useful tool that simplifies some of the subsequent work. We call this "equivariant reduction" or "G-reduction."

We begin by proving the known result that a homologically trivial action of a finite group G on a finite, connected 1-complex X is G-homotopy equvalent to a trivial action, provided $\chi(X) \neq 0$. Indeed the inclusion of the fixed point set $X^G \to X$ is a G-homotopy equivalence. Versions of this result are known, and some references are given in Secton 3. We then turn to the main result, a normalization theorem for semifree actions with nonempty fixed point set, on finite, connected

1-complexes, that permits an elementary description of the induced representation on homology, which is determined by the ranks of $H_1(X)$, $H_0(X^G)$, and $H_1(X^G)$. Such an action is classified up to G-homotopy equivalence by the ranks of $H_1(X)$, $H_0(X^G)$, and $H_1(C)$, where C ranges over the components of X^G. In this way one achieves an enumeration of the actions on such 1-complexes and of the actions inducing a given representation on homology.

We then briefly consider free actions and note that finite abelian groups induce a unique action on homology, which meshes well with our analysis of semifree actions in the case of cyclic groups. In particular, this allow us to give a complete accounting of all semifree actions of a cyclic group of prime order on finite, connected 1-complexes, up to equivariant homotopy equivalence, including the case of free actions.

We close with a bit of speculation about possible extensions of the results.

2. Equivariant reduction

Any 1-complex can be simplified, up to homotopy, by collapsing any edge that has distinct end points. one can repeat the process as long as possible, to show that any finite, connected 1-complex is homotopy equivalent to a wedge of circles, with a single vertex. In the equivariant case, if one collapses an edge, then one must also collapse the other edges in its orbit. In general this can change the homotopy type.

Suppose a finite group G acts on a finite, connected, 1-complex X. We will always assume that the action maps cells to cells and that if an element of the group sends a cell to itself, then the element is the identity on the cell. The latter property can always be achieved by subdivision if necessary. Such a complex is obtained from a discrete set X^0 by attaching equivariant 1-cells of the form $G/H \times I$ via maps $G/H \times \partial I \to X^0$.

Now suppose $E \subset X$ is an edge with distinct end points. If the full orbit $G(E)$ has acyclic components, then one can form the quotient $Y = X/\{G(E)\}$ in which each component of $G(E)$ is identified to a point. The action descends to one on Y, which has the same G-homotopy type as X. One can continue this process until one arrives at a point where no such edge E exists. If no such edge exists then we say that X is G-reduced.

The notion of G-reduction can help to simplify some arguments. Here are two elementary consequences of G-reduction.

LEMMA 2.1. *If X is a finite connected 1-complex on which a finite group G acts and if X is G-reduced, then the fixed point set X^G is a disjoint union of wedges of circles (and isolated points).*

PROOF SKETCH. Any edge with distinct end points in the fixed point set can be collapsed. □

LEMMA 2.2. *If X is a finite connected 1-complex on which a finite group G acts freely and if X is G-reduced, then the orbit space is a wedge of circles, with a single vertex.*

PROOF SKETCH. In this case, any edge with distinct end points that projects homeomorphically to the orbit space can be collapsed. □

3. Homologically trivial actions

As a first important special case we will show that almost all homologically trivial finite group actions on a 1-complexes are equivariantly homotopy equivalent to trivial actions. This is essentially a known result, but it makes a natural starting point for the present discussion. A similar result in slightly different form appears in work of B. Zimmerman, e.g. [**9**], Lemma 2.1, not unlike what we outline below. Also see the Proposition in his recent note [**10**]. A version of the result was recently published by Estelyi, Karaba, Mednykh, and Nedela [**3**], from more of a graph-theoretic point of view. The initial step of the proof begins with the well-known special case of finite group actions on trees as in [**7**], p. 20.

THEOREM 3.1. *If a finite group G acts homologically trivially on a finite, connected 1-complex X with $\chi(X) \neq 0$, then $X^G \neq \emptyset$ and the inclusion $X^G \to X$ induces an isomorphism $H_*(X^G) \to H_*(X)$.*

PROOF. We begin with the familiar case when $\chi(X) = 1$ and X is a tree, with trivial homology. It follows immediately from the G-reduction process applied to an edge with a free end point that the action is G-homotopy equivalent to an action on a one-point space. Henceforth we may assume $\chi(X) < 0$ and that the action is G-reduced.

Consider an element $g \in G$. Since g acts trivially on homology, the Lefschetz number $L(g) = \chi(X)$, which is nonzero by assumption. Therefore $X^g \neq \emptyset$ and, indeed, $\chi(X^g) = \chi(X)$. Thus $1 - b_1(X) = b_0(X^g) - b_1(X^g)$. Then

$$b_1(X^g) = b_1(X) + b_0(X^g) - 1 \geq b_1(X) \text{ since } X^g \neq \emptyset.$$

But $b_1(X^g) \leq b_1(X)$ (as is true of any subcomplex of X), so $b_1(X^g) = b_1(X)$. It then follows as well that $b_0(X^g) = 1$.

So X^g is a nonempty, connected subcomplex of X that carries the homology of X. In particular, each X^g contains the union of all primitive cycles. It follows that $X^G = \cap_{g \in G} X^g$ also contains the union of all primitive cycles, and is then nonempty and carries all the homology $H_1(X)$, as required.

It remains to show that $b_0(X^G) = 1$ so that X^G is connected. We know that inclusion induces an isomorphism $H_1(X^G) \to H_1(X)$. Let Y be the 1-complex obtained from X by identifying each component of X^G to a point. It follows that this kills all of $H_1(X)$ and does not create any new H_1. In particular Y is acyclic and hence a tree. In addition the action on X induces one on Y with the fixed point set Y^G being the collection of the vertices corresponding to the components of X^G. But since Y is a tree, the fixed point set Y^G is contractible and hence must consist of a single point. Therefore the original fixed point set X^G is connected, as required. □

In the context above, where the inclusion $X^G \to X$ induces an isomorphism $H_*(X^G) \to H_*(X)$, it follows by G-reduction that X actually G-equivariantly strong deformation retracts onto X^G.

COROLLARY 3.2. *If an action of a finite group G on a finite connected 1-complex X, with Euler characteristic $\chi(X) \neq 0$, induces the identity on homology $H_1(X; \mathbb{Z})$, then the action is G-homotopy equivalent to a trivial action.*

REMARK 1. In the case when $\chi(X) = 0$, X is G-homotopy equivalent to a circle, with either a trivial action or an action on the circle by rotations.

A slight elaboration of the argument above also proves the following, which allows one to reduce the studies here to the case of homologically faithful actions.

COROLLARY 3.3. *If a finite group G acts on a finite connected 1-complex X, with Euler characteristic $\chi(X) \neq 0$, and N is the normal subgroup of group elements inducing the identity on homology $H_1 X; \mathbb{Z})$, then the action is G-homotopy equivalent to an action for which N acts trivially.*

4. Semifree actions

Here we explain how to classify semifree actions of a finite group G on finite, connected 1-complexes X, assuming the fixed point set $X^G \neq \emptyset$. Recall that an action is said to be *semifree* if for any $x \in X$ either $gx = x$ for all $g \in G$ or $gx \neq x$ for all $g \in G, g \neq 1$.

It will be useful to describe such an action by its labelled orbit space. Each vertex and edge is labelled by its corresponding isotropy group, which is either 1 or G. Any such labelled 1-complex, with at least one label G gives rise to a semifree action of G, well-defined up to weak equivalence, where we we identify two actions if they differ by an automorphism of G.

We begin with a normalization theorem, building on the previous notion of reduction, to put any given action into a standard form. Then we draw out a number of applications and interpretations of the normalization.

4.1. Normalization.

THEOREM 4.1. *If a finite group G acts semifreely on a finite 1-complex X with nonempty fixed point set, then the action is G-homotopy equivalent to one in a certain normal form, described as follows:*

(1) *Each component of the fixed point set is a bouquet of circles or a single point. There is one vertex for each component of the fixed point set, and there are no non-fixed vertices.*
(2) *In the orbit complex, vertices are connected in a linear chain, with fixed point components in order of increasing betti number, by principal edges (i.e., with trivial isotropy group). There is one principal edge between each successive pair of fixed point components.*
(3) *All remaining principal edges are attached to a single vertex as a bouquet of freely permuted circles (except for the common fixed point) attached at the initial fixed point component of smallest betti number.*

PROOF. We begin by G-reducing the action. Then assertion (1) follows as previously observed.

For assertion (2) we may, in the orbit complex choose a minimal set of edges (with trivial isotropy group) that connect the components of X^G. We may change the way any one of these additional chosen edges is attached to the fixed point set by homotoping its end points within the fixed point set together with the other chosen edges. One extends such a move to a homotopy equivalence of the ambient orbit complex by the homotopy extension principle. Note also that any such move is covered by a G-homotopy equivalence. Similarly, one may homotope the attaching maps to interchange the order of any two fixed point components in the linear arrangement. In this way one may turn the minimal system of attached edges into a linear arrangement, in which the components of the fixed set are arranged in order of increasing betti number, achieving point (2).

Finally, the remaining edges with trivial isotropy group are now being attached to a connected space and can therefore be homotoped until they are all attached by a constant map at the end of smallest betti number. □

4.2. Interpretation of normalization. Here we record a number of consequences of the normalization theorem, together with explanatory comments, in lieu of formal proof.

COROLLARY 4.2 (Classification). *Two semifree G actions on finite connected 1-complexes with fixed points are weakly G-homotopy equivalent (i.e., up to conjugation) if and only if they have identical normal forms.*

PROOF SKETCH. We see that $b_1(X)$ together with the list of first betti numbers of the components of the fixed point set, form a complete set of invariants for normalized orbit diagrams and no two diagrams correspond to G-homotopy equivalent actions unless they have the same invariants. □

Below $\mathbb{Z}[G]$ denotes the integral group ring of G, $I[G]$ denotes the augmentation ideal of $\mathbb{Z}[G]$ viewed as a $\mathbb{Z}[G]$-module, and $\mathbb{Z}$ denotes the integers viewed as a $\mathbb{Z}[G]$-module with trivial G action.

COROLLARY 4.3 (Representation of a semifree action). *If a finite group G acts* semifreely *on a finite, connected, 1-complex X* with nonempty fixed point set, *then the representation on homology is of the form $\mathbb{Z}^t \oplus I[G]^a \oplus \mathbb{Z}[G]^r$.*

PROOF SKETCH. We use an implicit Mayer-Vietoris argument. Each circle in the fixed point set contributes a $\mathbb{Z}$ and $t = b_1(X^G)$. Each edge in the orbit complex chosen to connect distinct components of the fixed point set contributes $I[G]$, as one may check, and $a = b_0(X^G) - 1$. Finally r is the number of circles with trivial isotropy group attached at the end of the process. □

REMARK 2. For a cyclic group of prime order there are more abstract and theoretical ways to know that the representation has this form.

COROLLARY 4.4 (Realization of representations). *Any representation of the form $\mathbb{Z}^t \oplus I[G]^a \oplus \mathbb{Z}[G]^r$ can be realized by a semifree action with nonempty fixed point set on a finite 1-complex.*

PROOF SKETCH. One can construct a suitable labelled orbit complex with these invariants. And any such labelled orbit space is covered by an actual semifree action. □

COROLLARY 4.5 (Counting realizations). *The number of G-homotopy equivalence classes of realizations of a representation of the form $\mathbb{Z}^t \oplus I[G]^a \oplus \mathbb{Z}[G]^r$ by a semifree action with nonempty fixed point set on a finite 1-complex is given by $P_{a+1}(t)$, the partitions of t into at most $a+1$ parts.*

PROOF SKETCH. One just keeps track of how $t = b_1(X^G)$ is distributed among the $b_0(X^G) = a+1$ components. Distinct partitions correspond to distinct actions. □

5. Free actions

One would like to extend the analysis of semifree actions with nonempty fixed point set to the case of free actions, that is, semifree actions with no fixed points.

5.1. Representation on homology. We begin here with a brief aside.

THEOREM 5.1 (Gaschütz, [**4**], 1954). *If a finite group G acts freely on a finite, connected, 1-complex X, then the representation on rational homology is given by $H_1(X;\mathbb{Q}) = \mathbb{Q} \oplus \mathbb{Q}[G]^r$.*

This is a fairly direct consequence of the semisimplicity of the rational group ring, comparing homology groups with cellular chain groups. See Theorem 2.1 of [**5**] for a more recent proof.

One might ask whether or when the integral form of this identity holds. We offer a possibly new answer.

THEOREM 5.2. *If a finite group G acts freely on a finite, connected, 1-complex X, with the representation on integral homology given by $H_1(X;\mathbb{Z}) = \mathbb{Z} \oplus \mathbb{Z}[G]^r$, then G is cyclic.*

PROOF SKETCH. The idea is to create a periodic resolution of $\mathbb{Z}$ by free $\mathbb{Z}[G]$-modules and then use the known fact that finite groups with periodic cohomology of period 2 are necessarily cyclic. We refer to Ken Brown's *Cohomology of Groups* [**1**] for basics about groups with periodic cohomology.

Start with the equivariant chain complex of X, consisting of free module, which gives an exact sequence

$$0 \to H_1(X) \to C_1(X) \to C_0(X) \to H_0(X) \to 0$$

or

$$0 \to \mathbb{Z} \oplus \mathbb{Z}[G]^r \to C_1(X) \to C_0(X) \to \mathbb{Z} \to 0.$$

In the case when $r = 0$, it is standard to splice together copies of this sequence to obtain the required resolution. In general, after the first stage, we augment the sequence slightly so that kernel and cokernel are isomorphic, and splice copies of

$$0 \to \mathbb{Z} \oplus \mathbb{Z}[G]^r \to C_1(X) \to C_0(X) \oplus \mathbb{Z}[G]^r \to \mathbb{Z} \oplus \mathbb{Z}[G]^r \to 0.$$

This produces a free resolution that is eventually 2-periodic, which is enough to insure periodic cohomology of period 2. □

5.2. Classifying free actions. In general, one studies free actions via the associated regular G-covering $X \to Y$ where $Y = X/G$. By G-reduction we may assume that Y is a wedge of m circles with a single vertex $y_0 \in Y$. We identify $\pi_1(Y, y_0) = F_m$, the free group on m symbols corresponding to loops in Y. Then the covering is classified by a surjective homomorphism $F_m \to G$. In favorable cases, but not always, there is a unique such homomorphism up to automorphisms of F_m and of G.

Recall that the automorphism group of the free group F_m, with basis $e_1, \dots e_m$, is generated by Nielsen transformations: (1) replace one generator by its product (on the left, say) with another generator and (2) invert one generator, or (3) permute generators.

PROPOSITION 5.3. *If G is a finite cyclic group of order p, then up to automorphisms of F_m and G, there is a unique surjective homomorphism $F_m \to G$.*

PROOF SKETCH. This is a special case of a known result, so we just indicate the argument. Any surjective homomorphism $F_m \to C_p$ is given by a sequence $(g_1, g_s, \dots, g_m)$ of elements of G. Using Nielsen transformations, one can convert this into a sequence of the form $(g_0^i, 1, \dots, 1)$, where $g_0 \in G$ is a given generator. Then an automorphism of G converts this into $(g_0, 1, \dots, 1)$. $\square$

REMARK 3. The result holds more generally if G is a finite abelian group and $n \geq \operatorname{rank} G$, or if G is a dihedral group. But in general there can be more than one such homomorphism $F_m \to G$. See McCullough and Wanderley [**6**].

COROLLARY 5.4. *If G is a finite cyclic group of order p, then any free action of G on a 1-complex X induces representation on integral homology of the form $H_1(X;\mathbb{Z}) = \mathbb{Z} \oplus \mathbb{Z}[G]^r$.*

PROOF. Let the cyclic group $G = C_p$, have a chosen generator g_0. By covering space theory, the free action is given by a surjective homomorphism $F_m \to G$. By the preceding proposition this homomorphism is weakly equivalent to the homomorphism $F_m \to C_p$ given by $e_1 \mapsto g_0$ and $e_i \mapsto 1$ for $i > 1$, which clearly yields such a representation on H_1. $\square$

5.3. Counting actions. The discussion in this section provides the final piece to giving a complete count of the actions of the cyclic group G of prime order p on 1-complexes X of Euler characteristic $\chi(X) = 1 - n$: We already saw that for each way of writing $n = t + a(p-1) + rp$ there exist exactly $P_{a+1}(t)$ semifree actions with nonempty fixed point set, where $P_{a+1}(t)$ denotes the partitions of t into at most $a+1$ parts, describing how t loops are distributed among the $a+1$ components of the fixed point set. Now we know that, In addition, if $t = 1$ and $a = 0$, then there is one additional *free* action.

6. Discussion

One of the more interesting remaining questions arising from the present work, is whether an action of a finite group G on a finite, connected 1-complex X is necessarily G-homotopy equivalent to an action that is semifree, provided it induces on homology the sort of representation we have been considering.

CONJECTURE 1. *If a finite group G acts on a finite, connected 1-complex X inducing a representation of the form $H_1(X) = \mathbb{Z}^t \oplus I[G]^a \oplus \mathbb{Z}[G]^r$, then the action is G-homotopy equivalent to a semifree action.*

Here is one tentative result in this direction, where we assume the finite group is solvable in order to proceed by induction on the order of the group.

THEOREM 6.1. *If a finite* solvable *group G acts on a finite connected 1-complex X so that $H_1(X) = \mathbb{Z}[G]^r$, then the fixed point set X^G is acyclic, and the action is G-homotopy equivalent to a semifree action with a single fixed point.*

PROOF SKETCH. We proceed by induction on the order of the group to prove that X^G is acyclic. The result is clearly true for a cyclic group of prime order, by earlier work. Inductively suppose that N is a proper normal subgroup of G of index p. By induction X^N is acyclic. Moreover G/N acts on on the tree X^N, with $X^G = (X^N)^{G/N}$. It follows that X^G is acyclic. Actually, the full singular set of points with nontrivial isotropy group is acyclic, which one can see by observing that it is the union of acyclic pieces with acyclic intersections, since

$$X^{g_1} \cap X^{g_2} \cap \dots X^{g_k} = X^{\langle g_1, g_2, \dots, g_k \rangle}$$

It follows that if the action is G-reduced then the full singular set is a single point. □

If G is not solvable it seems that one can show this way only that the fixed point set consists of acyclic components, with only a bound on the number of components.

A weaker question would be whether a condition on integral homology could at least imply the existence of fixed points.

CONJECTURE 2. *If a finite group G acts on a finite, connected 1-complex X inducing a representation of the form $H_1(X) = \mathbb{Z}^t \oplus I[G]^a \oplus \mathbb{Z}[G]^r$, then $X^G \neq \emptyset$, provided $t - a \neq 1$.*

The condition $t - a \neq 1$ insures that each element of G has nonzero Lefschetz number and thus has nonempty fixed point set.

References

[1] Kenneth S. Brown, *Cohomology of groups*, Graduate Texts in Mathematics, vol. 87, Springer-Verlag, New York, 1994. Corrected reprint of the 1982 original. MR1324339

[2] Marc Culler, *Finite groups of outer automorphisms of a free group*, Contributions to group theory, Contemp. Math., vol. 33, Amer. Math. Soc., Providence, RI, 1984, pp. 197–207, DOI 10.1090/conm/033/767107. MR767107

[3] István Estélyi, Ján Karabáš, Roman Nedela, and Alexander Mednykh, *On a representation of the automorphism group of a graph in a unimodular group*, Discrete Math. **344** (2021), no. 12, Paper No. 112606, 4, DOI 10.1016/j.disc.2021.112606. MR4308002

[4] Wolfgang Gaschütz, *Über modulare Darstellungen endlicher Gruppen, die von freien Gruppen induziert werden* (German), Math. Z. **60** (1954), 274–286, DOI 10.1007/BF01187377. MR65564

[5] Fritz Grunewald, Michael Larsen, Alexander Lubotzky, and Justin Malestein, *Arithmetic quotients of the mapping class group*, Geom. Funct. Anal. **25** (2015), no. 5, 1493–1542, DOI 10.1007/s00039-015-0352-5. MR3426060

[6] Darryl McCullough and Marcus Wanderley, *Free actions on handlebodies*, J. Pure Appl. Algebra **181** (2003), no. 1, 85–104, DOI 10.1016/S0022-4049(02)00308-0. MR1971807

[7] Jean-Pierre Serre, *Trees*, Springer Monographs in Mathematics, Springer-Verlag, Berlin, 2003. Translated from the French original by John Stillwell; Corrected 2nd printing of the 1980 English translation. MR1954121

[8] Bruno Zimmermann, *Über Homöomorphismen n-dimensionaler Henkelkörper und endliche Erweiterungen von Schottky-Gruppen* (German), Comment. Math. Helv. **56** (1981), no. 3, 474–486, DOI 10.1007/BF02566224. MR639363

[9] Bruno Zimmermann, *Finite groups of outer automorphisms of free groups*, Glasgow Math. J. **38** (1996), no. 3, 275–282, DOI 10.1017/S0017089500031700. MR1417356

[10] Bruno P. Zimmermann, *A note on the Nielsen realization problem for connected sums of* $S^2 \times S^1$, Rend. Istit. Mat. Univ. Trieste **53** (2021), Art. No. 15, 7, DOI 10.1112/blms.12389. MR4442770

DEPARTMENT OF MATHEMATICS, INDIANA UNIVERSITY, BLOOMINGTON, INDIANA 47405
Email address: edmonds@indiana.edu

Contemporary Mathematics
Volume **811**, 2025
https://doi.org/10.1090/conm/811/16239

Affine structures on surfaces and the twisted cubic cone

William M. Goldman

Dedicated to Ravi S. Kulkarni, with admiration

ABSTRACT. We identify the deformation space of marked complete affine strctures on the 2-torus $\mathbb{T}^2$ with the cone over a twisted cubic curve in $\mathbb{R}\mathsf{P}^3$.

CONTENTS

Introduction
Outline of the paper
1. The cone on the twisted cubic
2. Affine structures on the torus
3. Commutative nilpotent algebras
4. Flat torsionfree equiaffine connections
Acknowledgments
References

Introduction

An affine structure on an n-dimensional manifold is defined by a system of local coordinates where the coordinate changes are locally defined by affine transformations of $\mathbb{R}^n$. In this way the manifold is locally modeled on an affine space A (corresponding to $\mathbb{R}^n$, but without the algebraic structure of a vector space). Equivalently, an affine structure on a manifold is a flat torsionfree affine connection. In [**12**], Kuiper described the (geodesically) complete affine structures on $\mathbb{T}^2$ and showed they are of two types: either Eucildean structures (*flat Riemannian structures*), or other structures defined by flat but *non-Riemannian* connections. Figures 1 and 2 depict these two types of structures. In this paper we show that they naturally form a *twisted cubic cone* in $\mathbb{R}^4$.

In a more geometric context, a *complete affine structure* on a manifold M is a representation of M as a quotient $\Gamma\backslash\mathsf{A}$, where A is an affine space, and $\Gamma < \mathsf{Aff}(\mathsf{A})$ a discrete group of affine transformations acting properly on A. More generally, Kuiper classifies affine manifolds covered by *convex domains* in A^2. He shows that

2020 *Mathematics Subject Classification.* Primary 57M50; Secondary 22F50, 57K20.

The author was partially supported through NSF grants DMS1709791 and DMS2203493, as well as a membership in the School of Mathematics at the Institute for Advanced Study in 2022-2023.

if a closed surface admits such a structure then $\chi(M) = 0$, and conjectures this is true without assuming convexity. This was later proved by Benzécri [**7**].

If Σ is a fixed surface, a *marked affine structure* on Σ is a pair (f, M) where M is an affine manifold and f is diffeomorphism $\Sigma \to M$. Marked affine structures (f, M) and (f', M') are *equivalent* if they differ (up to isotopy) by an affine isomorphism $M \xrightarrow{\phi} M'$ of affine manifolds such that f is isotopic to $f' \circ \phi$. The space of equivalence classes of marked structures is called the *deformation space.* It has a natural topology (see [**10**], §6 Classification) which in general can be quite pathological. However, in many cases it is locally homeomorphic to the "finite-dimensional" space $\mathsf{Hom}(\pi_1(\Sigma), G)/G$ where $G = \mathsf{Aff}(\mathsf{A})$. [1] This space is analogous to the Teichmüller space, consisting of equivalence classes of marked Riemann surfaces. The mapping class group $\mathsf{Mod}(\Sigma)$ acts simply transitively on the markings, and its quotient is the Riemann moduli space of Riemann surfaces homeomorphic to Σ.

The classification of affine structures on surfaces was completed by Nagano-Yagi [**13**] and Arrowsmith-Furness [**1**]. In particular the deformation space of *all* affine structures on $\mathbb{T}^2$ is not Hausdorff. In this paper we discuss the subspace $\mathcal{CA}(\mathbb{T}^2)$ corresponding to *complete* structures. Baues [**2**–**4**] showed that $\mathcal{CA}(\mathbb{T}^2)$ is homeomorphic [2] to $\mathbb{R}^2$.

A parametrization of this space by a "period map" is given in [**6**], and raises the question whether this space has a natural *singular* smooth structure. This paper resolves this question.

The deformation space of *marked* complete structures admits a natural action of the *mapping class group* $\mathsf{Mod}(\mathbb{T}^2) \cong \mathsf{GL}(2, \mathbb{Z})$, permuting the markings. Baues observed that this is the natural *linear* action of $\mathsf{GL}(2, \mathbb{Z})$ on $\mathbb{R}^2$. This action is chaotic, and the quotient $\mathsf{GL}(2, \mathbb{Z})\backslash\mathbb{R}^2$ is intractable. Indeed, this dynamical system is orbit-equivalent to the horocycle flow on the unit tangent bundle $\mathsf{SL}(2, \mathbb{Z})\backslash\mathsf{SL}(2, \mathbb{R})$ of the elliptic modular curve, which is known to be uniquely ergodic.

This dramatically contrasts with the action of $\mathsf{Mod}(\mathbb{T}^2)$ on the deformation space of Euclidean structures. That action is *proper,* with quotient $\mathbb{R}_+ \times \mathfrak{M}_0$ where $\mathfrak{M}_0$ the Riemann moduli space of elliptic curves and the $\mathbb{R}_+$ parameter corresponds to the area of a Euclidean structure on a torus. Of course, Euclidean structures on $\mathbb{T}^2$ are *complete affine* structures. However, since all Euclidean structures are affinely equivalent, the subspace

$$\mathcal{CA}(\mathbb{T}^2)_{\mathbf{Euc}} \subset \mathcal{CA}(\mathbb{T}^2)$$

corresponding to Eudclidean structures collapses to a single point in $\mathcal{CA}(\mathbb{T}^2)$. This point is the origin in $\mathbb{R}^2 \approx \mathcal{CA}(\mathbb{T}^2)$, which maps to the unique singular point in the twisted cubic cone.

The main theorem of the paper is the following:

THEOREM. *The deformation space of affine equivalence classes of marked complete affine structures on* $\mathbb{T}^2$ *naturally identifies with a twisted cubic cone* $\mathfrak{C} \subset \mathbb{R}^4$.

---

[1] $\mathsf{Hom}(\pi_1(\Sigma), G)$ has the natural structure as a real affine algebraic set, and $\mathsf{Hom}(\pi_1(\Sigma), G)/G$ is given the quotient topology by the action of $\mathsf{Inn}(G)$ by composition.

[2] Indeed, for a while it was believed [**9**] that the subspace comprising *complete* structures is not Hausdorff. Baues [**2**] addresses this error in the literature.

FIGURE 1. One example of a complete affine 2-manifold is a Euclidean flat torus, obtained as the quotient of the Eudlidean plane E^2 by a lattice $\Lambda \in \mathbb{R}^2$ of translations.

Outline of the paper

§1 describes the cone on the twisted cubic and its symmetries. §2 describes the theory of affine structures and reduces the classification to commutative nilpotent 2-dimensional $\mathbb{R}$-algebras, which are studied in §3. Such an algebra $\mathfrak{A}$, together with a basis, completely determines the marked structure. From the general theory, this already gives a proof of our main theorem. One key point is that completeness is equivalent to the connection being *equiaffine* (sometimes called *parallel volume*). This means its holonomy preserves volume, that is, its linear holonomy has determinant ± 1. In this particular case, it is equivalent to the stronger condition of unipotence of the linear holonomy. Compare the discussion of the Markus conjecture in [**10**], §11 and the proof of completeness of closed affine manifolds with unipotent holonomy in [**10**], §8.4.

§4 gives a direct proof of the main theorem, using the Christoffel symbols $\mathbf{\Gamma}^k_{ij}$ of the flat torsionfree equiaffine connection, that the deformation space $\mathcal{CA}(\mathbb{T}^2)$ of complete affine structures identifies with the cone $\mathfrak{C}$ on the twisted cubic. The Christoffel symbols are just the structure constants of the algebra $\mathfrak{A}$. The key point is that Kuiper's classification implies that the identity component of the affine automorphism group acts simply transitively on M. This implies that the Christoffel symbols are constant, enabling the identification of marked structures with nilpotent commutative associative 2-dimensional algebras over $\mathbb{R}$.

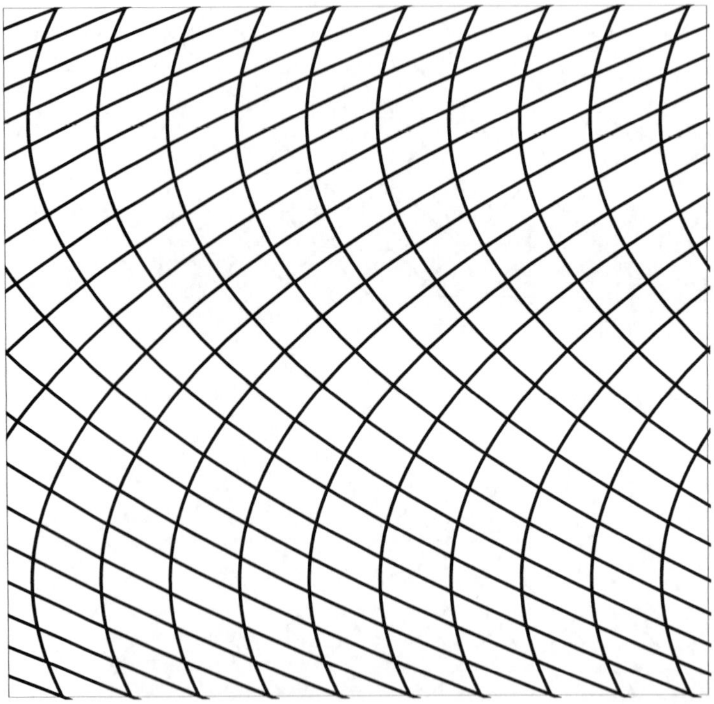

FIGURE 2. Applying a polynomial diffeomorphism of A^2 such as $(x, y) \overset{f}{\mapsto} (x + y^2, y)$ conjugates translations to a simply transitive affine action of $\mathbb{R}^2$ which is *not* Euclidean-isometric. The quotient $\mathsf{A}^2/f\Lambda f^{-1}$ is a complete affine 2-torus.

1. The cone on the twisted cubic

The *twisted cubic cone* $\mathfrak{C} \subset \mathbb{R}^4$ is the image of the map

$$\mathbb{R}^2 \hookrightarrow \mathsf{Sym}^3(\mathbb{R}^2) \cong \mathbb{R}^4 \tag{1.1}$$

$$\begin{bmatrix} A \\ B \end{bmatrix} \longmapsto \begin{bmatrix} A^3 \\ A^2B \\ AB^2 \\ B^3 \end{bmatrix} =: \begin{bmatrix} U \\ X \\ Y \\ Z \end{bmatrix} \in \mathbb{R}^4.$$

which embeds $\mathbb{R}^2$ topologically in $\mathbb{R}^4$. It is singular, and its unique singularity is the origin. It is the cone over the rational normal curve of degree 3 in $\mathbb{R}\mathsf{P}^3$.

Explicitly, $\mathfrak{C}$ is defined by the three homogeneous quadratic equations

$$X^2 = UY$$
$$UZ = XY$$
$$Y^2 = XZ$$

in the coordinates U, X, Y, Z. No two of these equations suffice to define $\mathfrak{C}$, despite $\mathfrak{C}$ having codimension two in $\mathbb{R}^4$.

The map (1.1) is $\mathsf{GL}(2,\mathbb{R})$-equivariant with respect to the standard action on $\mathbb{R}^2$ and the induced action on the symmetric power $\mathsf{Sym}^3(\mathbb{R}^2)$. The cone $\mathfrak{C} \subset \mathbb{R}^4$ is invariant under this action of $\mathsf{GL}(2,\mathbb{R})$. In particular this action restricts to $\mathsf{GL}(2,\mathbb{Z}) < \mathsf{GL}(2,\mathbb{R})$, and $\mathsf{GL}(2,\mathbb{Z})$ is isomorphic to the mapping class group $\mathsf{Mod}(\mathbb{T}^2)$.

Later in this paper we identify $\mathfrak{C}$ with the deformation space $\mathcal{CA}(\mathbb{T}^2)$ of marked complete affine structures on $\mathbb{T}^2$ and the $\mathsf{GL}(2,\mathbb{Z})$-action on $\mathfrak{C}$ with the $\mathsf{Mod}(\mathbb{T}^2)$-action on $\mathcal{CA}(\mathbb{T}^2)$.

2. Affine structures on the torus

Kuiper [**12**] showed that every complete affine structure on the torus $\mathbb{T}^2$ arises as a flat Euclidean torus $\Lambda\backslash\mathsf{E}^2$ where $\Lambda < \mathbb{R}^2$ is a lattice of translations or a quotient $\Gamma\backslash\mathsf{A}^2$ where $\Gamma < G$ is a lattice where G is the group of affine transformations

$$\begin{bmatrix} x \\ y \end{bmatrix} \longmapsto \begin{bmatrix} 1 & 2b \\ 0 & 1 \end{bmatrix} \begin{bmatrix} x \\ y \end{bmatrix} + \begin{bmatrix} a+b^2 \\ b \end{bmatrix}$$

of A^2, where $a, b \in \mathbb{R}$. The group G is isomorphic to $\mathbb{R}^2$; indeed is conjugate to the translation subgroup $\mathbb{R}^2 < \mathsf{Aff}(\mathsf{A}^2)$ by the diffeomorphism $(x,y) \longmapsto (x+y^2, y)$.

Since the action of G on A^2 is simply transitive, it induces a left-invariant[3] affine structure on G. Furthermore this passes down to an invariant affine structure on the Lie group $\Gamma\backslash G$, so the complete affine structures on $\mathbb{T}^2$ are all invariant structures on $\mathbb{T}^2$ with respect to its structure as an abelian Lie group.

Left-invariant affine structures on Lie groups form a rich algebraic theory described by (possibly non-associative) algebras, where the algebraic structure is defined by covariant differentiation of left-invariant vector fields. Commutator in this algebra is just the usual Lie bracket of left-invariant vector fields (because the connection is torsionfree). Flatness of the connection implies the defining condition that the associator is symmetric in its first two arguments; hence these algebras are called *left-symmetric algebras*. Bi-invariant affine structures correspond to the case when this algebra is associative.[4]

Completeness for these structures is equivalent to several conditions, such as the parallelism of right-invariant volume forms, or unipotence of the affine action by left-multiplication

The theory of affine structures on closed manifolds with nilpotent holonomy is studied in [**8**] (building on Smillie [**14**] for abelian holonomy). In particular the equivalence of geodesic completeness with parallel volume (or *equiaffinity*) and unipotent linear holonomy is shown there, and expounded in §11 of [**10**] in a broader context. For left-invariant affine structures on Lie groups — that is, for *affine Lie groups,* — completeness is equivalent to right-invariant volume forms being parallel [**11**]. Thus for *abelian* affine Lie groups, completeness is equivalent to equiaffinity.

As mentioned in the introduction, $\mathbb{T}^2$ is the only closed orientable surface supporting an affine structure. Furthermore, a complete affine structure on $\mathbb{T}^2$ is

[3]Since G is abelian, it is also right-invariant.

[4]See §10 of [**10**] for an extensive discussion of this theory.

necessarily invariant Equivalently, every complete affine structure on $\mathbb{T}^2$ is invariant under a structure of $\mathbb{T}^2$ as an *abelian Lie group.* [5] Baues [**3**, **4**] surveys the classification of affine structures on surfaces in detail; see also [**10**].

Since the connection is invariant, the covariant derivative of invariant vector fields is invariant, and the *Christoffel symbols,* the coefficients of the covariant derivatives of a basis of invariant vector fields are constant.

3. Commutative nilpotent algebras

THEOREM 3.1. *A marked complete affine structure on* $\mathbb{T}^2$ *corresponds to a based 2-dimensional vector space* V *together with a symmetric bilinear form* Γ *making* V *into an algebra* $\mathfrak{A}$*, such that* $\mathfrak{A}$ *is a commutative* $\mathbb{R}$*-algebra with* $\mathfrak{A}^3 = 0$.

For an extensive discussion, see [**10**],§8.4. For the reader's convenience we sketch the proof here.

PROOF. Let $\mathbb{T}^2 \longrightarrow M$ be a marked complete affine structure, that is a homeomorphism (defined up to isotopy) onto a complete affine 2-manifold M. Choose a universal covering $\widetilde{M} \to M$ and a developing pair (dev, h) where the developing map $tM \xrightarrow{\mathsf{dev}} \mathsf{A}^2$ is a diffeomorphism and the holonomy homomorphism h is an embedding of $\pi_1(M)$ onto a discrete subgroup $\Gamma < \mathsf{Aff}(\mathsf{A}^2)$ acting properly on A^2. By the general algebraic theory described above (or direct calculation as in Kuiper's original classification), Γ is conjugate to a lattice in a subgroup G_μ of $\mathsf{Aff}(\mathsf{A}^2)$ comprising elements having the form

$$\begin{bmatrix} x \\ y \end{bmatrix} \longmapsto \begin{bmatrix} 1 & 2\mu b \\ 0 & 1 \end{bmatrix} \begin{bmatrix} x \\ y \end{bmatrix} + \begin{bmatrix} a + \mu b^2/2 \\ b \end{bmatrix}$$

where $a, b \in \mathbb{R}$ and $\mu \in \mathbb{R}$ is a fixed parameter. The Lie algebra of G is generated by affine vector fields:

$$X := \partial/\partial x$$
$$Y_\mu := \mu y\, \partial/\partial x + \partial/\partial y.$$

Under covariant differentiation these vector fields span an algebra with multiplication given by Table 1.

TABLE 1. A commutative 2-dimensional algebra with $\mathfrak{A}^3 = 0$

	X	Y_μ
X	0	0
Y_μ	0	μX

Furthermore every 2-dimensional commutative associative algebra $\mathfrak{A}$ with $\mathfrak{A}^3 = 0$ has a basis as above: If $\mathfrak{A}^2 = 0$, the multiplication is trivial and we may take any basis X, Y (and $\mu = 0$). Otherwise nilpotence implies that $\dim \mathfrak{A}^2 = 1$. Let X base $\mathfrak{A}^2$, and extend $\{X\}$ to a basis $\{X, Y\}$ of $\mathfrak{A}$. □

[5]This was known to Kuiper [**12**], although he didn't state it in this form.

In practice V will be the abelian Lie algebra of left-invariant (right-invariant) vector fields, and Γ is the covariant derivative operator

$$\mathsf{V} \times \mathsf{V} \xrightarrow{\Gamma} \mathsf{V}$$

$$(\xi, \eta) \longmapsto \nabla_\xi \eta.$$

The above theory implies that $\mathfrak{A}$ is nilpotent, and indeed $\mathfrak{A}^3 = 0$ (so associativity is obvious).

Such a connection can be described in terms of the twisted cubic cone (1.1) as follows.

The connection (or the algebra structure) relates to the symmetric power $\mathsf{Sym}^3(\mathsf{V})$ as follows. For convenience choose a nonzero element ω in the line $\bigwedge^2(\mathsf{V})$. A vector $v \in \mathsf{V}$ determines a covector $\omega_v \in \mathsf{V}^*$ by

$$u \wedge v = \omega_v(u)\omega.$$

An endomorphism $\mathcal{E} \in \mathsf{End}(\mathsf{V})$ of rank one is determined by nonzero vectors k (generating the kernel) and j (generating the image) by

$$\mathcal{E} = \omega_k \otimes j.$$

The endomorphism $\mathcal{E}$ is nilpotent if and only if j and k are linearly dependent. In our case, unipotence of the linear action is equivalent to nilpotence of $\mathcal{E}$.

A connection is given by the tensor product $\psi \otimes \mathcal{E}$ where $\psi \in \mathsf{V}^*$ is a covector and $\mathcal{E}$ is an endomorphism. Flatness of the corresponding connection means that in the decomposition $\mathcal{E} = \omega_v \otimes w$, the covectors ω_v and ψ are linearly dependent. Hence a flat connection taking values in nilpotent endomorphisms corresponds to an element of $\mathsf{V}^* \otimes \mathsf{End}(\mathsf{V})$ of the form $\omega_v \otimes \mathcal{E}_v$ where $\mathcal{E}_v := \omega_v \otimes v$, that is in the image of

$$\mathsf{V} \longrightarrow \mathsf{V}^* \otimes \mathsf{End}(\mathsf{V})$$

$$v \longmapsto \omega_v \otimes \big(\omega_v \otimes v\big)$$

which is evidently equivalent to (1.1). The final section §4 gives a direct identification using only the assumption of area-preserving holonomy rather than unipotent holonomy.

4. Flat torsionfree equiaffine connections

Choose a coordinate system (x^1, x^2) on A^2 and coordinate vector fields

$$\partial_i := \frac{\partial}{\partial x^i}$$

for $i = 1, 2$. Write ∇_i for the covariant differential operator $\nabla_{\partial i}$ and $\mathbf{\Gamma}^k_{ij}$ for the *Christoffel symbols* for ∇ with respect to the frame ∂_i:

$$\nabla_i \partial_j = \mathbf{\Gamma}^k_{ij} \partial_k$$

(Einstein summation). If ∇ has zero torsion, $\nabla_i \partial_j = \nabla_j \partial_i$, and

$$\mathbf{\Gamma}^k_{21} = \mathbf{\Gamma}^k_{12}. \tag{4.1}$$

for $k = 1, 2$. Since ∇ is *equiaffine* $\nabla\big(dx^1 \wedge dx^2\big) = 0$. It follows that for every vector field ξ, the curvature operator $\eta \longmapsto \nabla_\xi \eta$ has trace zero. Take $\xi = \partial_i$ for $i = 1, 2$ to obtain:

$$\mathbf{\Gamma}^1_{i1} + \mathbf{\Gamma}^2_{i2} = 0 \tag{4.2}$$

for $i = 1, 2$. In the following calculations, (4.2) and (4.1) imply

$$\begin{aligned} \mathbf{\Gamma}^1_{21} &= \mathbf{\Gamma}^1_{12} = -\mathbf{\Gamma}^2_{22} \\ \mathbf{\Gamma}^2_{21} &= \mathbf{\Gamma}^2_{12} = -\mathbf{\Gamma}^1_{11} \end{aligned} \tag{4.3}$$

so we reduce our calculations to the four variables

$$\mathbf{\Gamma}^1_{11},\ \mathbf{\Gamma}^2_{11},\ \mathbf{\Gamma}^1_{22},\ \mathbf{\Gamma}^2_{22}.$$

Since the covariant derivatives are constant,

$$\begin{aligned} \nabla_1\nabla_2\partial_1 &= \nabla_1\big(\mathbf{\Gamma}^1_{21}\,\partial_1 + \mathbf{\Gamma}^2_{21}\,\partial_2\big) \\ &= \big(\mathbf{\Gamma}^1_{11}\mathbf{\Gamma}^1_{21} + \mathbf{\Gamma}^1_{12}\mathbf{\Gamma}^2_{21}\big)\ \partial_1 + \big(\mathbf{\Gamma}^2_{11}\mathbf{\Gamma}^1_{21} + \mathbf{\Gamma}^2_{12}\mathbf{\Gamma}^2_{21}\big)\,\partial_2 \end{aligned} \tag{4.4}$$

$$\begin{aligned} \nabla_2\nabla_1\partial_1 &= \nabla_2\big(\mathbf{\Gamma}^1_{11}\,\partial_1 + \mathbf{\Gamma}^2_{11}\,\partial_2\big) \\ &= \big(\mathbf{\Gamma}^1_{21}\mathbf{\Gamma}^1_{11} + \mathbf{\Gamma}^1_{22}\mathbf{\Gamma}^2_{11}\big)\,\partial_1 + \big(\mathbf{\Gamma}^2_{21}\mathbf{\Gamma}^1_{11} + \mathbf{\Gamma}^2_{22}\mathbf{\Gamma}^2_{11}\big)\,\partial_2 \end{aligned} \tag{4.5}$$

$$\begin{aligned} \nabla_1\nabla_2\partial_2 &= \nabla_1\big(\mathbf{\Gamma}^1_{22}\,\partial_1 + \mathbf{\Gamma}^2_{22}\,\partial_2\big) \\ &= \big(\mathbf{\Gamma}^1_{11}\mathbf{\Gamma}^1_{22} + \mathbf{\Gamma}^1_{12}\mathbf{\Gamma}^2_{22}\big)\,\partial_1 + \big(\mathbf{\Gamma}^2_{11}\mathbf{\Gamma}^1_{22} + \mathbf{\Gamma}^2_{12}\mathbf{\Gamma}^2_{22}\big)\,\partial_2 \end{aligned} \tag{4.6}$$

$$\begin{aligned} \nabla_2\nabla_1\partial_2 &= \nabla_2\big(\mathbf{\Gamma}^1_{12}\,\partial_1 + \mathbf{\Gamma}^2_{12}\,\partial_2\big) \\ &= \big(\mathbf{\Gamma}^1_{21}\mathbf{\Gamma}^1_{12} + \mathbf{\Gamma}^1_{22}\mathbf{\Gamma}^2_{12}\big)\,\partial_1 + \big(\mathbf{\Gamma}^2_{21}\mathbf{\Gamma}^1_{12} + \mathbf{\Gamma}^2_{22}\mathbf{\Gamma}^2_{12}\big)\,\partial_2 \end{aligned} \tag{4.7}$$

Flatness of ∇ implies $\nabla_1 \circ \nabla_2 = \nabla_2 \circ \nabla_1$. Subtracting (4.5) from (4.4), $[\nabla_1, \nabla_2]\,\partial_1 = 0$ implies:

$$\mathbf{\Gamma}^1_{12}\mathbf{\Gamma}^2_{21} = \mathbf{\Gamma}^1_{22}\mathbf{\Gamma}^2_{11} \tag{4.8}$$

for the ∂_1-component of $[\nabla_1, \nabla_2]\partial_1$

$$\mathbf{\Gamma}^2_{11}\mathbf{\Gamma}^1_{21} + \mathbf{\Gamma}^2_{12}\mathbf{\Gamma}^2_{21} = \mathbf{\Gamma}^2_{21}\mathbf{\Gamma}^1_{11} + \mathbf{\Gamma}^2_{22}\mathbf{\Gamma}^2_{11} \tag{4.9}$$

for the ∂_2-component of $[\nabla_1, \nabla_2]\partial_1$.

Subtracting (4.7) from (4.6), $[\nabla_1, \nabla_2]\,\partial_2 = 0$ implies:

$$\mathbf{\Gamma}^1_{11}\mathbf{\Gamma}^1_{22} + \mathbf{\Gamma}^1_{12}\mathbf{\Gamma}^2_{22} = \mathbf{\Gamma}^1_{21}\mathbf{\Gamma}^1_{12} + \mathbf{\Gamma}^1_{22}\mathbf{\Gamma}^2_{12} \tag{4.10}$$

for the ∂_1-component of $[\nabla_1, \nabla_2]\partial_2$.

$$\mathbf{\Gamma}^2_{11}\mathbf{\Gamma}^1_{22} = \mathbf{\Gamma}^2_{21}\mathbf{\Gamma}^1_{12} \tag{4.11}$$

for the ∂_2-component of $[\nabla_1, \nabla_2]\partial_2$.

Now apply (4.3) to rewrite (4.8), (4.9), (4.10), and (4.11) in terms of the four variables

$$\mathbf{\Gamma}^1_{11},\ \mathbf{\Gamma}^2_{11},\ \mathbf{\Gamma}^1_{22},\ \mathbf{\Gamma}^2_{22}.$$

First, (4.8) becomes

$$\mathbf{\Gamma}^1_{11}\mathbf{\Gamma}^2_{22} = \mathbf{\Gamma}^2_{11}\mathbf{\Gamma}^1_{22}, \tag{4.12}$$

Since ∇ is equiaffine, the curvature tensor takes values in traceless endomorphisms. Thus the ∂_1-component of $[\nabla_1, \nabla_2]\,\partial_2$ is the negative of ∂_2-component of $[\nabla_1, \nabla_2]\,\partial_1$. Thus (4.11) is equivalent to (4.8) and (4.11) provides nothing new.

Next, (4.9) and (4.10) respectively become:

$$\big(\mathbf{\Gamma}^1_{11}\big)^2 = \mathbf{\Gamma}^2_{11}\mathbf{\Gamma}^2_{22} \tag{4.13}$$

$$\big(\mathbf{\Gamma}^2_{22}\big)^2 = \mathbf{\Gamma}^1_{11}\mathbf{\Gamma}^1_{22}. \tag{4.14}$$

The three equations (4.12), (4.13) and (4.14) now describe the twisted cubic cone $\mathfrak{C}$ as in §1, by taking:

$$\begin{aligned} U &= \mathbf{\Gamma}_{11}^2 \\ X &= \mathbf{\Gamma}_{11}^1 \\ Y &= \mathbf{\Gamma}_{22}^2 \\ Z &= \mathbf{\Gamma}_{22}^1, \end{aligned}$$

thus identifying the deformation space $\mathcal{CA}(\mathbb{T}^2)$ of marked complete affine structures on $\mathbb{T}^2$ with $\mathfrak{C}$.

Acknowledgments

I thank Pierre Deligne for this result and many inspiring conversations. This research was performed during a membership at the School of Mathematics of the Institute for Advanced Study in 2022-2023. Indeed it was during a lecture there that Deligne suggested the main theorem of this paper. I wish to express my gratitude to the Institute for their hospitality and excellent working conditions there.

I am also grateful to the anonymous referee for a careful reading and suggesting many improvements.

References

[1] P. M. D. Furness and D. K. Arrowsmith, *Locally symmetric spaces*, J. London Math. Soc. (2) **10** (1975), no. 4, 487–499, DOI 10.1112/jlms/s2-10.4.487. MR467598

[2] Oliver Baues, *Gluing affine 2-manifolds with polygons*, Geom. Dedicata **75** (1999), no. 1, 33–56, DOI 10.1023/A:1005064409224. MR1680196

[3] Oliver Baues, *Varieties of discontinuous groups*, Crystallographic groups and their generalizations (Kortrijk, 1999), Contemp. Math., vol. 262, Amer. Math. Soc., Providence, RI, 2000, pp. 147–158, DOI 10.1090/conm/262/04172. MR1796130

[4] Oliver Baues, *Deformation spaces for affine crystallographic groups*, Cohomology of groups and algebraic K-theory, Adv. Lect. Math. (ALM), vol. 12, Int. Press, Somerville, MA, 2010, pp. 55–129. MR2655175

[5] Oliver Baues, *The deformations of flat affine structures on the two-torus*, Handbook of Teichmüller theory Vol. IV, European Math. Soc. 2014, pp. 461–537.

[6] Oliver Baues and William M. Goldman, *Is the deformation space of complete affine structures on the 2-torus smooth?*, Geometry and Dynamics, Contemp. Math., vol. 389, Amer. Math. Soc., Providence, RI, 2005, pp. 69–89, DOI 10.1090/conm/389/07272. MR2181958

[7] Jean-Paul Benzécri, *Variétés localement affines*, Seminaire de Topologie et Géom. Diff., Ch. Ehresmann (1958–60) (1959), no. 7, 1–34.

[8] David Fried, William Goldman, and Morris W. Hirsch, *Affine manifolds with nilpotent holonomy*, Comment. Math. Helv. **56** (1981), no. 4, 487–523, DOI 10.1007/BF02566225. MR656210

[9] William M. Goldman, *Geometric structures on manifolds and varieties of representations*, Geometry of group representations (Boulder, CO, 1987), Contemp. Math., vol. 74, Amer. Math. Soc., Providence, RI, 1988, pp. 169–198, DOI 10.1090/conm/074/957518. MR957518

[10] William M. Goldman, *Geometric structures on manifolds*, Graduate Studies in Mathematics, vol. 227, American Mathematical Society, Providence, RI, 2022 ©2022, DOI 10.1090/gsm/227. MR4500072

[11] William M. Goldman and Morris W. Hirsch, *Affine manifolds and orbits of algebraic groups*, Trans. Amer. Math. Soc. **295** (1986), no. 1, 175–198, DOI 10.2307/2000152. MR831195

[12] N. H. Kuiper, *Sur les surfaces localement affines* (French), Géométrie différentielle. Colloques Internationaux du Centre National de la Recherche Scientifique, Strasbourg, 1953, CNRS, Paris, 1953, pp. 79–87. MR60288

[13] Tadashi Nagano and Katsumi Yagi, *The affine structures on the real two-torus. I*, Osaka Math. J. **11** (1974), 181–210. MR377917

[14] John David Smillie, *Affinely flat manifolds*, ProQuest LLC, Ann Arbor, MI, 1977. Thesis (Ph.D.)–The University of Chicago. MR2611759

Department of Mathematics, University of Maryland, College Park, Maryland 20742

Email address: `wmg@umd.edu`

Contemporary Mathematics
Volume **811**, 2025
https://doi.org/10.1090/conm/811/16240

Computing mixed multiplicities, mixed volumes and sectional Milnor numbers

Kriti Goel, Vivek Mukundan, Sudeshna Roy, and J. K. Verma

Dedicated to Professor Ravi Kulkarni on the occasion of his 80th birthday

ABSTRACT. This is an expository version of our paper [J. Softw. Algebra Geom., 13 (2023), 1–12]. Our aim is to present recent Macaulay2 algorithms for computation of mixed multiplicities of ideals in a Noetherian ring which is either local or a standard graded algebra over a field. These algorithms are based on computation of the equations of multi-Rees algebras of ideals that generalises a result of Cox, Lin and Sosa [Proc. Amer. Math. Soc. 147 (2019), 4605–4616]. Using these equations we propose efficient algorithms for computation of mixed volumes of convex lattice polytopes and sectional Milnor numbers of hypersurfaces with an isolated singularity.

1. Introduction

In this expository article we describe the Macaulay2 package `MixedMultiplicity` which computes the mixed multiplicities of ideals having positive grade in a Noetherian ring, mixed volume of a collection of convex lattice polytopes and sectional Milnor numbers of hypersurfaces with an isolated singularity. One of the main steps of the algorithms is the computation of the defining equations of the multi-Rees algebra

$$\mathcal{R}(I_1,\ldots,I_s)=R[I_1t_1,\ldots,I_st_s]=\bigoplus_{a_1,\ldots,a_s\geq 0} I_1^{a_1}\cdots I_s^{a_s}t_1^{a_1}\cdots t_s^{a_s}\subseteq R[t_1,\ldots,t_s]$$

of ideals $I_1,\ldots,I_s$ in a Noetherian ring R. When R is a polynomial ring of finite Krull dimension over any field k and $I_1,\ldots,I_s$ are monomial ideals in R, D. Cox, K.-N. Lin, and G. Sosa give an explicit formula for the defining equations of $\mathcal{R}(I_1,\ldots,I_s)$ in [**5**]. In this article, we obtain an analogue of their result for any set of ideals $I_1,\ldots,I_s$ in a Noetherian ring R such that each I_i has positive grade. The latter condition is always satisfied when R is a domain or for any ideal of positive height in a reduced ring or in a Cohen-Macaulay ring. Several authors have proposed algorithms to determine the defining equations of the multi-Rees algebra. For example, J. Ribbe [**20**, Proposition 3.1, Proposition 3.4], K.-N. Lin and C. Polini [**17**, Theorem 2.4], G. Sosa [**21**, Lemma 2.1], and B. Jabarnejad [**15**, Theorem 1].

2020 *Mathematics Subject Classification.* Primary 13-04, 13A30, 13H15.

Key words and phrases. Multi-Rees algebras, mixed multiplicities, sectional Milnor numbers, mixed volume.

They assumed $I_i = \mathfrak{m}^{a_i}$ or $I_i = I^{a_i}$, where I is an ideal of linear type, or I_i's are monomial ideals. We show the following.

THEOREM 1.1. *Let R be a Noetherian ring and $I_1, \ldots, I_s \subseteq R$ be ideals of positive grade. For each i, consider some generating set $\{f_{ij} \mid j = 1, \ldots, n_i\}$ of I_i which contains at least one nonzerodivisor f_{ij_i}. We set $h = \prod_{i=1}^{s} f_{ij_i}$. Consider the indeterminates $\underline{Y} = \{Y_{ij} \mid i = 1, \ldots, s, j = 1, \ldots, n_i\}$ and $\underline{T} = (T_1, \ldots, T_s)$. Define an R-algebra homomorphism $\varphi : S = R[\underline{Y}] \to \mathcal{R}(I_1, \ldots, I_s) \subseteq R[\underline{T}]$ by $Y_{ij} \mapsto f_{ij}T_i$ for all $i = 1, \ldots, s$, and $j = 1, \ldots, n_i$. Then*

$$\Gamma := \langle Y_{ij}f_{ij'} - Y_{ij'}f_{ij} \mid i = 1, \ldots, s \text{ and } j, j' \in \{1, \ldots, n_i\} \text{ with } j \neq j' \rangle : h^\infty \subseteq R[\underline{Y}].$$

is the defining ideal of $\mathcal{R}(I_1, \ldots, I_s)$, that is, $\Gamma = \ker \varphi$.

Using this result, we write a function `multiReesIdeal` to compute the defining ideal of a multi-Rees algebra in Macaulay2. When the ring is not a domain, the algorithm follows an analogue of the `reesIdeal` function, albeit for the multiple ideal setting.

The algorithm for computing defining equations of $\mathcal{R}(I_1, \ldots, I_s)$ helps us construct algorithms to compute *mixed multiplicities, mixed volume* and *sectional Milnor numbers* in the general setting. To justify importance of such an algorithm, we now recall these invariants and their importance.

1.1. Mixed multiplicities of ideals. Let $I_1, \ldots, I_r$ be a sequence of ideals in a local ring $(R, \mathfrak{m})$ (or a standard graded algebra over a field and $\mathfrak{m}$ is the maximal graded ideal) and let I_0 be an $\mathfrak{m}$-primary ideal. In [**16**], the authors prove that $\ell\left(I_0^{u_0} I_1^{u_1} \cdots I_r^{u_r} / I_0^{u_0+1} I_1^{u_1} \cdots I_r^{u_r}\right)$ is a polynomial $P(\underline{u})$, for u_i large, where $\underline{u} = (u_0, \ldots, u_r)$. Write this polynomial in the form

$$P(\underline{u}) = \sum_{\substack{\alpha \in \mathbb{N}^{r+1} \\ |\alpha| = t}} \frac{1}{\alpha!} e_\alpha(I_0 \mid I_1, \ldots, I_r) u^\alpha + \text{ lower degree terms},$$

where $t = \deg P(\underline{u}), \alpha = (\alpha_0, \ldots, \alpha_r) \in \mathbb{N}^{r+1}, \alpha! = \prod_{i=0}^{r} \alpha_i!$ and $|\alpha| = \sum_{i=0}^{r} \alpha_i$. If $|\alpha| = t$, then $e_\alpha(I_0 \mid I_1, \ldots, I_r)$ are called the *mixed multiplicities* of the ideals $I_0, I_1, \ldots, I_r$. In [**23**], the authors prove the following result.

THEOREM 1.2 ([**23**, Theorem 1.2]). *Assume that $d = \dim R/(0 : I^\infty) \geq 1$, where $I = I_1 \cdots I_r$. Then $\deg P(\underline{u}) = d - 1$.*

We now briefly describe several applications of mixed multiplicities of ideals. Suppose that $(R, \mathfrak{m})$ is a local ring of positive dimension d and $I_0 \ldots, I_r \subseteq R$ are ideals of positive height.

(i) Let $t_0, \ldots, t_r$ be indeterminates. Consider the multi-form rings

$$\operatorname{gr}_R(I_0, \ldots, I_r; I_i) = \mathcal{R}(I_0, \ldots, I_r)/(I_i) \quad \text{for } i = 0, 1, \ldots, r. \tag{1}$$

Note that if I_0 is an $\mathfrak{m}$-primary ideal and $I_1, \ldots, I_r$ are ideals of positive height, then

$$\ell\left(\frac{I_0^{u_0} I_1^{u_1} \cdots I_r^{u_r}}{I_0^{u_0+1} I_1^{u_1} \cdots I_r^{u_r}}\right) = \ell([\operatorname{gr}_R(I_0, \ldots, I_r; I_0)]_{\underline{u}})$$

is a polynomial of total degree $d - 1$ for $u_i \gg 0$ and $i = 0, 1, \ldots, r$ as we have observed above. Let us rewrite the terms in total degree $d - 1$ in the form

$$\sum_{\alpha \in \mathbb{N}^{r+1},\ |\alpha| = d-1} \frac{1}{\alpha!} e(I_0^{[\alpha_0+1]} | I_1^{[\alpha_1]} | \cdots | I_r^{[\alpha_r]}) u_0^{\alpha_0} \cdots u_r^{\alpha_r}.$$

For this case, M. Herrmann and his coauthors **[12]** expressed the multiplicity of $\mathrm{gr}_R(I_0,\ldots,I_r;I_0)$ by means of mixed multiplicities in the following way:

$$e(\mathrm{gr}_R(I_0,\ldots,I_r;I_0)) = \sum_{\substack{\alpha\in\mathbb{N}^{r+1}\\ |\alpha|=d-1}} e(I_0^{[\alpha_0+1]}|I_1^{[\alpha_1]}|\cdots|I_r^{[\alpha_r]}).$$

(ii) In **[24]**, J. K. Verma proved that

$$\begin{aligned} e(\mathcal{R}(I_1\ldots,I_r)) &= e(\mathrm{gr}_R(\mathfrak{m},I_1\ldots,I_r;\mathfrak{m}))\\ &= \sum_{\alpha\in\mathbb{N}^{r+1},\ |\alpha|=d-1} e(\mathfrak{m}^{[\alpha_0+1]}|I_1^{[\alpha_1]}|\cdots|I_r^{[\alpha_r]}).\end{aligned}$$

(iii) In **[7]**, C. D'Cruz showed similar result for the multi-graded extended Rees algebra $T:=\mathcal{B}(I_1,\ldots,I_r)=R[I_1t_1,\ldots,I_rt_r,t_1^{-1},\ldots,t_r^{-1}]$. She showed that

$$e(T_N)=\frac{1}{2^d}\left[\sum_{j=0}^{r}\sum_{\substack{\beta+\beta_1+\cdots+\beta_j=d-1\\ 1\le i_1<\cdots<i_j\le r}} 2^{\beta_1+\cdots+\beta_j}e(L^{[\beta+1]}|I_{i_1}^{[\beta_1]}|\cdots|I_{i_j}^{[\beta_j]})\right],$$

where $N:=\mathcal{N}(I_1,\ldots,I_r)=(I_1t_1,\ldots,I_rt_r,\mathfrak{m},t_1^{-1},\ldots,t_r^{-1})$ and $L=\mathfrak{m}^2+I_1+\cdots+I_r$.

In view of the above, we can easily compute multiplicities of Rees algebras, extended Rees algebras and certain form rings, once mixed multiplicities are known.

1.2. Mixed volumes of lattice polytopes. We first recall the definition of mixed volumes. Let P,Q be two polytopes (not necessarily distinct) in $\mathbb{R}^n$. Their Minkowski sum is defined as the polytope

$$P+Q:=\{a+b \mid a\in P, b\in Q\}.$$

Let $Q_1,\ldots,Q_n$ be an arbitrary collection of lattice polytopes in $\mathbb{R}^n$ and $\lambda_1,\ldots,\lambda_n\in\mathbb{R}_+$. H. Minkowski proved that the function $\mathrm{vol}_n(\lambda_1Q_1+\cdots+\lambda_nQ_n)$ is a homogeneous polynomial of degree n in $\lambda_1,\ldots,\lambda_n$, see **[6**, Theorem 4.9, Chapter 7]. The coefficient of $\lambda_1\cdots\lambda_n$ is called *the mixed volume* of $Q_1,\ldots,Q_n$ and denoted by $MV_n(Q_1,\ldots,Q_n)$. We set $[n]:=\{1,\ldots,n\}$. Then

$$MV_n(Q_1,\ldots,Q_n):=\sum_{\emptyset\neq J\subseteq[n]}(-1)^{n-|J|}V_n\left(\sum_{j\in J}Q_j\right),$$

see **[6**, Theorem 4.12 d., Chapter 7]. Here V_n denotes the n-dimensional Euclidean volume.

In **[23]**, the authors explore the relationship between mixed multiplicities of multigraded rings and mixed volumes.

PROPOSITION 1.3 ([**23**, Corollary 2.5]). *Let $Q_1,\ldots,Q_n$ be lattice polytopes in $\mathbb{R}^n$. Suppose that $R=k[x_0,x_1,\ldots,x_n]$ is a polynomial ring and $\mathfrak{m}$ is its graded maximal ideal. Let M_i be any set of monomials of the same degree in R such that Q_i is the convex hull of the lattice points of their dehomogenized monomials in $k[x_1,\ldots,x_n]$. Let I_j be the ideal of R generated by the monomials of M_j. Then*

$$MV_n(Q_1,\ldots,Q_n)=e_{(0,1,\ldots,1)}(\mathfrak{m}|I_1,\ldots,I_n).$$

Mixed volumes of lattice polytopes have diverse applications. We describe a few of them.

(1) In 1975, D. N. Bernstein proved that the mixed volume of the Newton polytopes of Laurent polynomials $f_1, \ldots, f_n \in k[x_1^{\pm 1}, \ldots, x_n^{\pm 1}]$ is a sharp upper bound for the number of isolated common zeros in the torus $(k^*)^n$ where k is an algebraically closed field. Furthermore, this bound is attained for a generic choice of coefficients in these n polynomials if char $k = 0$.

(2) In 1993, W. Fulton proved that the mixed volume of the Newton polytopes of Laurent polynomials $f_1, \ldots, f_n \in k[x_1^{\pm 1}, \ldots, x_n^{\pm 1}]$ is an upper bound of $\sum_\alpha i(\alpha)$, where $\alpha \in (\overline{k}^*)^n$ is any isolated common zero and $i(\alpha)$ denotes the intersection multiplicity at α.

(3) Let P be a lattice polytope of dimension n and $|P \cap \mathbb{Z}^n|$ denote the number of lattice points. In 1962, E. Ehrhart proved that the function $L(t \cdot P) = |tP \cap \mathbb{Z}^n|$, for all non-negative integers t, is a polynomial of degree n denoted by $E_P(t) \in \mathbb{Q}[t]$. The polynomial $E_P(t)$ is called the *Ehrhart polynomial* of P. For lattice polytopes $P_1, \ldots, P_k \subseteq \mathbb{R}^n$ in the integer lattice $\mathbb{Z}^n$, we have the following *mixed Ehrhart polynomial* in one variable t:

$$ME_{P_1,\ldots,P_k}(t) = \sum_{\emptyset \neq J \subseteq [k]} (-1)^{k-|J|} L(t \cdot \sum_{j \in J} P_j) \in \mathbb{Q}[t].$$

If $\dim\left(\sum_{i=1}^k P_i\right) = n$, then by [**11**, Corollary 2.5], the leading coefficient of $ME_{P_1,\ldots,P_k}(t)$ is

$$\sum_{\sum_{i=1}^k s_i = n,\ s_i \geq 1} \binom{n}{s_1, \ldots, s_k} MV_n\left(P_1[s_1], \ldots, P_k[s_k]\right),$$

where $\binom{n}{s_1,\ldots,s_k} := \frac{n!}{s_1! \cdots s_k!}$ and $MV_n\left(P_1[s_1], \ldots, P_k[s_k]\right)$ means that the polytope P_j is taken s_j times. In particular, when $k = n$,

$$ME_{P_1,\ldots,P_n}(t) = n! MV_n(P_1, \ldots, P_n) t^n.$$

The coefficient of t^{n-1} in $ME_{P_1,\ldots,P_k}(t)$ is also described in [**11**, Corollary 2.6].

1.3. Sectional Milnor numbers. Suppose that the origin is an isolated singular point of a complex analytic hypersurface $H = V(f) \subset \mathbb{C}^{n+1}$. Let f_{z_i} denote the partial derivative of f with respect to z_i. Set

$$\mu = \dim_{\mathbb{C}} \frac{\mathbb{C}\{z_0, z_1, \ldots, z_n\}}{(f_{z_0}, f_{z_1}, \ldots, f_{z_n})}.$$

The number μ is called the *Milnor number* of the hypersurface H at the origin. In his Cargèse paper [**22**], B. Teissier refined the notion of Milnor number by replacing it with a sequence of Milnor numbers of intersections with general linear subspaces. Let $(X, 0)$ be a germ of a hypersurface in $\mathbb{C}^{n+1}$ with an isolated singularity. The Milnor number of $X \cap E$, where E is a general linear subspace of dimension i passing through the origin, is called the ith*-sectional Milnor number* of X at the origin. It is denoted by $\mu^{(i)}(X, 0)$. These are collected together in the sequence

$$\mu^*(X, 0) = (\mu^{(n+1)}(X, 0), \mu^{(n)}(X, 0), \ldots, \mu^{(0)}(X, 0)).$$

Let $R = \mathbb{C}[x_0, x_1, \ldots, x_n]$ be a polynomial ring in $n+1$ variables, $\mathfrak{m}$ be the maximal graded ideal and $f \in R$ be any polynomial having an isolated singularity at the origin. Recall that a point $a \in V(f)$ is said to be *singular* if the rank of the $1 \times n$ matrix $J(f)|_a$ is zero, where $J(f) = (f_{x_0}, f_{x_1}, \ldots, f_{x_n})$ denotes the *Jacobian matrix*. A polynomial f is said to have an isolated singularity at a point a if $\{a\}$

is a (connected) component of the variety $\text{Sing}(f) = \{a \in V(f) \mid J(f)|_a = 0\}$. We use the same notation $J(f)$ to denote the *Jacobian ideal* which is generated by the entries of the Jacobian matrix associated to f. Teissier proved that the ith-mixed multiplicity, denoted by $e_i(\mathfrak{m}|J(f))$ or $e(\mathfrak{m}^{[n+1-i]}, J(f)^{[i]})$ is equal to the ith-sectional Milnor number of the singularity. Note that the the $(n+1)^{th}$ sectional Milnor number of f is the Minor number of f, namely, $\dim_C R/J(f)$.

Let $f \in \mathbb{C}[x_0, x_1, \dots, x_n]$ be a homogeneous polynomial in positive degree. Set $D(f) = \mathbb{P}^n \backslash V(f)$. In [**13**, p. 6], J. Huh showed that *the Euler Characteristic* of $D(f)$ is given by

$$\chi(D(f)) = \sum_{i=0}^{n} (-1)^i e_i(\mathfrak{m} \mid J(f)).$$

We now describe the contents of the paper. In section 2, we generalize the result of Cox, Lin and Sosa to find the defining ideal of multi-Rees algebra of any collection of ideals having positive grade in a local ring or in a standard graded algebra over a field. For any ideal J in a Noetherian ring B, we define the *grade of* J to be the common length of the maximal B-sequences in J and denote it by grade(J). In view of our result, we write a Macaulay2 algorithm to compute the defining ideal. The computation time of these algorithms is compared with the computation time of the Macaulay2 function `reesIdeal` in [**9**].

In section 3, we present a Macaulay2 algorithm which calculates mixed multiplicities of ideals in a polynomial ring. We start with generalizing a result of D. Katz, S. Mandal and J. K. Verma, to give a precise formula for the Hilbert polynomial of the quotient of a multi-graded algebra over an Artinian local ring. This result helps to calculate the mixed multiplicities of a set of ideals $I_0, I_1, \dots, I_r$, where I_0 is primary to the maximal ideal and $I_0, I_1, \dots, I_r$ are ideals in a local ring or in a standard graded algebra over a field. The algorithm computes their mixed multiplicities, where grade$(I_j) > 0$ for all j.

Section 4 contains an in depth explanation of the codes which are used to write the function `mixedMultiplicity` in the package `MixedMultiplicity`. This is meant for the new users of Macaulay2.

In section 5, we give an algorithm which computes mixed volume of a collection of lattice polytopes in $\mathbb{R}^n$. We start with an algorithm which outputs the homogeneous ideal corresponding to the vertices of a lattice polytope. Given a collection of lattice polytopes in $\mathbb{R}^n$, N. V. Trung and the last author proved that their mixed volume is equal to a mixed multiplicity of a set of homogeneous ideals. The function `mMixedVolume` is written based on their results to calculate the mixed volume.

In the last section, we give an algorithm to compute the sectional Milnor numbers. We use Teissier's observation of identifying the sectional Milnor numbers with mixed multiplicities to achieve this task. Let f be a polynomial in the ring $k[x_1, \dots, x_n]$, where char$(k) = 0$ and $J(f)$ is $\mathfrak{m}$-primary. With an input of the polynomial f, the function `secMilnorNumbers` outputs $e_0(\mathfrak{m} \mid J(f)), \dots, e_n(\mathfrak{m} \mid J(f))$. Teissier [**22**] conjectured that invariance of the Milnor number of an isolated singularity in a family of hypersurfaces implies invariance of the sectional Milnor numbers. The conjecture was disproved by Joël Briançon and Jean-Paul Speder. We verify their example using our algorithm and also by explicitly calculating the mixed multiplicities.

The algorithms written for the computation of defining ideal, mixed multiplicities, mixed volume and sectional Milnor numbers are used in the Macaulay2 ([**10**]) package `MixedMultiplicity`. Macaulay2 is a software system devoted to support research in algebraic geometry and commutative algebra.

2. Defining equations of multi-Rees algebras of ideals

An explicit formula for the defining ideal of the multi-Rees algebra of a finite collection of monomial ideals in a polynomial ring was given by D. Cox, K.-N. Lin, and G. Sosa in [**5**]. In this section, we generalize their result to find the defining ideal of the multi-Rees algebra of a collection of ideals with positive grade in a Noetherian ring. We use this result to write a Macaulay2 algorithm to compute the defining ideal when the base ring is a domain. We provide another algorithm for the non-domain case.

Let R be a Noetherian ring and $I_1, \ldots, I_s \subseteq R$ be ideals. Suppose that $I_i = \langle f_{ij} \mid j = 1, \ldots, n_i \rangle$ for all $i = 1, \ldots, s$. Let $\mathcal{R}(I_1, \ldots, I_s)$ be the multi-Rees algebra of ideals $I_1, \ldots, I_s$. Consider the set of indeterminates $\underline{Y} = \{Y_{ij} \mid i = 1, \ldots, s, j = 1, \ldots, n_i\}$ and $\underline{T} = (T_1, \ldots, T_s)$ over R. Define an R-algebra homomorphism $R[\underline{Y}] \xrightarrow{\varphi} \mathcal{R}(I_1, \ldots, I_s) \subseteq R[\underline{T}]$ such that $\varphi(Y_{ij}) = f_{ij}T_i$, for all $i = 1, \ldots, s$, $j = 1, \ldots, n_i$ and $\varphi(r) = r$ for all $r \in R$. Then $\mathcal{R}(I_1, \ldots, I_s) \simeq R[\underline{Y}]/\ker(\varphi)$. The ideal $\ker\varphi$ is called the defining ideal of $\mathcal{R}(I_1, \ldots, I_s)$. We give an explicit description of $\ker(\varphi)$.

THEOREM 2.1. *Let R be a Noetherian ring and $I_1, \ldots, I_s \subseteq R$ be ideals of positive grade. For each i, consider a generating set $\{f_{ij} \mid j = 1, \ldots, n_i\}$ of I_i which contains at least one nonzerodivisor f_{ij_i}. We set $h = \prod_{i=1}^{s} f_{ij_i}$ and set*

$$\Gamma = \langle Y_{ij}f_{ij'} - Y_{ij'}f_{ij} \mid i = 1, \ldots, s \text{ and } j, j' \in \{1, \ldots, n_i\} \text{ with } j \neq j' \rangle : h^{\infty} \subseteq R[\underline{Y}].$$

Then $\Gamma \subseteq R[\underline{Y}]$ is the defining ideal of $\mathcal{R}(I_1, \ldots, I_s)$.

PROOF. Without loss of generality, we may assume that $j_i = 1$ for all $i = 1, \ldots, s$ and $h = \prod_{i=1}^{s} f_{i1}$. Consider the ring homomorphism

$$\phi : R \to R[f_{11}^{-1}, f_{21}^{-1}, \ldots, f_{s1}^{-1}] \cong R[h^{-1}],$$

which induces a natural map $\widetilde{\phi} : R[\underline{Y}] \to R[h^{-1}][\underline{Y}] \cong R[\underline{Y}]_h$. Note that $\phi(I_i) = (1, f_{i2}/f_{i1}, \ldots, f_{in_i}/f_{i1}) \subseteq R[h^{-1}]$ is the unit ideal for $i = 1, \ldots, s$. Consider the ring homomorphism

$$\theta : R[h^{-1}][\underline{Y}] \to \mathcal{R}(\phi(I_1), \ldots, \phi(I_s)) \subseteq R[h^{-1}][\underline{T}]$$

sending Y_{i1} to T_i and Y_{ij} to $\frac{f_{ij}}{f_{i1}}T_i$ for $i = 1, \ldots, s$ and $j = 1, \ldots, n_i$. Then we show that the defining ideal $\ker\theta$ of $\mathcal{R}(\phi(I_1), \ldots, \phi(I_s))$ is $J := J_1 + \cdots + J_s$, where $J_i := (Y_{i2} - \frac{f_{i2}}{f_{i1}}Y_{i1}, \ldots, Y_{in_i} - \frac{f_{in_i}}{f_{i1}}Y_{i1})$. Clearly, $J \subseteq \ker\theta$. On the other hand,

$$\frac{R[h^{-1}][\underline{Y}]}{J} \cong R[h^{-1}][Y_{i1} \mid i = 1, \ldots, s] \cong R[h^{-1}][\underline{T}].$$

This forces $\ker\theta = J$, as the last isomorphism is via θ. We further claim that $\widetilde{\phi}^{-1}(J) = \Gamma$. For all $j \neq j'$, and every i,

$$f_{ij}Y_{ij'} - f_{ij'}Y_{ij} = f_{ij}\left(Y_{ij'} - \frac{f_{ij'}}{f_{i1}}Y_{i1}\right) - f_{ij'}\left(Y_{ij} - \frac{f_{ij}}{f_{i1}}Y_{i1}\right)$$

is in J_i. So $\Gamma \subseteq \widetilde{\phi}^{-1}(J)$. Now let $r \in \widetilde{\phi}^{-1}(J)$. Then $\widetilde{\phi}(r) \in J$, i.e.,

$$\frac{r}{1} = \sum_{i=1}^{s}\sum_{j=2}^{n_i} \frac{a_{ij}}{h^{m_{ij}}}\left(Y_{ij} - \frac{f_{ij}}{f_{i1}}Y_{i1}\right)$$

for some $a_{ij} \in R[\underline{Y}]$. Thus we have

$$\begin{aligned} h^m r &\in (f_{i1}Y_{ij} - f_{ij}Y_{i1} \mid 1 \le i \le s \text{ and } 1 \le j \le n_i) \\ &\subseteq (f_{ij}Y_{ij'} - f_{ij'}Y_{ij} \mid 1 \le i \le s \text{ and } 1 \le j, j' \le n_i) \end{aligned}$$

for some $m \geq \max\{m_{ij} \mid 1 \le i \le s, 1 \le j \le n_i\}+1$. Therefore, $r \in \Gamma$ and hence the claim holds. Next, observe that h^t is a nonzerodivisor on R and hence on $R[\underline{Y}]/\mathcal{K} \cong \mathcal{R}(I_1, \dots, I_s)$ for every $t \geq 1$. So $\mathcal{K}$ is a contracted ideal by [**1**, Proposition 3.11 (iii)]. Besides, $\mathcal{R}(\phi(I_1), \dots, \phi(I_s)) \cong \mathcal{R}(I_1, \dots, I_s)_h$. Therefore,

$$R[h^{-1}][\underline{Y}]/J \cong \mathcal{R}(I_1, \dots, I_s)_h \cong (R[\underline{Y}]/\mathcal{K})_h \cong R[h^{-1}][\underline{Y}]/\mathcal{K}_h.$$

As $J = \ker\theta = \mathcal{K}_h = \langle\widetilde{\phi}(\mathcal{K})\rangle$ and $\mathcal{K}$ is a contracted ideal so we get that $\mathcal{K} = \widetilde{\phi}^{-1}(J) = \Gamma$. □

When R is a domain or when a list of nonzerodivisors (one each from the list of ideals with positive grades) is provided by the user, the function `multiReesIdeal` computes the defining ideal of the multi-Rees algebra using Theorem 2.1.

ALGORITHM (Version I: `multiReesIdeal`, set of ideals with positive grade). *Let $I_1, \dots, I_s$ be ideals of a Noetherian ring R with* grade $I_i > 0$ *for all i and let $a_1, \dots, a_s$ be a set of nonzerodivisors, where a_i belongs to the generating set of I_i for all i. When R is a domain, the function picks a_i to be the first element in the generating set of I_i for each i.*

Input: *The list $W = \{\{I_1, \dots, I_s\}, \{a_1, \dots, a_s\}\}$, or $W = \{I_1, \dots, I_s\}$ if R is a domain.*

1. *Define a polynomial ring S by attaching m indeterminates to the ring R, where m is the sum of the number of generators of all the ideals.*
2. *For each ideal I_i, construct a matrix $M(i)$ whose first row consists of the generators of the ideal and the second row consists of the indeterminates.*
3. *Add the ideals generated by 2×2 minors of these matrices to get an ideal L.*
4. *To get the defining ideal, saturate L with the product of a_i's.*

Output: *The defining ideal of the Rees algebra $\mathcal{R}(I_1, \dots, I_s)$.*

The elements of the defining ideal are assigned $\mathbb{N}^{s+1}$ degree by the function, where the first $\mathbb{N}^s$ coordinates point to the component of $\mathcal{R}(I_1, \dots, I_s)$ where the element lies and the last coordinate is the degree of the element. Observe that the defining ideal in the following example is generated in degrees $(1, 0, 5)$. The first two coordinates asserts that the generator lies in I_1t_1, whereas the last coordinate specifies that the element belongs to $(I_1)_5 \subseteq R_5$.

EXAMPLE 2.2.

```
i1 : R = QQ[w,x,y,z];
i2 : I = ideal(x^2-y*w, x^3-z*w^2);
i3 : J = ideal(w^2+x^2+y^2+z^2);
i4 : M = multiReesIdeal {I,J};
```

```
i5 : transpose gens M
o5 = {-1, 0, -5} | (x3-w2z)X_0+(-x2+wy)X_1 |
i6 : (first entries gens M)/degree
o6 : {{1, 0, 5}}
```

2.1. Routine for the non-domain case. In this section we present an algorithm to find the defining ideal of the Rees algebra using the definition of Rees algebra. This method does not have any requirements on the grade of the ideals or the domain property of the ring, but it seems to be comparably slower than the previous method.

We can construct the Rees algebra of I_i as the kernel of the map

$$\varphi_i : R[Y_{i1}, \ldots, Y_{in_i}] \to R[T_i]$$

where $\varphi_i(Y_{ij}) = f_{ij}T_i$ for $j = 1, \ldots, n_i$. Notice that the idea $(\ker \varphi_i)R[\underline{Y}] \subseteq \ker \varphi$. Suppose that ϕ_i is the presentation matrix of I_i for $i = 1, \ldots, s$. Then the symmetric algebra $\mathrm{Sym}(I_i)$ has a presentation $R[Y_{i1}, \ldots, Y_{in_i}]/\mathcal{L}_i$ where $\mathcal{L}_i = I_1([Y_{i1}, \ldots, Y_{in_i}] \cdot \phi_i)$. Clearly, $\mathcal{L}_i \subseteq \ker \varphi_i \subset \ker \varphi$. So the map φ_i factors through the symmetric algebra $\mathrm{Sym}(I_i)$. Now $\mathrm{Sym}(I_1) \otimes \cdots \otimes \mathrm{Sym}(I_s)$ has the presentation $R[\underline{Y}]/(\mathcal{L}_1 + \cdots + \mathcal{L}_s)$. Since each of $\mathcal{L}_i \subseteq \ker \varphi$, the map φ also factors through $\mathrm{Sym}(I_1) \otimes \cdots \otimes \mathrm{Sym}(I_s)$. Thus to find the defining ideal of the multi Rees algebra $\mathcal{R}(I_1, \ldots, I_s)$ it is enough to find the kernel of the surjective map $\mathrm{Sym}(I_1) \otimes \cdots \otimes \mathrm{Sym}(I_s) \to \mathcal{R}(I_1, \ldots, I_s)$.

ALGORITHM (Version II: `multiReesIdeal`, no assumptions). *Let $I_1, \ldots, I_s$ be ideals in the Noetherian ring R.*
Input: *The list $W = \{I_1, \ldots, I_s\}$.*

1. *For each I_i compute the presentation $F_i' \xrightarrow{\phi_i} F_i \to I_i \to 0$, where ϕ_i is the presentation matrix of I_i for $i = 1, \ldots, s$.*
2. *Now compute the source symmetric algebra $\mathrm{Sym}(F_1') \otimes \cdots \otimes \mathrm{Sym}(F_s')$ and the target symmetric algebra $\mathrm{Sym}(F_1) \otimes \cdots \otimes \mathrm{Sym}(F_s)$ of the map $\phi_1 \otimes \cdots \otimes \phi_s$.*
3. *Compute the map between the symmetric algebra of the source and target and return kernel of the above map.*

Output: *The defining ideal of the Rees algebra $\mathcal{R}(I_1, \ldots, I_s)$.*

In the following example the ring U is not a domain and hence the algorithm uses the above method. As expected, the computational time in the case a nonzerodivisor is given as an optional input is faster than the case where no optional input is given.

EXAMPLE 2.3.

```
i1 : R = QQ[w,x,y,z]/ideal(w*x, y*z);
i2 : m = ideal vars R;
i3 : time multiReesIdeal m
-- used 0.290619 seconds
o3 = ideal(y*X, z*X, z*X -x*X, y*X -x*X, w*X, z*X -w*X, y*X -w*X, x*X, X X, X X)
                3    2      1    3      1    2      1    0      3    0    2    0  2 3  0 1
i4 : time multiReesIdeal (m, w+x);
-- used 0.0283374 seconds
```

3. Computation of mixed multiplicities of multi-graded algebras

In [**25**], D. Katz, S. Mandal and the last author, found a precise formula for the Hilbert polynomial of the quotient of a bi-graded algebra over an Artinian local ring. We generalize this result for quotient of a multi-graded algebra over an Artinian local ring. This improvement helps in the calculation of the mixed multiplicities of a set of ideals $I_0, I_1, \ldots, I_r$, where I_0 is primary to the maximal ideal and $I_0, I_1, \ldots, I_r$ are ideals in a local ring or in a standard graded algebra over a field. We end the section by giving an algorithm to compute the mixed multiplicity of a sequence of ideals $I_0, I_1, \ldots, I_r$ of a polynomial ring, where I_0 is an primary to the maximal ideal and $\operatorname{ht}(I_j) > 0$ for all j.

Let S be an Artinian local ring and $A = S[X_1, \ldots, X_r]$ be an $\mathbb{N}^r$-graded ring over S, where for $1 \leq i \leq r$, $X_i = \{X_i(0), \ldots, X_i(s_i)\}$ is a set of indeterminates of degree e_i. Set $\underline{u} = (u_1, \ldots, u_r) \in \mathbb{N}^r$ and $|u| = u_1 + \cdots + u_r$. Then $A = \bigoplus_{\underline{u} \in \mathbb{N}^r} A_{\underline{u}}$, where $A_{\underline{u}}$ is the S-module generated by monomials of the form $P_1 \cdots P_r$, where P_i is a monomial of degree u_i in X_i. An element in $A_{\underline{u}}$ is called multi-homogeneous of degree $\underline{u}$. An ideal $I \subseteq A$ generated by multi-homogeneous elements is called a multi-homogeneous ideal. Then $R = A/I$ is an $\mathbb{N}^r$-graded algebra with $\underline{u}$-graded component $R_{\underline{u}} = A_{\underline{u}}/I_{\underline{u}}$. The Hilbert function of R is defined as $H(\underline{u}) = \lambda(R_{\underline{u}})$, where λ denotes the length as an S-module. Set $\underline{t}^{\underline{u}} = t_1^{u_1} \cdots t_r^{u_r}$. The Hilbert series of R is given by $HS(R, \underline{t}) = \sum_{\underline{u} \in \mathbb{N}^r} \lambda(R_{\underline{u}}) \underline{t}^{\underline{u}}$. Then there exists a polynomial $N(t_1, \ldots, t_r) \in \mathbb{Z}[t_1, \ldots, t_r]$ so that

$$HS(R, \underline{t}) = N(t_1, \ldots, t_r) / \left((1-t_1)^{s_1+1} \cdots (1-t_r)^{s_r+1}\right). \tag{2}$$

Note that the rational function on the right hand side need not be reduced.

THEOREM 3.1. *Write the Hilbert polynomial of R as*

$$P(\underline{u}, R) = \sum_{\alpha=\underline{0}}^{\underline{s}} c_\alpha \binom{u_1+\alpha_1}{\alpha_1} \cdots \binom{u_r+\alpha_r}{\alpha_r}. \tag{3}$$

Then

$$c_\alpha = \frac{(-1)^{|\underline{s}-\alpha|}}{(s_1-\alpha_1)! \cdots (s_r-\alpha_r)!} \cdot \frac{\partial^{|\underline{s}-\alpha|} N}{\partial t_1^{s_1-\alpha_1} \cdots \partial t_r^{s_r-\alpha_r}} \Bigg|_{(t_1,\ldots,t_r)=\underline{1}}.$$

PROOF. We write

$$N^{(\underline{\alpha})} = \frac{\partial^{|\alpha|} N(t_1, \ldots, t_r)}{\partial t_1^{\alpha_1} \cdots \partial t_r^{\alpha_r}} \Bigg|_{(t_1,\ldots,t_r)=\underline{1}}.$$

Then

$$HS(R, \underline{t}) - \sum_{\underline{\alpha}=\underline{0}}^{\underline{s}} \frac{N^{(\underline{\alpha})}(-1)^{|\alpha|}}{\alpha_1! \cdots \alpha_r! (1-t_1)^{s_1+1-\alpha_1} \cdots (1-t_r)^{s_r+1-\alpha_r}}$$
$$= \frac{D(t_1, \ldots, t_r)}{(1-t_1)^{s_1+1} \cdots (1-t_r)^{s_r+1}}$$

where

$$D(t_1, \ldots, t_r) = N(t_1, \ldots, t_r) - \sum_{\underline{\alpha}=\underline{0}}^{\underline{s}} \frac{N^{\underline{\alpha}}}{\alpha_1! \cdots \alpha_r!} (t_1-1)^{\alpha_1} \cdots (t_r-1)^{\alpha_r}.$$

Thus $D(t_1,\dots,t_r)$ is the remainder of the Taylor series of $N(t_1,\dots,t_r)$ about the point $(1,\dots,1)\in\mathbb{N}^r$ having terms of degree $\geq s_i+1$ in t_i-1, for all $1\leq i\leq r$. So $D(t_1,\dots,t_r)$ is divisible by $(1-t_1)^{s_1+1}\cdots(1-t_r)^{s_r+1}$. Therefore, for all large u_i's, $\lambda(R_{\underline{u}})$ is the coefficient of $t^{\underline{u}}$ in the power series expansion of

$$E(t_1,\dots,t_r)=\sum_{\underline{\alpha}=\underline{0}}^{\underline{s}}\frac{N^{(\underline{\alpha})}(-1)^{|\underline{\alpha}|}}{\alpha_1!\cdots\alpha_r!(1-t_1)^{s_1+1-\alpha_1}\cdots(1-t_r)^{s_r+1-\alpha_r}}. \tag{4}$$

As the coefficient of $t^{\underline{u}}$ in $E(t_1,\dots,t_r)$ is given by a polynomial for all $\underline{u}$ so

$$E(t_1,\dots,t_r)=\sum_{\underline{u}\in\mathbb{N}^r}P(\underline{u};R)t^{\underline{u}}. \tag{5}$$

Here we are using the fact that two polynomials in $\mathbb{Z}[y_1,\dots,y_r]$ coinciding at $\underline{u}$, for all u_i large, are equal. Now expanding the rational function in (4), we get

(6)

$$\begin{aligned}E(t_1,\dots,t_r)&=\sum_{\underline{\alpha}=\underline{0}}^{\underline{s}}\frac{N^{(\underline{\alpha})}(-1)^{|\underline{\alpha}|}}{\alpha_1!\cdots\alpha_r!}\sum_{\underline{u}\in\mathbb{N}^r}\binom{s_1-\alpha_1+u_1}{u_1}\cdots\binom{s_r-\alpha_r+u_r}{u_r}t^{\underline{u}}\\&=\sum_{\underline{u}\in\mathbb{N}^r}\left(\sum_{\underline{\alpha}=\underline{0}}^{\underline{s}}\frac{N^{(\underline{\alpha})}(-1)^{|\underline{\alpha}|}}{\alpha_1!\cdots\alpha_r!}\right)\binom{s_1-\alpha_1+u_1}{u_1}\cdots\binom{s_r-\alpha_r+u_r}{u_r}t^{\underline{u}}.\end{aligned}$$

Comparing (6) with (5), we get the result. □

Note that

$$\binom{u_i+\alpha_i}{\alpha_i}=\frac{1}{\alpha_i!}u_i^{\alpha_i}+\text{ lower degree terms.}$$

So if we write $P(\underline{u})$ as in (3), then $c_\alpha=e_\alpha$ for all $\alpha\in\mathbb{N}^{r+1}$ with $|\alpha|=d-1$, see Theorem 1.2. Therefore, Theorem 3.1 gives an expression for e_α.

Remark 3.2. Let $I_0',I_1',\dots,I_r'$ denote the images of ideals $I_0,I_1,\dots,I_r$ in $A':=A/(0:I^\infty)$, where $I=I_1\cdots I_r$. Following the notations in **[23]**, we set $R:=R(I_0|I_1,\dots,I_r)=\mathcal{R}(I_0,\dots,I_r)/(I_0)$ and put $R':=R(I_0'|I_1',\dots,I_r')$. Notice $R=\operatorname{gr}_A(I_0,I_1,\dots,I_r;I_0)$, defined in (1). Then for $\underline{u}$ large, $P_R(\underline{u})=P_{R'}(\underline{u})$ (see **[23**, Theorem 1.2] for details). Therefore, in case grade $I_i=0$ for some i, the user needs to work in the quotient ring A' and input the images of the ideals in the quotient ring.

Algorithm. *The algorithm for the function* `mixedMultiplicity` *uses the above ideas to calculate the mixed multiplicity. Let $I_0,I_1,\dots,I_r$ be a set of ideals of a Noetherian ring R of dimension $d\geq 1$, where I_0 is primary to the maximal ideal and* grade$(I_i)>0$ *for all i; $\underline{a}=(a_0,a_1,\dots,a_r)\in\mathbb{N}^{r+1}$ with $|\underline{a}|=d-1$.*

Input: *The sequence $W=((I_0,I_1,\dots,I_r),(a_0,a_1,\dots,a_r))$.*

1. *Compute the defining ideal of the multi-Rees algebra using the function* `multiReesIdeal` *and use it to find the Hilbert series of $R(I_0\mid I_1,\dots,I_r)$.*
2. *Extract the powers of $(1-T_i)$ in the denominator of the Hilbert series.*
3. *Calculate $e_{\underline{a}}$ using the formula given in Theorem* 3.1.

Output: *The mixed multiplicity $e_{\underline{a}}(I_0\mid I_1,\dots,I_r)$.*

Example 3.3.

```
i1 : R = QQ[w,x,y,z];                       o4 = 1
i2 : I = ideal(x^2-y*w, x^3-z*w^2);         i5 : mixedMultiplicity((m,I),(2,1))
i3 : m = ideal vars R;                      o5 = 2
i4 : mixedMultiplicity((m,I),(3,0))
```

When some ideal has grade zero, the following example explains how to compute the mixed multiplicity.

EXAMPLE 3.4. Let $R = \mathbb{Q}[w,x,y,z]/(wx,yz)$, $\mathfrak{m} = (w,x,y,z)$, and $I = (w,y)$. Notice that $\operatorname{grade} I = 0$, since $I \in \operatorname{Ass} R$.

```
i1 : R = QQ[w,x,y,z]/ideal(w*x, y*z); i7 : n = substitute(m, T);
i2 : I = ideal(w,y);                   i8 : dim T
i3 : m = ideal vars S;                 o8 = 2
i4 : L = saturate(sub(ideal 0, R), I); i9 : mixedMultiplicity ((n,J),(1,0))
i5 : T = S/L;                          o9 = 3
i6 : J = substitute(I, T);
```

To calculate mixed multiplicities, the function `mixedMultiplicity` computes the Hilbert polynomial of the graded ring $\bigoplus I_0^{u_0} I_1^{u_1} \cdots I_r^{u_r} / I_0^{u_0+1} I_1^{u_1} \cdots I_r^{u_r}$. In particular, if $I_1, \ldots, I_r$ are also $\mathfrak{m}$-primary ideals, then $e_{(a_0,a_1,\ldots,a_r)}(I_0 \mid I_1, \ldots, I_r) = e(I_0^{[a_0+1]}, I_1^{[a_1]}, \ldots, I_r^{[a_r]})$ (see [**14**, Definition 17.4.3]). Therefore, to compute the $(a_0+1, a_1, \ldots, a_r)$-th mixed multiplicity of $I_0, I_1, \ldots, I_r$, one needs to enter the sequence $(a_0, a_1, \ldots, a_r)$ in the function. The same is illustrated in Example 3.5.

EXAMPLE 3.5. Let $R = \mathbb{Q}[w,x,y,z]$ and $\mathfrak{m}$ be the maximal homogeneous ideal of R. Let $I = (x^2 - yw, x^3 - zw^2)$ and $J = \mathfrak{m}^4 + I$. We calculate the mixed multiplicities $e(\mathfrak{m}) = e(\mathfrak{m}^{[4]}, J^{[0]})$ and $e(\mathfrak{m}^{[3]}, J^{[1]})$.

```
i1 : R = QQ[w,x,y,z];                      i5 : mixedMultiplicity ((m,J),(3,0))
i2 : m = ideal vars R;                     o5 = 1
i3 : I = ideal(x^2-y*w, x^3-z*w^2);        i6 : mixedMultiplicity ((m,J),(2,1))
i4 : J = m^4 + I;                          o6 = 2
```

4. Writing the function `mixedMultiplicity` in Macaulay2

In this section, we illustrate a detailed overview of the function `mixedMultiplicity` based on the algorithm discussed in the previous section. One can refer to the package `MixedMultiplicity` for the Macaulay2 code of this function.

We start with two sequences `W1`$=(I_0, I_1, \ldots, I_r)$ and `W2`$=(a_0, a_1, \ldots, a_r)$. Then the command `W1#i` returns the i-th element of `W1`, that is, I_i. All ideals are expected to be in the same ring. To capture the ambient of these ideals we use `ring W1#0`. We now check that the ambient ring has dimension at least one, the two sequences `W1`, `W2` have same length, `W1` consists of ideals of positive grade with same ambient ring, and `W2` is a sequence of natural numbers. Now we set 'n' to be the cardinality of `W1`, that is, $n = r+1$. The defining ideal of the multi-Rees algebra $\mathcal{R}(I_0, \ldots, I_r)$ is computed using the function `multiReesIdeal` and named as `L1`. For $i = 1, \ldots, n$, `V(i)` gives us the unit vector $(0,..,0,1,0,..,0) \in \mathbb{N}^n$, where 1 is at the ith place. Next, `Q` is a list of `V(i)`'s where `V(i)` occurs as many times as the number of generators of `W1#i` (i.e., I_i) for $i = 0, \ldots, r$. We use `compress gens` to remove zero entries, if any, from the input generating set of ideals. As we use `compress gens` in the function `multiReesIdeal` too so it is compatible. Since `gens` gives us the generators of an ideal as a matrix, we could also use `#(first entries compress gens W1#i)`. The command `first entries` extracts the first row of a matrix in the form of a list. Other possible commands are `numgens trim W1#i` and `rank source mingens W1#i`.

Then we define a ring `S`, which is the same ring as the ambient ring of `L1`, but is now multigraded with degrees given by `Q`. By `sub(W1#0, S)`, we view the ideal `W1#0` as an ideal in `S` and call it `p`. Next, we view `L1` as an ideal in `S/p` and call it `L2`. Recall that `W1#0`$= I_0$ is primary to the maximal ideal in its ambient ring. `hilbertSeries` computes the Hilbert series of `L2` and reduce it using the command `reduceHilbert`. Thus we get an expression `H` as in (2). To use Theorem 3.1, we extract the numerator `B0` and denominator `B1` of `H`. Factoring out `B1`, we capture the powers. For instance, if `B1` $= T_0T_1^2 - 2T_0T_1 - T_1^2 + T_0 + 2T_1 - 1$, then `facB1 = factor B1` $= (T_1-1)^2(T_0-1)$, and `facB1#0` $= (T_1-1)^2$. The command `facB1#0` executes because `factor` outputs an expression of the class `Product`, which is a type of a basic list. Since `Power` is also a type of a basic list, we get `facB1#0#1`$=2$. We now extract a_i's with the command `W2#i` and verify that $a_i \leq$ power of $(1-T_i)$. At this stage the function has the value of all the variables in the expression given in Theorem 3.1. So we just compute the value.

5. Mixed volume of lattice polytopes

Given a collection of lattice polytopes in $\mathbb{R}^n$, Trung and the last author proved that their mixed volume is equal to a mixed multiplicity of a set of homogeneous ideals, see Proposition 1.3. We use this result to construct an algorithm which calculates the mixed volume of a collection of lattice polytopes. We also give an algorithm which outputs the homogeneous ideal corresponding to the vertices of a lattice polytope.

Let Q be a lattice polytope in $\mathbb{R}^n$ with the set of vertices $\{p_1, \ldots, p_r\} \subseteq \mathbb{N}^n$. We first compute the corresponding homogeneous ideal I in the polynomial ring $R = k[x_1, \ldots, x_{n+1}]$ such that Q is the convex hull of the lattice points of the dehomogenization of a monomial generating set of I in $k[x_1, \ldots, x_n]$. For this purpose, we write a function `homIdealPolytope` which requires as an input the list of vertices of Q.

ALGORITHM. *Suppose that $p_1, \ldots, p_r \subseteq \mathbb{N}^n$ are vertices of a polytope Q.*
Input*: The list $W = \{p_1, p_2, \ldots, p_r\}$.*

1. *Define a polynomial ring R in $(n+1)$-variables over $\mathbb{Q}$, where the vertices are in $\mathbb{R}^n$.*
2. *For each vertex p_i, associate a monomial in the first n-variables, where the j-th component of p_i is the power of the j-th variable and add them up.*
3. *Add the monomials and homogenize it with respect to the last variable.*
4. *Define an ideal generated by the terms of the homogeneous element.*

Output*: The corresponding homogeneous ideal in $\mathbb{Q}[x_1, \ldots, x_{n+1}]$.*

We now write a function `mMixedVolume` to calculate the mixed volume of a collection of n lattice polytopes in $\mathbb{R}^n$. Let $Q_1, \ldots, Q_n$ be an arbitrary collection of lattice polytopes in $\mathbb{R}^n$. Let I_i be the homogeneous ideal of $R = k[x_1, \ldots, x_{n+1}]$ associated to Q_i for $i = 1, \ldots, n$. The function `mMixedVolume` takes the list $\{I_1, \ldots, I_n\}$ as an input and produces the mixed volume of $Q_1, \ldots, Q_n$ as an output. The function can also take the list of lists of vertices of the polytope as an input to compute their mixed volume. Since calculating the mixed volume is same as calculating a mixed multiplicity, the algorithm of the function `mMixedVolume` is similar to that of `mixedMultiplicity`.

EXAMPLE 5.1. Let Q_1, Q_2, Q_3 be the same tetrahedron with vertices $(1,1,0)$, $(2,1,0), (1,3,0)$ and $(1,1,3)$. Using the formula, the mixed volume can be calculated as follows:

$$\begin{aligned} MV_3(Q_1,Q_2,Q_3) = & V_3(Q_1+Q_2+Q_3) - V_3(Q_1+Q_2) - V_3(Q_2+Q_3) - V_3(Q_3+Q_1) \\ & + V_3(Q_1) + V_3(Q_2) + V_3(Q_3), \end{aligned}$$

where V_3 denotes the 3-dimensional Euclidean volume. One can check that $V_3(Q_i) = 1$, for all $i = 1, 2, 3$; $V_3(Q_i + Q_j) = 8$, for all $i \neq j$ and $V_3(Q_1 + Q_2 + Q_3) = 27$. Hence $MV_3(Q_1, Q_2, Q_3) = 27 - (3 \times 8) + (3 \times 1) = 6$. The following Macaulay2 working verifies this calculation.

```
i1 : A = {(1,1,0),(2,1,0),(1,3,0),(1,1,3)};
i2 : mMixedVolume{A,A,A}
o2 = 6
```

EXAMPLE 5.2. Consider a generic vertical parabola f_1 and a generic horizontal parabola f_2 given by

$$\begin{aligned} f_1 &= c_{1,(0,0)} + c_{1,(1,0)}x + c_{1,(2,0)}x^2 + c_{1,(0,1)}y, \\ f_2 &= c_{2,(0,0)} + c_{2,(1,0)}x + c_{2,(0,1)}y + c_{2,(0,2)}y^2. \end{aligned}$$

We compute the mixed volume of the Newton polytopes associated to the supports Q_1, Q_2 of f_1, and f_2 respectively.

```
i1 : Q1 = {(0,0),(1,0),(2,0),(0,1)};
i2 : Q2 = {(0,0),(1,0),(0,1),(0,2)};
i3 : mMixedVolume {Q1, Q2}
o3 = 4
```

This computation verifies [**2**, Example 1.2] and also gives a bound for the number of common zeros (in the complex torus) of f_1 and f_2 (see Bernstein's theorem in subsection 1.2).

EXAMPLE 5.3. We compute and verify the mixed volumes of the two pairs of convex polytopes, which appeared in [**4**, Example 1.1] and [**4**, Section 2] (take $\lambda = 1$), respectively.

```
i1 : Q1 = {(1, 1),(3, 0),(4, 0),(4, 1),(3, 3),(1, 4),(0, 4),(0, 3)};
i2 : Q2 = {(0, 1),(0, 0),(3, 0),(4, 1),(4, 4),(3, 4)};
i3 : mMixedVolume {Q1, Q2}
o3 = 32
i4 : P1 = {(0, 0),(0, 2),(2, 0),(2, 2)};
i5 : P2 = {(0, 0),(1, 2),(2, 1)};
i6 : mMixedVolume {P1, P2}
o6 = 8
```

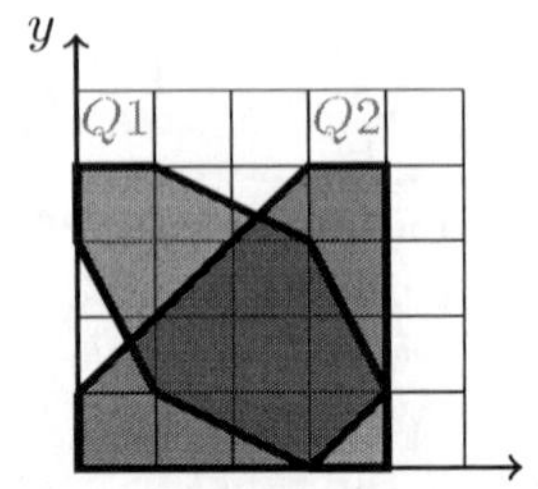

6. Sectional Milnor number

In this section, we give an algorithm to compute the sectional Milnor numbers. We use Teissier's observation of identifying the sectional Milnor numbers with mixed multiplicities to achieve this task. Teissier ([**22**]) conjectured that invariance of Milnor number implies invariance of the sectional Milnor numbers. The conjecture was disproved by Joël Briançon and Jean-Paul Speder. We verify their example using our algorithm and also by explicitly calculating the mixed multiplicities.

Let $R = \mathbb{C}[x_1, \ldots, x_n]$ be a polynomial ring in n variables, $\mathfrak{m}$ be the maximal graded ideal and $f \in R$ be any polynomial with an isolated singularity at the origin. Using Theorem 3.1, one can now calculate the mixed multiplicities of $\mathfrak{m}$ and $J(f)$. We use the ideas in the previous section to write a function `secMilnorNumbers` for computing the first $n-1$ sectional Milnor numbers. With a polynomial f given as an input, the algorithm calculates the Jacobian ideal of f and then using the function `multiReesIdeal`, it finds the defining ideal of $\mathcal{R}(\mathfrak{m}, J(f))$. This helps to compute the Hilbert series of the special fiber $\mathcal{F}(\mathfrak{m}, J(f)) = \mathcal{R}(\mathfrak{m}, J(f)) \otimes_R R/\mathfrak{m}$. Using the formula given in Theorem 3.1, it then calculates the mixed multiplicities. Note that the nth-sectional Milnor number is the Milnor number of the hypersurface f at the origin. So under the extra assumption that the ideal $J(f)$ is $\mathfrak{m}$-primary, we have $\mu^{(n)}(X,0) = \dim_{\mathbb{C}} R/J(f)$. Together, the function `secMilnorNumbers` outputs $\left(\mu^{(0)}(X,0), \mu^{(1)}(X,0), \ldots, \mu^{(n)}(X,0)\right)$.

EXAMPLE 6.1. Let $R = \mathbb{Q}[x, y, z]$ and $f = x^4 + y^4 + z^4$ be a polynomial in the ring.

```
i1 : R = QQ[x,y,z];
i2 : f = x^4+y^4+z^4;
i3 : secMilnorNumbers f
o3 = HashTable{0 => 1, 1 => 3, 2 => 9, 3 => 27}
```

REMARKS 6.2.

(1) We set $I := (f, J(f))$. If f has an isolated singularity at the origin, then by definition $IR_\mathfrak{m}$ is an $\mathfrak{m}R_\mathfrak{m}$-primary ideal. In view of [**14**, Theorem 7.1.5, Proposition 1.6.2], $f/1 \in \overline{J(f)R_\mathfrak{m}}$ and hence $J(f)R_\mathfrak{m} \subset IR_\mathfrak{m}$ is a reduction. Thus

$$IR_\mathfrak{m} \subset \overline{IR_\mathfrak{m}} = \overline{J(f)R_\mathfrak{m}} \subseteq \sqrt{J(f)R_\mathfrak{m}} \subset \sqrt{IR_\mathfrak{m}},$$

see [**14**, Remark 1.1.3 (3)]. So $\sqrt{J(f)R_\mathfrak{m}} = \sqrt{IR_\mathfrak{m}} = \mathfrak{m}R_\mathfrak{m}$. Therefore, $J(f)R_\mathfrak{m}$ is $\mathfrak{m}R_\mathfrak{m}$-primary. The example in (2) says that $J(f)$ need not be $\mathfrak{m}$-primary even when I is $\mathfrak{m}$-primary. However, if f is homogeneous, then by the *classical Euler formula* $(\deg f) \cdot f = \sum_{i=1}^n x_i \cdot \partial f/\partial x_i \in J(f)$. Hence $J(f) = I$. Consequently, $J(f)$ is $\mathfrak{m}$-primary if I is so.

(2) If $J(f)$ is not $\mathfrak{m}$-primary, then the Milnor number is $\dim_{\mathbb{C}} \frac{R_{\mathfrak{m}}}{J(f)R_{\mathfrak{m}}}$, the multiplicity of $R/J(f)$ at $\mathfrak{m}$. It can be computed using Macaulay2 by means of the R-module isomorphism

$$\frac{R_{\mathfrak{m}}}{J(f)R_{\mathfrak{m}}} \overset{\overline{\varphi}}{\cong} \frac{R}{\varphi^{-1}(J(f)R_{\mathfrak{m}})} = \frac{R}{(J(f) : (J(f) : \mathfrak{m}^{\infty}))} \tag{7}$$

induced from the natural inclusion $\varphi : R \to R_{\mathfrak{m}}$, see [**8**, Page 61]. Since $\varphi^{-1}(J(f)R_{\mathfrak{m}})R_{\mathfrak{m}} = J(f)R_{\mathfrak{m}}$ by [**1**, Proposition 1.17], $\overline{\varphi}_{\mathfrak{m}}$ is an isomorphism. So from [**1**, Proposition 3.9] it follows that $\overline{\varphi}$ is an isomorphism. It is now enough to show

$$\varphi^{-1}(J(f)R_{\mathfrak{m}}) = (J(f) : (J(f) : \mathfrak{m}^{\infty})).$$

Suppose that $J(f) = (\cap_{i=1}^{r} Q_i) \cap Q$ is a minimal primary decomposition, where $\sqrt{Q} = \mathfrak{m}$ and $\sqrt{Q_i} = P_i \neq \mathfrak{m}$ for $i = 1, \ldots, r$. Since $J(f)R_{\mathfrak{m}}$ is $\mathfrak{m}R_{\mathfrak{m}}$-primary, $\mathfrak{m}$ is a minimal prime of $J(f)$ and $P_i \not\subseteq \mathfrak{m}$ for $i = 1, \ldots, r$. By [**1**, Theorem 4.10], the $\mathfrak{m}$-primary component of $J(f)$ is unique. We now show that

$$\varphi^{-1}(J(f)R_{\mathfrak{m}}) = Q = (J(f) : (J(f) : \mathfrak{m}^{\infty})).$$

By [**1**, Proposition 4.9], $J(f)R_{\mathfrak{m}} = QR_{\mathfrak{m}}$ and $\varphi^{-1}(J(f)R_{\mathfrak{m}}) = Q$. On the other hand, $\widetilde{J(f)} := (J(f) : \mathfrak{m}^{\infty}) = \cap_{i=1}^{r} Q_i$, the intersection of all non $\mathfrak{m}$-primary components of $J(f)$. Due to [**1**, Proposition 1.11], we can pick $x \in (\cap_{i=1}^{r} Q_i) \setminus \mathfrak{m}$. Then

$$(J(f) : \widetilde{J(f)}) \subseteq (J(f) : x) = (\cap_{i=1}^{r}(Q_i : x)) \cap (Q : x) = Q,$$

by [**1**, Lemma 4.4]. Besides, as $\widetilde{J(f)}, Q$ are co-maximal so $Q \cdot \widetilde{J(f)} = Q \cap \widetilde{J(f)} = J(f)$. Therefore, $Q \subseteq (J(f) : \widetilde{J(f)})$ and hence $(J(f) : \widetilde{J(f)}) = Q$. Note that the ideal $J(f)$ is $\mathfrak{m}$-primary if and only if $J(f) : \mathfrak{m}^{\infty} = R$.

(3) If $J(f)$ is not $\mathfrak{m}$-primary, then one can use the isomorphism in (7) to compute the Milnor number of f. However, computing 'saturation' is time-consuming. Besides, the Macaulay2 commands for working in a local ring are not fully developed. So we decided not to use either of these in the function `secMilnorNumbers` and restricted ourselves to the case when $J(f)$ is $\mathfrak{m}$-primary. This assumption forces $\mathrm{Sing}(f) = \{\text{origin}\}$, as the ideal $(f, J(f))$ is also $\mathfrak{m}$-primary. Nevertheless, users can use the succeeding commands to compute the Milnor number of f directly, whether $J(f)$ is $\mathfrak{m}$-primary or not. We take $f = x^2 + y^2 + z^2 + xyz \in \mathbb{Q}[x, y, z]$. By [**18**, Example 2.4], the Milnor number of f is 1. Notice $(f, J(f))$ is an $\mathfrak{m}$-primary ideal.

```
i1 : R = QQ[x,y,z];
i2 : f = x^2+y^2+z^2+x*y*z;
i3 : J = ideal jacobian f;
i4 : m= ideal vars R;
i5 : J1= saturate(J, m);
i6 : J1 == sub(ideal 1, R)
o6 : false
i7 : I= ideal f + J;
i8 : saturate(I, m) == sub(ideal 1, R)
```

```
o8 : true
i9 : J2 = J : J1;
i10 : degree J2
o10 : 1
-- the Milnor number of f
```

Let h be a non-constant homogeneous polynomial in $\mathbb{C}[z_0, \ldots, z_n]$ and let $\Delta_h \subseteq \mathbb{R}^n$ be the convex hull of exponents of dehomogenized monomials in $\mathbb{C}[z_1, \ldots, z_n]$ appearing in one of the partial derivatives of h. Note that such h need not have an isolated singularity at the origin. We set $\mathfrak{m} = (z_0, \ldots, z_n)$. In [**13**], June Huh extended the notion of sectional Milnor numbers and defined a sequence $\{\mu^i(h)\}_{i=0}^n$ by setting $\mu^i(h) = e_i(\mathfrak{m}, J(h))$ for $i = 0, \ldots, n$. We want to remark that once can compute $\mu^i(h)$ using the command `mixedMultiplicity`. Huh compared $\mu^i(h)$ with the mixed volume of standard n-dimensional simplex Δ and Δ_h in $\mathbb{R}^n$.

THEOREM 6.3 ([**13**, Theorem 15]). *Suppose that h is a non-constant homogeneous polynomial in $\mathbb{C}[z_0, \ldots, z_n]$. For $i = 0, \ldots, n$, we have*

$$\mu^{(i)}(h) \leq MV_n(\underbrace{\Delta, \ldots, \Delta}_{n-i}, \underbrace{\Delta_h, \ldots, \Delta_h}_{i}).$$

We use our algorithms to verify the above result through examples.

EXAMPLE 6.4. Let $h = z_0^3 + z_1^3 + z_2^3 + z_0z_1z_2 \in \mathbb{Q}[z_0, z_1, z_2]$. Then the Jacobian ideal $J(h) = (h_{z_0}, h_{z_1}, h_{z_2})$ is $(3z_0^2 + z_1z_2, 3z_1^2 + z_0z_2, 3z_2^2 + z_0z_1)$. We set $\mathfrak{m} = (z_0, z_1, z_3)$. Using Macaulay2, we verify that $\operatorname{ht} J(h) = 3$. This ensures that $J(h)$ is an $\mathfrak{m}$-primary ideal, since $J(h)$ is homogeneous. Consequently, h has an isolated singularity at the origin. Notice that the vertices of Δ_h are $\{(0,0), (1,1), (2,0), (0,1), (1,0), (0,2)\}$. The following session in Macaulay2 supports Theorem 6.3. In fact, we get equality for all i.

```
i1 : R = QQ[z_0..z_2];
i2 : h = z_0^3+z_1^3+z_2^3+z_0*z_1*z_2;
i3 : J = ideal jacobian ideal h;
i4 : codim J
o4 : 3
i5 : I = homIdealPolytope {(0,0),(1,1),(2,0),(0,1),(1,0),(0,2)};
i6 : W = ring I;
i7 : n = ideal vars W;
i8 : mMixedVolume {n,n}
o8 : 1
i9 : mMixedVolume {n,I}
o9 : 2
i10 : mMixedVolume {I,I}
o10 : 4
i11 : secMilnorNumbers(h)
o11 : HashTable{0=>1, 1=>2, 2=>4, 3=>8}
```

In [**13**, Example 17], J. Huh produced an example of a polynomial h which gives a strict inequality for some i. Although, this h does not have isolated singularity at the origin. The following Macaulay2 session verifies the example.

```
i1 : R = QQ[z_0..z_2];
i2 : h = z_1*(z_0*z_1 - z_2^2);
i3 : J = ideal jacobian ideal h;
i4 : codim J
o4 : 2
i5 : m = ideal vars R;
i6 : mixedMultiplicity ((m,J),(2,0))
o6 : 1
i7 : mixedMultiplicity ((m,J),(1,1))
o7 : 2
i8 : mixedMultiplicity ((m,J),(0,2))
o8 : 1
i9 : I = homIdealPolytope
        {(2,0),(1,0),(0,2),(1,1)};
i10 : W = ring I;
i11 : n = ideal vars W;
i12 : mMixedVolume {n,n}
o12 : 1
i13 : mMixedVolume {n,I}
o13 : 2
i14 : mMixedVolume {I,I}
o14 : 2
```

Note that for $i = 2$, we are getting the inequality $1 < 2$ as it was claimed by Huh. One can also consider the homogeneous polynomial $h = z_0^2 z_1 + z_0^2 z_2 + z_1^2 z_2 + z_1 z_2^2 \in \mathbb{Q}[z_0, z_1, z_2]$. Running a similar session as above, we get that $\mu^{(2)}(h) = 2$ whereas $MV_2(\Delta_h, \Delta_h) = 4$.

6.1. Verifying example of Joël Briançon and Jean-Paul Speder. We first recall Teissier's conjecture which claims that the invariance of Milnor number implies invariance of sectional Milnor numbers.

Teissier's Conjecture. [**22**] If (X, x) and (Y, y) have same topological type, then

$$\mu^*(X, x) = \mu^*(Y, y).$$

In [**3**], Joël Briançon and Jean-Paul Speder disproved the conjecture by giving a counter-example. They considered the family of hypersurfaces $\mathbf{X}_t \in \mathbb{C}^3$ defined by

$$F_t(x, y, z) = z^5 + ty^6 z + xy^7 + x^{15} = 0.$$

This family $\mathbf{X}_t$ provides a counter example to Teissier's conjecture. We verify the example by using Teissier's observation of identifying the sectional Milnor numbers with mixed multiplicities of ideals.

Consider the ideals $\mathfrak{m} = (x, y, z)$ and $J(F_t) = (\partial F_t/\partial x, \partial F_t/\partial y, \partial F_t/\partial z)$ in $\mathbb{C}[x, y, z]$, where

$$\frac{\partial F_t}{\partial x} = y^7 + 15x^{14}, \quad \frac{\partial F_t}{\partial y} = 6ty^5 z + 7xy^6 \quad \text{and} \quad \frac{\partial F_t}{\partial z} = 5z^4 + ty^6.$$

We show that while $e(J(F_t))$ in independent of t, $e_2(\mathfrak{m} \mid J(F_t))$ depends on t. Recall that

$$e_3(\mathfrak{m} \mid J(F_t)) = e(J(F_t)) = \ell\left(\frac{\mathbb{C}[x, y, z]}{J(F_t)}\right).$$

Let $t = 0$. Since $J(F_0)$ is generated by a system of parameters,

$$\begin{aligned} e(J(F_0)) = e(y^7 + 15x^{14}, xy^6, z^4) &= 4e(y^7 + 15x^{14}, xy^6, z) \\ &= 4e(y^7 + 15x^{14}, x, z) + 4e(y^7 + 15x^{14}, y^6, z) \\ &= 28 + 24e(x^{14}, y, z) = 28 + 336 = 364. \end{aligned}$$

Here, we use a fact that in a Cohen-Macaulay ring, if (a, b, c) and (a, b, d) are ideals generated by system of parameters such that c and d are co-prime and a, b, cd is a system of parameter, then $e(a, b, cd) = e(a, b, c) + e(a, b, d)$. Now let $t \neq 0$. Using the above observation and [**19**, Theorem 14.11], we get

$$\begin{aligned}
&e(J(F_t))\\
&\quad = e(y^7 + 15x^{14}, y^5(6tz + 7xy), 5z^4 + ty^6)\\
&\quad = 5e(y^7 + 15x^{14}, y, 5z^4 + ty^6) + e(y^7 + 15x^{14}, 6tz + 7xy, 5z^4 + ty^6)\\
&\quad = 5e(x^{14}, y, z^4) + e(y^7 + 15x^{14}, 5x^4y^4\alpha^4 + ty^6) \quad \text{(by putting } z = xy\alpha \text{ where } \alpha = -\frac{7}{6t})\\
&\quad = 280 + e(y^7 + 15x^{14}, y^4(5x^4\alpha^4 + ty^2))\\
&\quad = 280 + 4e(y^7 + 15x^{14}, y) + e(y^7 + 15x^{14}, 5(\alpha x)^4 + ty^2)\\
&\quad = 280 + 56 + e(y^7 + 15x^{14}, \sqrt{5}\alpha^2x^2 + i\sqrt{t}y) + e(y^7 + 15x^{14}, \sqrt{5}\alpha^2x^2 - i\sqrt{t}y)\\
&\quad = 336 + 14 + 14 = 364.
\end{aligned}$$

Hence $e(J(F_t)) = 364$ and is independent of t. We now calculate $e_2(\mathfrak{m} \mid J(F_t))$.

In order to calculate $e_2(\mathfrak{m} \mid J(F_t))$, we find a joint reduction of $(\mathfrak{m}, J(F_t), J(F_t))$. Let $t = 0$. Consider the set of elements $\{x, \partial F_0/\partial x, \partial F_0/\partial z\}$. Using Macaulay2, we could check that

$$\mathfrak{m}^{10}J^2 = (x)\mathfrak{m}^9J^2 + (\partial F_0/\partial x, \partial F_0/\partial z)\mathfrak{m}^{10}J$$

in $\mathbb{Q}[x, y, z]$. This implies

$$\{x, \partial F_0/\partial x, \partial F_0/\partial z\}$$

is a joint reduction of $(\mathfrak{m}, J(F_0), J(F_0))$ and hence

$$e_2(\mathfrak{m} \mid J(F_0)) = e(x, \partial F_0/\partial x, \partial F_0/\partial z) = e(x, y^7 + 15x^{14}, z^4) = e(x, y^7, z^4) = 28.$$

If $t \neq 0$, we consider the set of elements $\{x, \partial F_t/\partial y, \partial F_t/\partial z\}$. Using Macaulay2, we could check that

$$\mathfrak{m}^{11}J^2 = (x)\mathfrak{m}^{10}J^2 + (\partial F_t/\partial y, \partial F_t/\partial z)\mathfrak{m}^{11}J$$

in the ring $\mathbb{Q}(t)[x, y, z]$. So

$$\{x, \partial F_t/\partial y, \partial F_t/\partial z\}$$

is a joint reduction of $(\mathfrak{m}, J(F_t), J(F_t))$ and hence

$$\begin{aligned}
e_2(\mathfrak{m} \mid J(F_t)) = e(x, \partial F_t/\partial y, \partial F_t/\partial z) &= e(x, 6ty^5z + 7xy^6, 5z^4 + ty^6)\\
&= e(x, y^5z, 5z^4 + ty^6)\\
&= e(x, y^5, z^4) + e(x, z, y^6) = 20 + 6 = 26.
\end{aligned}$$

This proves that the mixed multiplicities are dependent on t, verifying the example given by Briançon and Speder. The following displays the working in Macaulay2.

```
i1 : k = frac(QQ[t]);
i2 : R = k[x,y,z];
i3 : f = z^5 + t*y^6*z + x*y^7 + x^15;
i4 : secMilnorNumbers (f)
o4 = HashTable{0=> 1, 1=> 4, 2=> 26, 3=> 364}
i5 : g = z^5 + x*y^7 + x^15;
i6 : secMilnorNumbers (g)
o6 = HashTable{0=> 1, 1=> 4, 2=> 28, 3=> 364}
```

References

[1] M. F. Atiyah and I. G. Macdonald, *Introduction to commutative algebra*, Student economy edition, Addison-Wesley Series in Mathematics, Westview Press, Boulder, CO, 2016. For the 1969 original see [MR0242802]. MR3525784

[2] Gennadiy Averkov, Christopher Borger, and Ivan Soprunov, *Classification of triples of lattice polytopes with a given mixed volume*, Discrete Comput. Geom. **66** (2021), no. 1, 165–202, DOI 10.1007/s00454-020-00246-4. MR4270639

[3] Joël Briançon and Jean-Paul Speder, *La trivialité topologique n'implique pas les conditions de Whitney* (French, with English summary), C. R. Acad. Sci. Paris Sér. A-B **280** (1975), no. 6, Aiii, A365–A367. MR425165

[4] Tianran Chen, *Unmixing the mixed volume computation*, Discrete Comput. Geom. **62** (2019), no. 1, 55–86, DOI 10.1007/s00454-019-00078-x. MR3959922

[5] David A. Cox, Kuei-Nuan Lin, and Gabriel Sosa, *Multi-Rees algebras and toric dynamical systems*, Proc. Amer. Math. Soc. **147** (2019), no. 11, 4605–4616, DOI 10.1090/proc/14579. MR4011498

[6] David A. Cox, John Little, and Donal O'Shea, *Using algebraic geometry*, 2nd ed., Graduate Texts in Mathematics, vol. 185, Springer, New York, 2005. MR2122859

[7] Clare D'Cruz, *A formula for the multiplicity of the multi-graded extended Rees algebra*, Comm. Algebra **31** (2003), no. 6, 2573–2585, DOI 10.1081/AGB-120021882. MR1986206

[8] David Eisenbud, Daniel R. Grayson, Michael Stillman, and Bernd Sturmfels (eds.), *Computations in algebraic geometry with Macaulay 2*, Algorithms and Computation in Mathematics, vol. 8, Springer-Verlag, Berlin, 2002, DOI 10.1007/978-3-662-04851-1. MR1949544

[9] Kriti Goel, Vivek Mukundan, Sudeshna Roy, and Jugal Verma, *Algorithms for computing mixed multiplicities, mixed volumes, and sectional Milnor numbers*, J. Softw. Algebra Geom. **13** (2023), no. 1, 1–12, DOI 10.2140/jsag.2023.13.1. MR4633631

[10] Daniel R. Grayson and Michael E. Stillman, *Macaulay2, a software system for research in algebraic geometry*, Available at `http://www.math.uiuc.edu/Macaulay2/`.

[11] Christian Haase, Martina Juhnke-Kubitzke, Raman Sanyal, and Thorsten Theobald, *Mixed Ehrhart polynomials*, Electron. J. Combin. **24** (2017), no. 1, Paper No. 1.10, 14, DOI 10.37236/5815. MR3609180

[12] Manfred Herrmann, Eero Hyry, Jürgen Ribbe, and Zhongming Tang, *Reduction numbers and multiplicities of multigraded structures*, J. Algebra **197** (1997), no. 2, 311–341, DOI 10.1006/jabr.1997.7128. MR1483767

[13] June Huh, *Milnor numbers of projective hypersurfaces and the chromatic polynomial of graphs*, J. Amer. Math. Soc. **25** (2012), no. 3, 907–927, DOI 10.1090/S0894-0347-2012-00731-0. MR2904577

[14] Craig Huneke and Irena Swanson, *Integral closure of ideals, rings, and modules*, London Mathematical Society Lecture Note Series, vol. 336, Cambridge University Press, Cambridge, 2006. MR2266432

[15] Babak Jabarnejad, *Equations defining the multi-Rees algebras of powers of an ideal*, J. Pure Appl. Algebra **222** (2018), no. 7, 1906–1910, DOI 10.1016/j.jpaa.2017.08.013. MR3763290

[16] Daniel Katz and J. K. Verma, *Extended Rees algebras and mixed multiplicities*, Math. Z. **202** (1989), no. 1, 111–128, DOI 10.1007/BF01180686. MR1007742

[17] Kuei-Nuan Lin and Claudia Polini, *Rees algebras of truncations of complete intersections*, J. Algebra **410** (2014), 36–52, DOI 10.1016/j.jalgebra.2014.03.022. MR3201047

[18] Yongqiang Liu, *Milnor and Tjurina numbers for a hypersurface germ with isolated singularity* (English, with English and French summaries), C. R. Math. Acad. Sci. Paris **356** (2018), no. 9, 963–966, DOI 10.1016/j.crma.2018.07.004. MR3849082

[19] Hideyuki Matsumura, *Commutative ring theory*, 2nd ed., Cambridge Studies in Advanced Mathematics, vol. 8, Cambridge University Press, Cambridge, 1989. Translated from the Japanese by M. Reid. MR1011461

[20] J. Ribbe, *On the defining equations of multi-graded rings*, Comm. Algebra **27** (1999), no. 3, 1393–1402, DOI 10.1080/00927879908826501. MR1669107

[21] Gabriel Sosa, *On the Koszulness of multi-Rees algebras of certain strongly stable ideals*, 2014, preprint: `arXiv:1406.2188v1`.

[22] Bernard Teissier, *Cycles évanescents, sections planes et conditions de Whitney* (French), Singularités à Cargèse (Rencontre Singularités Géom. Anal., Inst. Études Sci., Cargèse, 1972), Astérisque, Nos. 7 et 8, Soc. Math. France, Paris, 1973, pp. 285–362. MR374482

[23] Ngo Viet Trung and Jugal Verma, *Mixed multiplicities of ideals versus mixed volumes of polytopes*, Trans. Amer. Math. Soc. **359** (2007), no. 10, 4711–4727, DOI 10.1090/S0002-9947-07-04054-8. MR2320648

[24] J. K. Verma, *Multigraded Rees algebras and mixed multiplicities*, J. Pure Appl. Algebra **77** (1992), no. 2, 219–228, DOI 10.1016/0022-4049(92)90087-V. MR1149023

[25] J. K. Verma, D. Katz, and S. Mandal, *Hilbert functions of bigraded algebras*, Commutative algebra (Trieste, 1992), World Sci. Publ., River Edge, NJ, 1994, pp. 291–302. MR1421092

Department of Mathematics, University of Utah, Salt Lake City, Utah 84112
Email address: kritigoel.maths@gmail.com

Department of Mathematics, Indian Institute of Technology Delhi, New Delhi, 110016, India
Email address: vmukunda@iitd.ac.in

Department of Mathematics, Chennai Mathematical Institute, Siruseri, Kelambakkam 603103, India
Email address: sudeshnaroy.11@gmail.com

Department of Mathematics, Indian Institute of Technology Gandhinagar Gujarat, 382355, India
Email address: jugal.verma@iitgn.ac.in

Contemporary Mathematics
Volume **811**, 2025
https://doi.org/10.1090/conm/811/16241

Differential central simple algebras

Parul Gupta, Yashpreet Kaur, and Anupam Singh

ABSTRACT. Differential central simple algebras are the main object of study in this survey article. We recall some crucial notions such as differential subfields, differential splitting fields, tensor products etc. Our main focus is on differential splitting fields which connects these objects to the classical differential Galois theory. We mention several known results and raise some questions along the line.

1. Introduction

Let R be a ring with identity (not necessarily commutative). A ring R equipped with an additive map $d_R : R \to R$ satisfying the Leibnitz rule $d_R(\alpha\beta) = \alpha d_R(\beta) + d_R(\alpha)\beta$ for all $\alpha, \beta \in R$, is called a **differential ring**, and the map d_R is called a **derivation**. We denote a differential ring by an ordered pair (R, d_R). Any ring can be made into a differential ring in a trivial way by taking the zero derivation on it. A differential ring which is a field (or algebra) will be called a **differential field (or algebra)**. Here are some examples.

EXAMPLE 1.1. The zero derivation is the only derivation on $\mathbb{Q}$.

EXAMPLE 1.2. Let $k = \mathbb{Q}(t)$ be the rational function field over $\mathbb{Q}$. For $a \in k$, setting $d_k(t) = a$ we get a derivation on k. For $a = 1$, we get the usual derivation $\frac{d}{dt}$.

EXAMPLE 1.3 (Inner derivations). Let R be a ring. For any element $a \in R$, the map $\partial_a : R \to R$ given by $\partial_a(x) = xa - ax$ for $x \in R$ is a derivation on R. We say that ∂_a is the **inner derivation** with respect to a. Note that ∂_a restricts to the zero derivation on the center of R. In particular, if R is commutative then ∂_a is essentially the zero derivation on R for any $a \in R$. Inner derivations were introduced and studied in the context of Lie algebras in [**14**].

EXAMPLE 1.4 (Matrix algebras and coordinate-wise derivation). Let (k, d_k) be a differential field. The map $\delta^c : M_m(k) \to M_m(k)$ defined by $\delta^c((a_{ij})) = (d_k(a_{ij}))$ is a derivation on $M_m(k)$ and δ^c is called the **coordinate-wise derivation on** $M_m(k)$. Then $(M_m(k), \delta^c)$ is a differential algebra over (k, d_k).

2020 *Mathematics Subject Classification.* Primary 12H05, 16H05, 16W25.

Key words and phrases. Derivations, Differential algebras, Differential splitting fields, Differential Brauer group.

The second author is supported by NBHM (DAE, Govt. of India): 0204/16(33)/2020/R&D-II/26. The third author is funded by SERB through CRG/2019/000271 for this research.

EXAMPLE 1.5 (Quaternion algebras and standard derivation). Let (k, d_k) be a differential field and $\alpha, \beta \in k^\times$. Let $Q = (\alpha, \beta)_k$ be a quaternion algebra and let $u, v \in Q$ such that $u^2 = \alpha$, $v^2 = \beta$ and $vu = -uv$. A derivation d on Q such that $d(u) \in k(u)$ and $d(v) \in k(v)$ is called a **standard derivation on** Q, and is denoted by $d_{(u,v)}$. This is defined in [**19**].

Let (R, d_R) be a differential ring. Elements $c \in R$ such that $d_R(c) = 0$ are called **constants.** The kernel of the derivation, $\{c \in R : d_R(c) = 0\}$ forms a ring (indeed a differential ring with the zero derivation). We call the kernel the **ring of constants** of (R, d_R) and denote it by $C_{(R,d_R)}$. A ring homomorphism $\phi : (R, d_R) \to (S, d_S)$ is called a **differential homomorphism** if $d_S(\phi(\alpha)) = \phi(d_R(\alpha))$ for all $\alpha \in R$. A ring $S \supseteq R$ is called a **differential ring extension** of R if there is a differential ring structure on S which is compatible with the differential ring structure on R, i.e., $d_S|_R = d_R$. A differential ring extension of a differential field (k, d_k) is a **differential k-algebra**.

The main objects of interest in this article are differential central simple algebras. We have already seen some examples of differential central simple algebras (Examples 1.4, 1.5). Classifications of the derivations on certain central simple algebras are noted in Section 3. A crucial aspect of differential central simple algebras is their differential splitting fields. These splitting fields can be obtained as Picard-Vessiot extensions of a matrix differential equation. This connects differential splitting fields to the differential Galois theory (see [**17**]). In Section 2, we recall the definitions and results related to Picard-Vessiot extensions of a matrix differential equation and their differential Galois groups. In Section 4, we have gathered several open questions concerning differential splitting fields. Section 5 explores the tensor products of differential central simple algebras and related properties.

2. Picard-Vessiot extensions

For an introduction to differential Galois theory we refer the interested readers to [**16**], [**17**] and [**20**]. For the results in this section, our main references are [**22**, Chapter 1] and [**5**].

Let (k, d_k) be a differential field with its field of constants C_k an algebraically closed field. Let $P \in M_m(k)$. A **Picard-Vessiot ring** (R, d_R) over (k, d_k) for the matrix differential equation $\delta^c(Y) = PY$ is a ring satisfying the following:

(a) The only ideals I of R satisfying $d_R(I) \subseteq I$ are (0) and R.
(b) There exists a matrix $F \in \mathrm{GL}_m(R)$ satisfying $\delta^c(F) = PF$.
(c) The ring R is generated as a ring by k, the components of F and the inverse of the determinant of F.

For a matrix differential equation $\delta^c(Y) = PY$ over (k, d_k), there exists a Picard-Vessiot ring and is unique up to k-differential isomorphism. A Picard-Vessiot ring is a domain and the fraction field of a Picard-Vessiot ring is called a **Picard-Vessiot extension field** over k and is abbreviated as **PVE** (see [**22**, Proposition 1.20, Definition 1.21]). The next proposition describes the properties of a PVE:

PROPOSITION 2.1 ([**22**, Proposition 1.22]). *Let (k, d_k) be a differential field and $P \in M_m(k)$. A differential field extension $(K, d_K) \supset (k, d_k)$ is a PVE for a matrix differential equation $\delta^c(Y) = PY$ if and only if the following hold.*

(a) *There exists $F \in \mathrm{GL}_m(K)$ such that $\delta^c(F) = PF$.*
(b) *The field K is generated over k by the entries of F.*

(c) *The field of constants* $C_K = C_k$.

Let $(K, d_K) \supset (k, d_k)$ be a PVE for matrix equation $\delta^c(Y) = PY$, where $P \in M_m(k)$. The group of differential k-automorphisms of K is called the **differential Galois group** and is denoted by $\mathsf{DGal}(K/k)$. The differential Galois group can be viewed as a subgroup of $\mathrm{GL}_m(C_k)$ (see [**22**, Observations 1.26]), and hence is a linear algebraic group. The transcendence degree of K over k is equal to the dimension of $\mathsf{DGal}(K/k)$.

Similar to the classical Galois correspondence, there exists a differential Galois correspondence. For a subgroup $H \subset \mathsf{DGal}(K/k)$, let K^H denote the set of elements of K that are fixed by every differential automorphism in H. For a differential field $k \subset L \subset K$, let $\mathrm{DAut}(L/k)$ denote the set of differential automorphisms in $\mathsf{DGal}(K/k)$ that fixes every element in L. We state the differential Galois correspondence in the theorem below.

THEOREM 2.2 ([**5**, Theorem 1]). *For $P \in M_m(k)$, let (K, d_K) be a PVE for a matrix equation $\delta^c(Y) = PY$ over (k, d_k). Then*

(a) *For every (Zariski) closed subgroup $H \subset \mathsf{DGal}(K/k)$, there exists a differential subfield $L \subset K$ containing k such that $K^H = L$ and vice versa.*
(b) *A differential subfield $L \subset K$ containing k is a PVE over k if and only if $\mathsf{DGal}(K/L)$ is a normal subgroup of $\mathsf{DGal}(K/k)$.*

In [**5**], authors established the Galois correspondence for general PVEs, i.e. without the assumption that C_k is algebraically closed.

3. Differential central simple algebras

Let k be a field. By a **central simple k-algebra** we always mean a finite-dimensional k-algebra that has no two-sided non-trivial ideal and whose center is k. We refer to [**6**],[**7**] or [**4**], for the standard theory of central simple algebras.

Let A be a central simple k-algebra. Let d_k be a derivation on k and d be an extension of d_k to A. Then (A, d) is a **differential central simple algebra over** (k, d_k). The set of all derivations on A that are extensions of d_k is denoted by $Der(A/(k, d_k))$. Derivations on simple algebras are studied in [**1**], [**12**] and [**14**]. Initially questions were asked about the existence of extensions of derivations, which were dealt with in [**11**], [**3**],[**23**] and [**2**].

THEOREM 3.1 ([**2**, Theorem, Proposition 1]). *A derivation d_k on the field k can always be extended to a central simple k-algebra A. If $d, d' \in Der(A/(k, d_k))$, then $d' = d + \partial_a$ for some $a \in A$.*

In particular, the above result exhibits that one derivation on a central simple k-algebra A determines the whole set $Der(A/(k, d_k))$.

EXAMPLE 3.2. Let d_k be the zero derivation on k. Then $Der(A/(k, d_k))$ consists of inner derivations on A.

In some cases we can define a special element in $Der(A/(k, d_k))$, which we may call "standard", depending on the parameter or structure of the algebra. In the following example, we see the structure of $Der(M_m(k)/(k, d_k))$ for martix algebras $M_m(k)$.

EXAMPLE 3.3. For the matrix algebra $M_m(k)$, we have $Der(M_m(k)/(k, d_k)) = \{d_P = \delta^c + \partial_P \mid P \in M_m(k), \mathrm{tr}(P) = 0\}$ (see [**2**, Theorem 2]).

Symbol algebras are another important example of central simple algebras where a specific derivation is known depending on the defining parameters of the algebra. For $m \geq 2$, assume that k contains a primitive mth root of unity ζ. For $\alpha, \beta \in k^{\times}$, the **symbol algebra** $A = (\alpha, \beta)_{k,\zeta}$ is an m^2-dimensional k-algebra generated by $u, v \in A$ satisfying the relations $u^m = \alpha, v^m = \beta$ and $vu = \zeta uv$. Any 4-dimensional central simple k-algebras is isomorphic to a symbol algebra with $m = 2$; these are also called **quaternion algebras**. We denote the set of trace 0 elements in A by A^0.

EXAMPLE 3.4 (Symbol algebras and standard derivation). Let A be a m^2-dimensional symbol algebra over k. A derivation $d \in Der(A/(k, d_k))$ is called a **standard derivation on** A if there exists $u, v \in A^0$ such that $u^m, v^m \in k^{\times}$, $vu = \zeta uv$, $d(u) \in k(u)$ and $d(v) \in k(v)$, and we denote this derivation by $d_{(u,v)}$. For every $d \in Der(A/(k, d_k))$, there exists a unique element $a \in A^0$ such that $d = d_{(u,v)} + \partial_a$ (see [**8**, Proposition 3.1]).

We now generalize the above definition to an arbitrary central division algebra. Let D be a finite-dimensional central division k-algebra. As an easy application of the Double Centraliser Theorem, one gets that D is generated by two elements $D = k[u, v]$ where u, v satisfies certain relations (see [**10**, Corollary 3]). We say $d \in Der(D/(k, d_k))$ is a **standard derivation on** D if there exists $u, v \in D$ such that $D = k[u, v]$, $d(u) \in k(u)$ and $d(v) \in k(v)$, and we denote this derivation by $d_{(u,v)}$.

In the context of Examples 3.3 and 3.4, it is natural to ask the following question:

QUESTION 3.5. *For a central simple algebra A over a differential field (k, d_k), can one completely determine the set of derivations $Der(A/(k, d_k))$ using a standard derivation?*

In the context of the above question, one can study the behavior of derivations on subfields of A. As we have seen in Example 3.4 a standard derivation on a symbol algebra A stabilises two maximal subfields of A. This motivates the study of subfields of A that are stable under d. A subfield $L \subseteq A$ is called a **differential subfield** of (A, d) if $d(L) \subseteq L$.

Note that, in this case, d restricts to the unique derivation d_L on L. In [**2**, Theorem], it was shown that given a central simple k-algebra A and a maximal subfield L of A, there exists $d \in Der(A/(k, d_k))$ such that L is a differential subfield of (A, d). However, there are differential central simple algebras (A, d) having no differential subfields (see [**19**] and [**8**]).

QUESTION 3.6. *For a central simple algebra A over a differential field (k, d_k), classify the set of derivations $d \in Der(A/(k, d_k))$ such that every maximal subfield of A is a differential subfield.*

We say that two differential central simple algebras (A_1, d_1) and (A_2, d_2) over (k, d_k) are **differentially isomorphic** if there exists a k-algebra isomorphism $\phi : A_1 \to A_2$ such that $\phi \circ d_1 = d_2 \circ \phi$; in this case ϕ is called a **differential isomorphism of** (A_1, d_1) **and** (A_2, d_2).

QUESTION 3.7. *What is the relation between two differential central simple algebras having the same differential subfields? Do there exist non-isomorphic differential algebras (A_1, d_1) and (A_2, d_2) over (k, d_k) having same differential subfields?*

4. Differential splitting field

Let A be an m^2-dimensional central simple k-algebra and let K/k be a field extension. Then $A \otimes_k K$ is again a central simple K-algebra. We say that A **is split over** K if $A \otimes_k K \simeq M_m(K)$. It is well-known that a central simple algebra splits over its maximal subfields. For every subfield $E \subseteq A$, we have $[E : k]$ divides $\sqrt{\dim_k(A)}$. Thus, in particular, there exists a field extension E/k with $[E : k] \leq \sqrt{\dim_k(A)}$ and the algebra A is split over E. We refer the reader to [**21**, Chapter 13] for the statements related to maximal subfields and splitting fields.

Analogously, Juan and Magid [**15**] gave a notion of a differential splitting field of a differential central simple algebra. A differential central simple algebra (A, d) over (k, d_k) **is split over a differential field** $(K, d_K) \supseteq (k, d_k)$ if $(A \otimes_k K, d^* = d \otimes d_K)$ and $(M_m(K), \delta^c)$ are differentially isomorphic algebras over (K, d_K); we say that (K, d_K) is a **differential splitting field** id (A, d).

Let (A, d) be a differential central simple algebra over (k, d_k). Let L/k be a splitting field of A. Then there exists a trace zero matrix $P \in M_m(L)$ such that $(A \otimes_k L, d^*) \simeq (M_m(L), d_P)$ (see Example 3.3). Therefore, in order to understand the splitting of differential central simple algebras, the first step is to understand the splitting of differential matrix algebras. In [**15**, Proposition 2], a necessary and sufficient condition was obtained for a differential matrix algebra to be split.

PROPOSITION 4.1. *Let $P, Q \in M_m(k)$. Then differential algebras $(M_m(k), d_P)$ and $(M_m(k), d_Q)$ are differentially isomorphic if and only if there exists $F \in GL_m(k)$ such that $F^{-1}\delta^c(F) + F^{-1}QF = P$. In particular, $(M_m(k), d_P)$ is split if and only if there exists $F \in GL_m(k)$ such that $\delta^c(F) = FP$.*

Note that, for F^{-1} we have $\delta^c(F^{-1}) = PF^{-1}$ and hence we can as well say that $(M_m(k), d_P)$ is split if and only if there exists $F \in GL_m(k)$ such that $\delta^c(F) = PF$. This leads to the following result for arbitrary differential central simple algebras.

THEOREM 4.2 ([**15**, Corollary 1]). *A differential central simple algebra over (k, d_k) splits over a PVE of (k, d_k).*

Differential splitting fields of a differential central simple algebra are not necessarily algebraic extensions (see [**19**, Example 4.3, Theorem 4.5], [**8**, Theorem 5.5] and [**9**, Example 5.2, Theorem 6.5]). It is intriguing to determine the transcendence degree of a differential splitting field for a given differential central simple algebra. We set

$$\mathsf{Ddeg}_k(A, d) := \min\{\operatorname{tr.deg}_k(E) \mid (E, d_E) \supseteq (k, d_k) \text{ splits } (A, d)\}$$

and call $\mathsf{Ddeg}_k(A, d)$ the **differential splitting degree of** (A, d).

It is clear from Proposition 4.1 that $\mathsf{Ddeg}_k(A, d)$ is bounded by $\dim_k(A)$. Let (K, d_K) be a PVE of (k, d_k) splitting (A, d) and let $\mathsf{DGal}(K/k)$ be its differential Galois group. Since we can choose the matrix P to be a trace zero matrix, $\mathsf{DGal}(K/k)$ can be viewed as a subgroup of $\mathrm{SL}_m(C_k)$, by [**22**, Exercise 1.35 (5a)]. Then

$$\operatorname{tr.deg}_k(K) = \dim_k(\mathsf{DGal}(K/k)) \leq m^2 - 1.$$

Thus $\mathsf{Ddeg}_k(A, d)$ is bounded by $m^2 - 1$.

QUESTION 4.3. *Let A be a central simple k-algebra. Which numbers i between $0 \leq i \leq m^2 - 1$ can occur as $\mathsf{Ddeg}_k(A, d)$ for derivations $d \in Der(A/(k, d_k))$?*

Examples of differential quaternion algebras (Q, d) over (k, d_k) such that $\mathsf{Ddeg}_k(Q, d) = i$ for each $0 \leq i \leq 3$ were produced in [**9**].

EXAMPLE 4.4 ([**9**, Example 5.2 and Theorem 6.5]). Let $(k = \mathbb{Q}(t), d_k)$ be a differential field with $d_k(t) = 1$. Let $Q = (1, t)_k$ be a quaternion algebra and $u, v \in Q^0$ such that $u^2 = 1$, $v^2 = t$ and $vu = -uv$. Consider a derivation $d = d_{(u,v)} + \partial_a$ on Q, where $a \in Q$. Then

(a) For $a = 0$, $\mathsf{Ddeg}_k(Q, d) = 0$.
(b) For $a = \frac{1}{t}v$, $\mathsf{Ddeg}_k(Q, d) = 1$.
(c) For $a = \frac{1}{4t}(-u - 2v + 2uv)$, $\mathsf{Ddeg}_k(Q, d) = 2$.
(d) For $a = -\frac{1}{4t}u - v$, $\mathsf{Ddeg}_k(Q, d) = 3$.

Using [**9**, Theorem 6.5], one can produce several examples of differential quaternion algebras with $\mathsf{Ddeg}_k(Q, d) = 0, 1$. On the other hand, for $\mathsf{Ddeg}_k(Q, d) = 2, 3$ we only know examples of split quaternion algebras. Therefore, we ask the following question.

QUESTION 4.5. *Does there exist a differential division quaternion algebra (Q, d) such that $\mathsf{Ddeg}_k(Q, d)$ equals 2 or 3?*

Since differential Galois groups of a PVE over k can be viewed as a subgroup of $\mathrm{SL}_m(C_k)$, there may exist a relation between differential splitting fields of differential central simple algebras and subgroups of $\mathrm{SL}_m(C_k)$.

QUESTION 4.6. *Which linear algebraic subgroups of $\mathrm{SL}_m(C_k)$ can occur as Galois groups of PVE over k that splits a differential central simple algebra over k?*

It is also interesting to relate differential subfields to differential splitting fields For example, it was shown in [**8**, Theorem 5.4], that a symbol algebra $(\alpha, \beta)_{k,\zeta}$ with a standard derivation (where we have differential subfields that are also maximal) is split by a finite extension of k.

QUESTION 4.7. *Let (L, d_L) be a PVE of (k, d_k) splitting (A, d). Explore the relation between differential subfields of (A, d) and (L, d_L). More precisely we can ask: Assuming that there is a maximal subfield $M \subseteq A$ which is also a differential subfield of A, what can we say about $\mathsf{Ddeg}_k(A, d)$?*

QUESTION 4.8. *Let A be a central simple k-algebra and let $d_1, d_2 \in \mathsf{Der}(A/(k, d_k))$. Asssume that (A, d_1) and (A, d_2) have the same differential splitting fields. Is $(A, d_1) \simeq (A, d_2)$?*

The following characterises differential quaternion algebras over (k, d_k) that are split over a finite differential field extension of (k, d_k).

THEOREM 4.9 ([**9**, Theorem 7.3]). *A differential quaternion algebra splits over a finite field extension of k if and only if the derivation on the quaternion algebra becomes standard over a finite extension of k.*

QUESTION 4.10. *Does Theorem 4.9 hold for symbol algebras and division algebras?*

5. Tensor products of differential central simple algebras

Given two differential central simple algebras (A_1, d_1) and (A_2, d_2) over (k, d_k), we obtain the tensor differential algebra as follows. The algebra $A_1 \otimes_k A_2$ is again a central simple k-algebra and the map $d_1 \otimes d_2 : A_1 \otimes_k A_2 \to A_1 \otimes_k A_2$ defined by $(d_1 \otimes d_2)(a_1 \otimes a_2) = d_1(a_1) \otimes a_2 + a_1 \otimes d_2(a_2)$ is a derivation on $A_1 \otimes_k A_2$. We write $(A_1, d_1) \otimes_{(k,d_k)} (A_1, d_1)$ for the differential central simple algebra $(A_1 \otimes_k A_2, d_1 \otimes d_2)$ over (k, d_k).

QUESTION 5.1. *Let A be a central simple k-algebra and let $d_1, d_2 \in \mathsf{Der}(A/(k, d_k))$. Asssume that $(A, d_1) \otimes_k (A, d_2)$ is split over (k, d_k). Are (A, d_1) and (A, d_2) differentially isomorphic?*

The Brauer group of a field k, denoted by $\mathrm{Br}(k)$, is the quotient group of the group generated by central simple algebras modulo the subgroup generated by matrix algebras. Due to Weddurburn's Theorem, every central simple k-algebra A is Brauer equivalent to a unique division algebra up to isomorphism.

Analogous to the usual definition of the Brauer group of a field, a definition of the differential Brauer group was given in [**13**]. However, with the definition in [**13**] it turned out that when the base field is of characteristic zero the differential Brauer group is the same as the usual Brauer group and thus does not contain any extra information. In [**15**], the **differential Brauer group** denoted by $Br^{diff}(k, d_k)$ is defined as the quotient group of the group generated by differential matrix algebras over (k, d_k) modulo the subgroup generated by matrix algebras with coordinate-wise derivation. Note that $Br^{diff}(k, d_k)$ encompasses information only about differential matrix algebras. When k is algebraically closed (or a C_1-field) then all central simple algebras are matrix algebras.

It is well-known that $\mathrm{Br}(k)$ is a torsion group, that is every element in $\mathrm{Br}(k)$ has finite order. It is well-known for a central division k-algebra D that the order of the class of D in $\mathrm{Br}(k)$ divides $\sqrt{\dim_k(D)}$. However, it is possible to construct a differential matrix algebra $(M_m(k), d)$ such that for every $n \in \mathbb{N}$, $(M_m(k), d) \otimes_k \cdots \otimes_k (M_m(k), d)$ (n-times) in a non-split differential algebra. A differential central simple algebra (A, d) over (k, d_k) is said to be of **finite order** if there exists an integer $n \in \mathbb{N}$ such that $(A, d) \otimes_k \cdots \otimes_k (A, d)$ (n-times) is split differential algebra over (k, d_k). The following question is motivated by the work of K. Singla and A. Kulshrestha [**18**, Theorem 5.1], where for certain matrix differential algebras, they have shown that finite order implies splitting by a finite extension, i.e. $\mathsf{Ddeg} = 0$.

QUESTION 5.2. *Let (A, d) be a differential central simple algebra over (k, d_k) of finite order. Does it imply that $\mathsf{Ddeg}(A, d) = 0$?*

Since every element in $\mathrm{Br}(k)$ has finite order, it is enough to consider the above question for matrix algebras. One can also try to deal with the case of quaternion algebras or symbol algebras separately.

QUESTION 5.3. *Find relation between differential splitting degrees of two Brauer equivalent differential central simple algebras.*

Acknowledgments

Authors wish to express their gratitude to Varadharaj R. Srinivasan and Amit Kulshrestha for several insightful discussions on this topic which has given rise to many questions in this article. We also thank Kanika Singla for discussions happened on this topic during a seminar series organised by Varadharaj R. Srinivasan. We thank the referee(s) for valuable comments which helped to improve the manuscript.

References

[1] S. A. Amitsur, *Derivations in simple rings*, Proc. London Math. Soc. (3) **7** (1957), 87–112, DOI 10.1112/plms/s3-7.1.87. MR88480

[2] S. A. Amitsur, *Extension of derivations to central simple algebras*, Comm. Algebra **10** (1982), no. 8, 797–803, DOI 10.1080/00927878208822750. MR651973

[3] Michael Barr and Max-Albert Knus, *Extensions of derivations*, Proc. Amer. Math. Soc. **28** (1971), 313–314, DOI 10.2307/2037807. MR272829

[4] Grégory Berhuy and Frédérique Oggier, *An introduction to central simple algebras and their applications to wireless communication*, Mathematical Surveys and Monographs, vol. 191, American Mathematical Society, Providence, RI, 2013. With a foreword by B. A. Sethuraman, DOI 10.1090/surv/191. MR3086869

[5] Teresa Crespo, Zbigniew Hajto, and Elżbieta Sowa-Adamus, *Galois correspondence theorem for Picard-Vessiot extensions*, Arnold Math. J. **2** (2016), no. 1, 21–27, DOI 10.1007/s40598-015-0029-z. MR3460041

[6] P. K. Draxl, *Skew fields*, London Mathematical Society Lecture Note Series, vol. 81, Cambridge University Press, Cambridge, 1983, DOI 10.1017/CBO9780511661907. MR696937

[7] Philippe Gille and Tamás Szamuely, *Central simple algebras and Galois cohomology*, Cambridge Studies in Advanced Mathematics, vol. 101, Cambridge University Press, Cambridge, 2006, DOI 10.1017/CBO9780511607219. MR2266528

[8] Parul Gupta, Yashpreet Kaur, and Anupam Singh, *Splitting fields of differential symbol algebras*, J. Pure Appl. Algebra **227** (2023), no. 5, Paper No. 107280, 16, DOI 10.1016/j.jpaa.2022.107280. MR4513883

[9] Parul Gupta, Yashpreet Kaur, and Anupam Singh, *Splitting of differential quaternion algebras*, J. Algebra **633** (2023), 43–55, DOI 10.1016/j.jalgebra.2023.06.022. MR4610779

[10] I. N. Herstein and A. Ramer, *A note on division algebras*, Canadian J. Math. **24** (1972), 734–736, DOI 10.4153/CJM-1972-069-4. MR297804

[11] G. Hochschild, *Restricted Lie algebras and simple associative algebras of characteristic p*, Trans. Amer. Math. Soc. **80** (1955), 135–147, DOI 10.2307/1993009. MR72428

[12] Klaus Hoechsmann, *Simple algebras and derivations*, Trans. Amer. Math. Soc. **108** (1963), 1–12, DOI 10.2307/1993822. MR152548

[13] R. T. Hoobler, *The differential Brauer group*, `arXiv:1003.1421` (2010).

[14] Nathan Jacobson, *Abstract derivation and Lie algebras*, Trans. Amer. Math. Soc. **42** (1937), no. 2, 206–224, DOI 10.2307/1989656. MR1501922

[15] Lourdes Juan and Andy R. Magid, *Differential central simple algebras and Picard-Vessiot representations*, Proc. Amer. Math. Soc. **136** (2008), no. 6, 1911–1918, DOI 10.1090/S0002-9939-08-09165-X. MR2383496

[16] Irving Kaplansky, *An introduction to differential algebra*, Publications de l'Institut Mathématique de l'Université de Nancago [Publications of the Mathematical Institute of the University of Nancago], No. 5, Hermann, Paris, 1957. Actualités Scientifiques et Industrielles [Current Scientific and Industrial Topics], No. 1251. MR93654

[17] E. R. Kolchin, *Algebraic matric groups and the Picard-Vessiot theory of homogeneous linear ordinary differential equations*, Ann. of Math. (2) **49** (1948), 1–42, DOI 10.2307/1969111. MR24884

[18] Amit Kulshrestha and Kanika Singla, *Finite splittings of differential matrix algebras*, J. Algebra **634** (2023), 74–96, DOI 10.1016/j.jalgebra.2023.06.019. MR4620670

[19] Amit Kulshrestha and Varadharaj R. Srinivasan, *Quaternion algebras with derivations*, J. Pure Appl. Algebra **226** (2022), no. 2, Paper No. 106805, 14, DOI 10.1016/j.jpaa.2021.106805. MR4276482

[20] Andy R. Magid, *Lectures on differential Galois theory*, University Lecture Series, vol. 7, American Mathematical Society, Providence, RI, 1994, DOI 10.1090/ulect/007. MR1301076

[21] Richard S. Pierce, *Associative algebras*, Studies in the History of Modern Science, vol. 9, Springer-Verlag, New York-Berlin, 1982. Graduate Texts in Mathematics, 88. MR674652

[22] Marius van der Put and Michael F. Singer, *Galois theory of linear differential equations*, Grundlehren der mathematischen Wissenschaften [Fundamental Principles of Mathematical Sciences], vol. 328, Springer-Verlag, Berlin, 2003, DOI 10.1007/978-3-642-55750-7. MR1960772

[23] A. Roy and R. Sridharan, *Derivations in Azumaya algebras*, J. Math. Kyoto Univ. **7** (1967), 161–167, DOI 10.1215/kjm/1250524275. MR222073

Department of Mathematics, Shiv Nadar Institution of Eminence, Gautam Buddh Nagar, Uttar Pradesh 201314, India
Email address: parulgupta1211@gmail.com

Department of Mathematics, Thapar Institute of Engineering and Technology, Bhadson Rd, Patiala Punjab 147004, India
Email address: yashpreetkm@gmail.com

Department of Mathematics, Indian Institute of Science Education and Research Pune, Dr. Homi Bhabha Road, Pashan, Pune, 411008, India
Email address: anupamk18@gmail.com

Contemporary Mathematics
Volume **811**, 2025
https://doi.org/10.1090/conm/811/16242

z-Classes and rational conjugacy classes

Vikas S. Jadhav

This paper is dedicated to Professor Ravi S. Kulkarni on the occasion of his eightieth birthday

Abstract. For a group G, we say that $x, y \in G$ are in the same z-class if their centralizers are conjugate in G. Two elements x and y in a group G are rational conjugate if the cyclic subgroups generated by x and y are conjugate subgroups of G. In this paper we obtain an upper bound on the number z-classes of finite groups in terms of number of rational conjugacy classes of certain associated quotient group, and give a necessary condition on groups attaining this bound. We prove the group $U(n, p)$ of upper triangular matrices for $2 \leq n \leq 4$, p-prime attains the upper bound. We also characterize the non-abelian p-groups of order $\leq p^5$, attaining this upper bound.

1. Introduction

For a group G, we say that $x, y \in G$ are in the same z-class if their centralizers in G are conjugate. Two elements x and y in a group G are rational conjugate or $\mathbb{Q}$-conjugate if the cyclic subgroups generated by x and y are conjugate subgroups of G. The notion of z-class has origin in a connection between geometry and groups. However, as the notion is purely group theoretic, in this paper, we focus on connection between number of rational conjugacy classes and number of z-classes, as the number of rational conjugacy classes is equal to number of rational irreducible representations of the group.

This paper is organized as follows: In Section 2, we obtain upper bound on the number of z-classes for general class of finite groups, and give a necessary condition on groups attaining this bound. In Section 3, we prove that the upper bound for the number of z-classes is attained for the the group $U(n, p)$ of upper triangular matrices for $2 \leq n \leq 4$, $p-$prime. In Section 4, we characterize the non-abelian p-groups of order $\leq p^5$, attaining the upper bound for the number of z-classes.

Notations. For a group G, $Z(G)$ denotes the center and $\gamma_2(G) = G'$ denotes the commutator subgroup of G. If H is a subgroup of G, we write $H \leq G$, and if H is proper subgroup, we write $H < G$. For $x \in G$ and $H \leq G$, we denote by $Z_G(x)$ and $Z_G(H)$ respectively, the centralizer of x, respectively H, in G (cf. [**3**], [**4**], [**5**], [**11**], [**12**]). Some authors use the notation $C_G(x)$ and $C_G(H)$ respectively for

2020 *Mathematics Subject Classification.* Primary 20D15.

Key words and phrases. z-classes, rational conjugacy classes, p-groups.

The author acknowledges support from UGC-Teacher fellowship received through Nowrosjee Wadia College, Pune to work at Bhaskarcharya Prathisthana, Pune, during which a part of this work was done.

centralizers, but we do not use it in this paper. Here, $N_G(H)$ denotes the normalizer of H in G; $[x,y]$ is the commutator $x^{-1}y^{-1}xy$; and C_n denotes the cyclic group of order n. The nilpotency class of G is denoted by $c(G)$; and $Z_2(G)$ is the subgroup of G such that $Z_2(G) \setminus Z(G) = Z(G \setminus Z(G))$. We write $x \sim_z y$ if x, y are in the same z-class.

2. An upper bound on number of z-classes in groups

Let G be a group acting on a set X. For $x \in X$, let G_x denote the stabilizer subgroup of G at x. We say that $x, y \in X$ are in *same orbit class* if G_x and G_y are conjugate in G. We shall denote this equivalence relation by $x \sim_0 y$. Let $R(x)$ denote the equivalence class of x with respect to $\sim_0$. When $X = G$ and G acts on itself by conjugation then the stabilizer group of an element is its centralizer. In this case we call $R(x)$ as the *z-class* of $x \in G$ (see [**11**]). Thus for a group G, we say that $x, y \in G$ are in the same z-class if their centralizers are conjugate in G.

Two elements x and y in a group G are rational conjugate if the cyclic subgroups generated by x and y are conjugate subgroups of G.

Here we will see connection between rational conjugacy classes (representation theory) and z-classes.

For any group G, let $\mathcal{C}_G$ be the set of conjugacy classes of cyclic subgroups of G, and $\mathcal{Z}_G$ be the set of z-classes of G. Let $\varphi\colon G \to \overline{G} = G/Z(G)$ be the natural homomorphism. For $H \leq G$ (resp. $\overline{H} \leq \overline{G}$), let $[H]$ (resp. $[\overline{H}]$) denote the conjugacy class of the subgroup H (resp. $\overline{H}$). Define,

$$h\colon \mathcal{C}_{\overline{G}} \to \mathcal{Z}_G, \ [\langle \bar{x} \rangle] \mapsto R(x),$$

where $R(x)$ is the z-class of x in G.

PROPOSITION 1. *The map h is well defined and surjective.*

PROOF. $R(x)$ is a $Z(G)$-set (action by multiplication) as well as a G-set (action by conjugation). Therefore, for any $t \in Z(G)$, $xt \in R(x)$ and $gxg^{-1} \in R(x)$, and so h is well defined and surjective. □

COROLLARY 1 (Upper bound). *If G is a group with $[G\colon Z(G)] < \infty$, the number of z-classes in G is at most equal to the number of rational irreducible representations of $G/Z(G)$.*

PROOF. This follows from surjectivity of h in Proposition 1 and the fact that in a finite group, the number of conjugacy classes of cyclic subgroups of the group is equal to the number of rational irreducible representations of the group (see [**15**, §13.1, p. 102]). □

LEMMA 2.1. *Let G be a non-abelian finite p-group and $[G\colon Z(G)] = p^2$. Then the number of z-classes in G is equal to the number of conjugacy classes of cyclic subgroups in $G/Z(G)$.*

PROOF. Let G be a non-abelian finite p-group and $[G\colon Z(G)] = p^2$, then $G/Z(G) \cong C_p \times C_p$. Here $G/Z(G)$ is abelian and has $p+2$ cyclic subgroups (including the trivial subgroup), hence $G/Z(G)$ has $p+2$ conjugacy classes of cyclic subgroups. Also by Lemma 7 in [**10**], G has $p+2$ z-classes. □

THEOREM 2.2. *Let G be a non-abelian finite p-group with $[G\colon Z(G)] = p^k$. If the number of z-classes in G is equal to the number of rational irreducible representations of $G/Z(G)$, then either $G/Z(G) \cong C_p \times C_p$ or G has no abelian subgroup of index p.*

PROOF. Suppose that $G/Z(G) \not\cong C_p \times C_p$. Note that, by Proposition 1, if $xZ(G)$, $yZ(G)$ are $\mathbb{Q}$-conjugate in $G/Z(G)$, then x, y are z-equivalent in G, and by hypothesis, converse holds for G. That is, in this case the map h is a bijection.

If possible, let A be an abelian subgroup of index p. Then $Z(G) < A$, otherwise $AZ(G) = G$ will be abelian. Thus

$$Z(G) < A < G \text{ and } [A\colon Z(G)] \geq p^2.$$

Recall Lemma 6 in [**10**] which says that in a non-abelian finite p-group a abelian maximal subgroup H of G is a union of two z-classes in G, viz. $Z(G)$ and $H \backslash Z(G)$.

CASE 1 ($G/Z(G)$ is abelian).

(i) If $A/Z(G)$ is cyclic (of order $\geq p^2$), let $aZ(G)$ be a generator of $A/Z(G)$. Then $aZ(G)$ and $a^pZ(G)$ are not $\mathbb{Q}$-conjugate in $G/Z(G)$ as their orders are different, but a and a^p are z-equivalent in G by Lemma 6 in [**10**], as both a and $a^p \notin Z(G)$. A contradiction.

(ii) If $A/Z(G)$ is non-cyclic, $\exists\, a, b \in A \setminus Z(G)$, such that $\langle aZ(G)\rangle \neq \langle bZ(G)\rangle$. Then $aZ(G)$, $bZ(G)$ are not $\mathbb{Q}$-conjugate in $G/Z(G)$, but a, b are z-equivalent in G (by Lemma 6 in [**10**]).

CASE 2 ($G/Z(G)$ is non-abelian). Then $Z(G) < Z_2(G)$. Since $A/Z(G)$ is an abelian subgroup of index p,

$$Z(G/Z(G)) = Z_2(G)/Z(G) < A/Z(G), \text{ hence } Z_2(G) < A.$$

For any $a \in A \setminus Z_2(G)$ and $b \in Z_2(G) \setminus Z(G)$, $Z_G(a) = A = Z_G(b)$, hence a, b are z-equivalent in G (by Lemma 6 in [**10**]), but $aZ(G), bZ(G)$ are not $\mathbb{Q}$-conjugate (since only $bZ(G)$ is central), a contradiction.

□

3. z-classes and rational conjugacy classes in the matrix group $U(n, p), 2 \leq n \leq 4$

There are two most commonly used non-abelian finite p-groups in the literature. One is the group of unitriangular matrices over a field and the other is the extra-special p-groups. In this section we prove that the upper bound for the number of z-classes given in Corollary 1 is attained for the group $U(n, p)$ of upper unitriangular matrices for $2 \leq n \leq 4$, p-prime. For $n \geq 2$, let $U(n, p)$ be the group of $n \times n$ upper triangular matrices with diagonal entries 1, over $\mathbb{F}_p$, where p is a prime. The group $U(2, p) \cong \mathbb{Z}_p$ has exactly one z-class and the result holds. The group $U(3, p)$ is non-abelian group of order p^3; it has $p + 2$ z-classes, whose representatives are given by

$$I = \begin{pmatrix} 1 & & \\ & 1 & \\ & & 1 \end{pmatrix}, \begin{pmatrix} 1 & & \\ & 1 & 1 \\ & & 1 \end{pmatrix}, \begin{pmatrix} 1 & 1 & \\ & 1 & \alpha \\ & & 1 \end{pmatrix}_{\alpha \in \mathbb{F}_p}.$$

$U(3,p)/Z(U(3,p)) \cong C_p \times C_p$, and the images of these representatives in the central quotient generate $p+2$ distinct cyclic subgroups. Therefore, *the number of z-classes in $U(3,p)$ is equal to the number of conjugacy classes of cyclic subgroups in $U(3,p)/Z(U(3,p))$.*

We now prove that $x, y \in G = U(4,p)$ are z-equivalent if and only if $\langle xZ(G)\rangle$ and $\langle yZ(G)\rangle$ are conjugate in $G/Z(G)$. Before that, we obtain some structural information about G and $G/Z(G)$. (Blank entries in the matrix represent 0).

1. $|G| = p^6$ and the subgroup

$$H = \left\{ \begin{pmatrix} 1 & * & * & * \\ & 1 & & * \\ & & 1 & * \\ & & & 1 \end{pmatrix} \right\}$$

is normal in G; it is an extra-special p-group of order p^5. Therefore, $\overline{H} = H/Z(G)$ is elementary abelian subgroup of index p in $\overline{G} = G/Z(G)$.
If $\bar{x} \in \overline{H} \setminus Z(\overline{G})$, then $Z_{\overline{G}}(\bar{x}) = \overline{H}$, hence $|Z_{\overline{G}}(\bar{x})| = p^4$.
If $\bar{x} \in \overline{G} \setminus \overline{H}$, then $Z_{\overline{G}}(\bar{x}) = \langle \bar{x}, Z(\overline{G})\rangle$, hence $|Z_{\overline{G}}(\bar{x})| = p^3$.
2. *For $x \in G$, the subgroup $Z_G(x)$ is maximal if and only if $x \in Z_2(G) \setminus Z(G)$:*
First note that if $x \in Z_2(G) \setminus Z(G)$ then $Z_G(x) < G$, hence

$$p^6 > |Z_G(x)| \geq |Z_{\overline{G}}(\bar{x})| = p^5,$$

hence $Z_G(x)$ is a maximal subgroup. Now,

$$Z_2(G) = \gamma_2(G) = \left\{ \begin{pmatrix} 1 & 0 & * & * \\ & 1 & 0 & * \\ & & 1 & 0 \\ & & & 1 \end{pmatrix} \right\}.$$

If $M \leq G$ is maximal, $M \supseteq \gamma_2(G)$, hence $Z_G(M) \subseteq Z_G(\gamma_2(G))$. For $i < j$, $E_{ij}(\alpha)$ be the upper triangular matrix with (i,j)th entry α, ones on the diagonal and remaining non-diagonal entries 0. Then $\gamma_2(G) = \langle E_{13}(1), E_{24}(1)\rangle$, and

$$Z_G(E_{13}(1)) = \left\{ \begin{pmatrix} 1 & * & * & * \\ & 1 & * & * \\ & & 1 & \\ & & & 1 \end{pmatrix} \right\}, \quad Z_G(E_{24}(1)) = \left\{ \begin{pmatrix} 1 & & * & * \\ & 1 & * & * \\ & & 1 & * \\ & & & 1 \end{pmatrix} \right\},$$

hence

$$K = Z_G(\gamma_2(G)) = Z_G(E_{13}(1)) \cap Z_G(E_{24}(1)) = \left\{ \begin{pmatrix} 1 & & * & * \\ & 1 & * & * \\ & & 1 & \\ & & & 1 \end{pmatrix} \right\}.$$

Therefore, if $Z_G(x)$ is maximal, then

$$\langle x, Z(G)\rangle \subseteq Z(Z_G(x)) \subseteq Z_G(\gamma_2(G)) = K, \text{ hence } x \in K.$$

Consider an element of K of the type

$$A = \begin{pmatrix} 1 & & a_{13} & a_{14} \\ & 1 & \alpha & a_{24} \\ & & 1 & \\ & & & 1 \end{pmatrix}, \quad \alpha \in \mathbb{F}_p^*.$$

By conjugating A by $E_{12}(1)$, we get $E_{12}(1)^{-1}AE_{12}(1)$. Thus by conjugating, part of the second row of A after diagonal gets added to first row. Also conjugating A by $E_{34}(1)$, the part of the third column above the diagonal gets substracted from the fourth column. Hence, conjugating A by suitable power of $E_{12}(1)$ and $E_{34}(1)$ if necessary, we can assume that $a_{13} = a_{24} = 0$. Hence,

$$A \sim A_1 = \begin{pmatrix} 1 & & & a_{14} \\ & 1 & \alpha & \\ & & 1 & \\ & & & 1 \end{pmatrix}, \quad \alpha \in \mathbb{F}_p^* \ (\sim \text{ stands for conjugacy})$$

If $\alpha \neq 1$, for some $1 \leq i < p$,

$$A_1^i = \begin{pmatrix} 1 & & & * \\ & 1 & 1 & \\ & & 1 & \\ & & & 1 \end{pmatrix}, \quad \alpha \in \mathbb{F}_p^*,$$

and $Z_G(A_1) = Z_G(A_1^i) = Z_G(E_{23}(1))$ (since we can ignore central part of element in finding its centralizer). Then,

$$Z_G(A) \sim Z_G(A_1) = Z_G(E_{24}(1)) = \begin{pmatrix} 1 & & * & * \\ & 1 & * & * \\ & & 1 & \\ & & & 1 \end{pmatrix},$$

which is not a maximal subgroup. Therefore, $Z_G(x)$ is a maximal subgroup only if $x \in Z_2(G) \setminus Z(G)$.
3. *If $u, v \in Z_2(G) \setminus Z(G)$ with $\langle u, Z(G)\rangle \neq \langle v, Z(G)\rangle$ then $Z_G(u) \neq Z_G(v)$*:
If $Z_G(u) = Z_G(v)$ $(= M)$, then $u, v \in Z(Z_G(u)) \subseteq Z_2(G)$; by hypothesis $Z_2(G) = \langle u, v, Z(G)\rangle$ and it is in the center of a maximal subgroup (M), and hence centralizer of all elements of $Z_2(G)\setminus Z(G)$ is the same maximal subgroup. But as shown above, $Z_G(E_{13}(1))$ and $Z_G(E_{24}(1))$ are distinct maximal subgroups.

THEOREM 3.1. *$x, y \in U(4,p)$ are z-equivalent if and only if their images in the central quotient generate conjugate cyclic subgroups.*

PROOF. Let $x, y \in G = U(4,p)$ be z-equivalent. If x and y are conjugate, then their images in the central quotient will generate conjugate cyclic subgroups; therefore, we assume that x, y are not conjugate. Since, for some $g \in G$, $Z_G(x) = gZ_G(y)g^{-1} = Z_G(gyg^{-1})$, replacing y by the conjugate $g^{-1}yg$, we can assume that $Z_G(x) = Z_G(y)$ and prove that $\langle xZ(G)\rangle$, $\langle yZ(G)\rangle$ are conjugate in $G/Z(G)$.

Clearly, if $x \in Z(G)$ and $Z_G(x) = Z_G(y)$, then $\langle \bar{x}\rangle = \langle \bar{y}\rangle = 1$.

If $x \in Z_2(G) \setminus Z(G)$ and $Z_G(x) = Z_G(y)$ then by (2), $y \in Z_2(G) \setminus Z(G)$ and by (3), $\langle x, Z(G)\rangle = \langle y, Z(G)\rangle$, hence $\langle \bar{x}\rangle = \langle \bar{y}\rangle$.

Let $x \in H \setminus Z_2(G)$ and $Z_G(x) = Z_G(y)$, where

$$H = \left\{ \begin{pmatrix} 1 & * & * & * \\ & 1 & & * \\ & & 1 & * \\ & & & 1 \end{pmatrix} \right\}.$$

Then $p^5 > |Z_G(x)| \geq |Z_{\overline{G}}(\bar{x})| = p^4$, hence $|Z_G(x)| = p^4 = |Z_G(y)|$. Since H is extra-special group of order p^5, and $x \notin Z(H)$, hence $Z_H(x)$ is maximal in H (hence of order p^4). Therefore, $Z_G(x) = Z_H(x)$. Also $Z_H(x)$ is non-abelian (otherwise, $|H| = p.|Z(H)|.|H'| = p^3$, contradiction). As center of $Z_G(x) = Z_H(x)$ contains x and $Z(G)$, we must have $Z(Z_G(x)) = \langle x, Z(G)\rangle$. Therefore,

$$\langle x, Z(G)\rangle = Z(Z_G(x)) = Z(Z_G(y)) = \langle y, Z(G)\rangle, \text{ hence } \langle \bar{x}\rangle = \langle \bar{y}\rangle.$$

Let $x \in G \setminus H$ and $Z_G(x) = Z_G(y)$. Then $\bar{x} \notin \overline{H}$, hence by (1), $Z_{\overline{G}}(\bar{x}) = \langle \bar{x}, Z(\overline{G})\rangle$. Also, $xy = yx$, hence $\bar{x}\bar{y} = \bar{y}\bar{x}$, i.e. $\bar{y} \in \langle \bar{x}, Z(\overline{G})\rangle$. Therefore,

$$\langle \bar{x}, Z(\overline{G})\rangle = \langle \bar{y}, Z(\overline{G})\rangle.$$

Since $\langle \bar{x}\rangle$ and $\langle \bar{y}\rangle$ are non-central cyclic subgroups of order p in $\overline{G}$ and they are contained in the normal subgroup $\langle \bar{x}, Z(\overline{G})\rangle$, hence all conjugates of $\langle \bar{x}\rangle$ are in this normal subgroup. The number of conjugates of $\langle \bar{x}\rangle$ in $\overline{G}$ is the index of the normalizer of $\langle \bar{x}\rangle$ in $\overline{G}$. Since $\langle \bar{x}\rangle \cong C_p$, normalizer of $\langle \bar{x}\rangle$ is same as its centralizer, which is equal to $\langle \bar{x}, Z(\overline{G})\rangle$, and has index p^2. Therefore, $\langle \bar{x}\rangle$ has p^2 conjugates in G, all are contained in the subgroup $\langle \bar{x}, Z(\overline{G})\rangle$. It is easy to see that the number of subgroups of order p contained in $\langle \bar{x}, Z(\overline{G})\rangle$ and non-central in $\overline{G}$ is equal to p^2. Therefore, $\langle \bar{x}\rangle$ and $\langle \bar{y}\rangle$ must be conjugate in $\overline{G}$. □

THEOREM 3.2. *The number of conjugacy classes of cyclic subgroups in* $U(4,p)/Z(U(4,p))$ *is* $2(p^2+p+1)$.

PROOF. Let $G = U(4,p)$ and $\overline{G} = G/Z(G)$. $|\overline{G}| = p^5$, and $\overline{G}$ has an elementary abelian subgroup $\overline{H}$ of index p, which is the image of the maximal subgroup

$$H = \left\{ \begin{pmatrix} 1 & * & * & * \\ & 1 & & * \\ & & 1 & * \\ & & & 1 \end{pmatrix} \right\}.$$

Also $Z(\overline{G}) = \gamma_2(\overline{G}) \cong C_p \times C_p$.

Clearly, $Z(\overline{G})$ has exactly $p+2$ distinct cyclic subgroups (including trivial) and each of these forms a singleton conjugacy class in the collection of cyclic subgroups of $\overline{G}$. Let $N \leq \overline{H}$ be a cyclic subgroup such that $N \cap Z(\overline{G}) = 1$. Since $\overline{H} \trianglelefteq \overline{G}$, all conjugates of N are in $\overline{H}$. The number of conjugates of N in $\overline{G}$ is equal to the index of normalizer of N, which is equal to the index of centralizer of N (since $N \cong C_p$). As centralizer of N is $\overline{H}$, N has p conjugates in $\overline{G}$, and they are contained in $\overline{H}$. If $\overline{H}$ contains l conjugacy classes of non-central cyclic subgroups of $\overline{G}$, then

$$p^4 - p^2 = |\overline{H} \setminus Z(\overline{G})| = lp.(p-1), \text{ hence } l = p^2 + p.$$

Let $L \leq \overline{G}$ be a cyclic subgroup such that $L \not\subseteq \overline{H}$. Then we can see that the centralizer (or normalizer) of L in $\overline{G}$ is $\langle L, Z(\overline{G})\rangle$, which has index p^2 in $\overline{G}$. Therefore, L has p^2 conjugates in $\bar{G}$. If there are t conjugacy classes of cyclic subgroups of $\bar{G}$ not contained in $\bar{H}$, then

$$p^5 - p^4 = |\bar{G} \setminus \bar{H}| = tp^2(p-1), \text{ hence } t = p^2.$$

Hence $\overline{G}$ has $(p+2)+l+t = 2p^2+2p+2$ conjugacy classes of cyclic subgroups. □

By the Theorem 3.1 and the Theorem 3.2, the number of z-classes in $U(4,p)$ is equal to the number of conjugacy classes of cyclic subgroups in $U(4,p)/Z(U(4,p))$ which is equal to $2p^2+2p+2$.

CONJECTURE. *For a prime p, and natural number $n \geq 2$, $x, y \in U(n,p)$ are z-equivalent if and only if their images in the central quotient generate conjugate cyclic subgroups.*

REMARK 3.3. In general for $U(n,q)$, $q = p^k$, $k > 1$, the above conjecture is not true. For example, let $G = U(3,4)$. Here the number of z-classes in G is 6 and the number of rational conjugacy classes in $G/Z(G)$ is 16.

4. Non-abelian p-groups of order $\leq p^5$ attaining an upper bound

In this section we will find the non-abelian p-groups of order $\leq p^5$, attaining the upper bound for the number of z-classes given in Corollary 1 viz. the number of rational conjugacy classes of $G/Z(G)$. We have the following observations:
(O_1) If G is a non-cyclic group of order p^2, then the number of irreducible rational representations of G, which is equal to the number of rational conjugacy classes in G is $p+2$.
(O_2) If $G \cong C_p \times C_p \times \cdots \times C_p$ (r times) is an elementary abelian p-group, then G has exactly $\dfrac{p^r - 1}{p-1} + 1$ distinct cyclic subgroups (including the trivial one) and each of these forms a singleton conjugacy class in the collection of cyclic subgroups of G. Thus the number of rational conjugacy classes in G is $\dfrac{p^r-1}{p-1} + 1$.
(O_3) If G is a non-abelian group of order p^3, then G is of class 2 and $G/Z(G) \cong C_p \times C_p$. By (O_1) above, the number of irreducible rational representations of $G/Z(G)$ is $p+2$, and hence by Corollary 3.4.7 in [**9**] (If G is a finite p-group of class 2, then G has at most one faithful irreducible $\mathbb{Q}$-representation.), the number of irreducible rational representations of G is $p+3$.

Let G be a non-abelian p-group of order $\leq p^5$, p-prime.

(1) If $[G\colon Z(G)] = p^2$, then by Lemma 7 in [**10**] and (O_1), the number of z-classes in G = the number of irreducible rational representations of $G/Z(G) = p+2$.
(2) If $[G\colon Z(G)] = p^3$, then either $|G| = p^4$ or $|G| = p^5$ and $\mathrm{cl}(G) = 2$ or 3.
    (a) If $|G| = p^4$, then $\mathrm{cl}(G) = 3$ i.e. $G/Z(G)$ is a non-abelian group of order p^3, since $Z(G) \subset \gamma_2(G)$. By Theorem 8 in [**10**], the number of z-classes in G are $p+2$, whereas by (O_3), we have seen that the number of irreducible rational representations of $G/Z(G)$ is $p+3$.
    (b) If $|G| = p^5$, then either $\mathrm{cl}(G) = 2$ or 3.
        (i) If $\mathrm{cl}(G) = 2$ (i.e. $G/Z(G)$ is abelian), then by Theorem 9(1) in [**10**], G has p^2+2 z-classes. Also in the proof of Theorem 9(in [**10**],we have seen that G has an abelian subgroup of index p, and $G/Z(G) \cong C_p \times C_p \times C_p$. Thus by (O_2), $G/Z(G)$ has p^2+p+2 irreducible rational representations.
        (ii) If $\mathrm{cl}(G) = 3$ (i.e. $G/Z(G)$ is non-abelian) and G has an abelian subgroup of index p, then by Theorem 9(2(a)) in [**10**], G has $p+2$ z-classes, whereas by (O_3), $G/Z(G)$ has $p+3$ irreducible rational representations.
        (iii) If $\mathrm{cl}(G) = 3$ (i.e. $G/Z(G)$ is non-abelian) and G has no abelian subgroup of index p, then by Theorem 9(2(b)) in [**10**] and (O_3), the number of z-classes in G = the number of irreducible rational representations of $G/Z(G) = p+3$.

(3) If $[G\colon Z(G)] = p^4$, then $|G| = p^5$ and $\mathrm{cl}(G) = 2, 3$ or 4.

(a) If $\mathrm{cl}(G) = 2$ (i.e. $G/Z(G)$ is abelian), then in the proof of Theorem 10(1) in [**10**], it is proved that there is a bijection between z-classes of G and the cyclic subgroups of $G/Z(G)$ (including $\{1\}$); these are $\dfrac{p^4-1}{p-1}+1$ in number. Hence the number of z-classes in G = the number of irreducible rational representations of $G/Z(G) = \dfrac{p^4-1}{p-1}+1$.

(b) If $\mathrm{cl}(G) = 3$ (i.e. $G/Z(G)$ is non-abelian), then by Theorem 10(2) in [**10**], G has $3p+2$ z-classes. Also in the proof of Theorem 10(2) in [**10**], it is proved that G has no abelian subgroup of index p and $\overline{G} = G/Z(G)$ is a non-abelian group of order p^4, of class 2. Also $Z(\overline{G})$ is not cyclic.

For $p = 2$, we can see the following using GAP Software: Either $G/Z(G) = \langle x, y, z | x^4 = 1 = y^2 = z^2, xy = yx, yz = zy, z^{-1}xz = xy\rangle \cong (C_4 \times C_2) \rtimes C_2$ or $G/Z(G) \cong C_2 \times D_8$.

(i) If $G/Z(G) \cong (C_4 \times C_2) \rtimes C_2$, then $G/Z(G)$ has 8 irreducible rational representations.

(ii) If $G/Z(G) \cong C_2 \times D_8$, then $G/Z(G)$ has 10 irreducible rational representations.

If $p \geq 3$, then using the classification of groups of order p^4 given in Appendix in [**9**], we see that as $Z(\overline{G})$ is not cyclic, $\overline{G} = G/Z(G)$ is isomorphic to one of the following groups:

$$
\begin{aligned}
K_1 &= \langle x, y | x^{p^2} = 1 = y^{p^2}, y^{-1}xy = x^{1+p}\rangle \cong C_{p^2} \rtimes C_{p^2} \\
K_2 &= \langle x, y, z | x^{p^2} = 1 = y^p = z^p, xy = yx, yz = zy, z^{-1}xz = x^{1+p}\rangle \\
&\cong (C_{p^2} \rtimes C_p) \times C_p \\
K_3 &= \langle x, y, z | x^{p^2} = 1 = y^p = z^p, xy = yx, yz = zy, z^{-1}xz = xy\rangle \\
&\cong (C_{p^2} \times C_p) \rtimes C_p \\
K_4 &= \langle x, y, z, w | x^p = 1 = y^p = z^p = w^p, w^{-1}zw = zx, \\
&\qquad wy = yw, wx = xw, zy = yz, zx = xz, xy = yx\rangle \\
&\cong ((C_p \times C_p) \rtimes C_p) \times C_p
\end{aligned}
$$

CLAIM 1. *$G/Z(G) \ncong K_2$ and $G/Z(G) \ncong K_3$.*

Suppose $G/Z(G) \cong K_2$. Here $o(x) = p^2$, and $o(xy) = p^2$ as x and y commute and $o(y) = p$. Also $(xz)^p = x^p z^p [z, x]^{\binom{p}{2}} = x^p \neq 1$, hence $o(xz) = p^2$. Thus the subgroups $\overline{H_1} = \langle x\rangle$, $\overline{H_2} = \langle xy\rangle$ and $\overline{H_3} = \langle xz\rangle$ has index p^2 in $\overline{G}$. Let $\varphi\colon G \to \overline{G} = G/Z(G)$ be the natural homomorphism and $\varphi^{-1}(\overline{H_i}) = H_i$, $i = 1, 2, 3$. We see that $\overline{H} = \cap_{i=1}^3 \overline{H_i} = \langle x^p\rangle$. Let $H = \varphi^{-1}(\overline{H})$. For each $i = 1, 2, 3$, H_i is an abelian subgroup of G (since $H_i/Z(G)$ is a cyclic group) and $[G\colon H_i] = p^2$. Note that H is central in each H_i and $\langle H_1, H_2, H_3\rangle = G$, hence H lies in the center of G. This is a contradiction, since $[G\colon H] = [\overline{G}\colon \overline{H}] = p^2$, whereas $[G\colon Z(G)]$. Hence $G/Z(G) \ncong K_2$. Similary we can see that $G/Z(G) \ncong K_3$.

By Lemma 2.4.8 in [**9**], $\overline{G}$ has an abelian subgroup $\overline{K}$ of index p. As $Z(\overline{G}) \cong C_p \times C_p$ and $Z(\overline{G}) \leq \overline{K}$, we have that either $\overline{K}$ is of the type $C_p \times C_p \times C_p$ or of the type $C_{p^2} \times C_p$. Note that the group K_1 has $p+1$ abelian subgroups of index p and each of them is of the type $C_{p^2} \times C_p$. Also the group K_4 has $p+1$ abelian subgroups of index p and each of them is of the type $C_p \times C_p \times C_p$. Thus if the abelian subgroup of index p in $\overline{G}$ is of the type $C_{p^2} \times C_p$, then $\overline{G} \cong K_1$, and if the abelian subgroup of index p in $\overline{G}$ is of the type $C_p \times C_p \times C_p$, then $\overline{G} \cong K_4$.

Thus for $p \geq 3$, either $G/Z(G) = \langle x, y | x^{p^2} = 1 = y^{p^2}, y^{-1}xy = x^{1+p}\rangle \cong C_{p^2} \rtimes C_{p^2}$ or $G/Z(G) = \langle x, y, z, w | x^p = 1 = y^p = z^p = w^p, w^{-1}zw = zx, wy = yw, wx = xw, zy = yz, zx = xz, xy = yx\rangle \cong C_p \times ((C_p \times C_p) \rtimes C_p)$.

(i) If $G/Z(G) \cong C_{p^2} \rtimes C_{p^2}$, then $G/Z(G)$ has $3p+2$ irreducible rational representations.

(ii) If $G/Z(G) \cong C_p \times ((C_p \times C_p) \rtimes C_p)$, then the $G/Z(G)$ has p^2+2p+2 irreducible rational representations.

(c) If $\mathrm{cl}(G) = 4$ i.e. G is a p-group of maximal class, then $G/Z(G)$ is a p-group of order p^4 and $\mathrm{cl}(G/Z((G)) = 3$.

(i) For $p = 2$, $G/Z(G) \cong D_{16}$, the Dihedral group of order 16. $G/Z(G) \cong D_{16}$ has 6 irreducible rational representations (using GAP software, see [**6**]).

(ii) For $p = 3$, $G/Z(G) = \langle x, y, z | x^9 = 1 = y^3 = z^3, xy = yx, z^{-1}yz = x^{-3}y, z^{-1}xz = xy\rangle \cong (C_9 \times C_3) \rtimes C_3$, and has 7 irreducible rational representations (using GAP software, see [**6**]).

(iii) For $p \geq 5$, the exponent of $G/Z(G)$ is p (see [**1**, §4.3 Theorem 4.7]). But a group of order p^4 of class 3 and exponent p is isomorphic to $(C_p \times C_p \times C_p) \rtimes C_p$. Thus $G/Z(G) \cong (C_p \times C_p \times C_p) \rtimes C_p$. The number of rational conjugacy classes in $G/Z(G)$ is $2p+3$. Hence $G/Z(G)$ has $2p+3$ irreducible rational representations. By Theorem 12 in [**10**], G has $p+2$ z-classes, if G has abelian subgroup of index p, and G has $p+4$ z-classes, if G has no abelian subgroup of index p.

From the above we have Proposition 2:

PROPOSITION 2. *Let G be a non-abelian p-group of order $\leq p^5$, p-prime. Then the number of z-classes in G is equal to the number of irreducible rational representations, if and only if G satisfies one of the following:*

(1) $[G\colon Z(G)] = p^2$.

(2) $|G| = p^5, [G\colon Z(G)] = p^3$, *$cl(G) = 3$ and G has no abelian subgroup of index p.*

(3) $|G| = p^5, [G\colon Z(G)] = p^4$, $cl(G) = 2$.

(4) $|G| = 2^5, [G\colon Z(G)] = 2^4$, $cl(G) = 3$ *and* $G/Z(G) = \langle x, y, z | x^4 = 1 = y^2 = z^2, xy = yx, yz = zy, z^{-1}xz = xy\rangle \cong (C_4 \times C_2) \rtimes C_2$.

(5) $|G| = p^5, [G\colon Z(G)] = p^4$, $cl(G) = 3$ *and* $G/Z(G) = \langle x, y | x^{p^2} = 1 = y^{p^2}, y^{-1}xy = x^{1+p}\rangle \cong C_{p^2} \rtimes C_{p^2}$, $p > 2$.

(6) $|G| = 3^5, [G\colon Z(G)] = p^4$, $cl(G) = 4$ *and* G *has no abelian subgroup of index* p.

Acknowledgment

The author wishes to thank Dr. Ravi Kulkarni, Dr. S. A. Katre and Dr. R. D. Kitture for their helpful input in the preparation of this article. The author thanks Dr. S. A. Katre and Dr. Rahul D. Kitture for their helpful input.

References

[1] Gustavo A. Fernández-Alcober, *An introduction to finite p-groups: regular p-groups and groups of maximal class*, Mat. Contemp. **20** (2001), 155–226. 16th School of Algebra, Part I (Portuguese) (Brasília, 2000). MR1868828

[2] Yakov Berkovich, *Groups of prime power order. Vol. 1*, De Gruyter Expositions in Mathematics, vol. 46, Walter de Gruyter GmbH & Co. KG, Berlin, 2008. With a foreword by Zvonimir Janko, DOI 10.1515/9783110208238.512. MR2464640

[3] Krishnendu Gongopadhyay and Ravi S. Kulkarni, *z-classes of isometries of the hyperbolic space*, Conform. Geom. Dyn. **13** (2009), 91–109, DOI 10.1090/S1088-4173-09-00190-8. MR2491719

[4] Krishnendu Gongopadhyay and Ravi S. Kulkarni, *The Z-classes of isometries*, J. Indian Math. Soc. (N.S.) **81** (2014), no. 3-4, 245–258. MR3308688

[5] Krishnendu Gongopadhyay, *The z-classes of quaternionic hyperbolic isometries*, J. Group Theory **16** (2013), no. 6, 941–964, DOI 10.1515/jgt-2013-0013. MR3198727

[6] The GAP Group, *GAP – Groups, Algorithms, and Programming*, Version 4.4; 2005, `http://www.gap-system.org`.

[7] P. Hall, *The classification of prime-power groups*, J. Reine Angew. Math. **182** (1940), 130–141, DOI 10.1515/crll.1940.182.130. MR3389

[8] Chenggong Hao and Zhuxuan Jin, *Finite p-groups which contain a self-centralizing cyclic normal subgroup*, Acta Math. Sci. Ser. B (Engl. Ed.) **33** (2013), no. 1, 131–138, DOI 10.1016/S0252-9602(12)60199-4. MR3003748

[9] Vikas S. Jadhav, *Arithmetic Aspects of Group Representation, z-Classes, Jacobi Sums and Codes*, Ph.D Thesis, Savitribai Phule Pune University.

[10] Vikas S. Jadhav and Rahul Dattatraya Kitture, *z-classes in p-groups of order* $\leq p^5$, Bull. Allahabad Math. Soc. **29** (2014), no. 2, 173–194. MR3331311

[11] Kulkarni Ravi S., *Dynamical types and conjugacy classes of centralizers in groups*, J. Ramanujan Math. Soc., 22, No. 1 (2007) 35-36.

[12] Kulkarni Ravi S., *Dynamics of Linear and Affine Maps*, Asian Journal of Mathematics, 12, No. 3, pp. 321-344.

[13] Ravindra Kulkarni, Rahul Dattatraya Kitture, and Vikas S. Jadhav, *z-classes in groups*, J. Algebra Appl. **15** (2016), no. 7, 1650131, 16, DOI 10.1142/S0219498816501310. MR3528559

[14] Derek J. S. Robinson, *A course in the theory of groups*, 2nd ed., Graduate Texts in Mathematics, vol. 80, Springer-Verlag, New York, 1996, DOI 10.1007/978-1-4419-8594-1. MR1357169

[15] Jean-Pierre Serre, *Linear representations of finite groups*, Graduate Texts in Mathematics, Vol. 42, Springer-Verlag, New York-Heidelberg, 1977. Translated from the second French edition by Leonard L. Scott. MR450380

[16] Anupam Singh, *Conjugacy classes of centralizers in* G_2, J. Ramanujan Math. Soc. **23** (2008), no. 4, 327–336. MR2492571

Department of Mathematics, Nowrosjee Wadia College, SPPU, Pune, India

Current address: Department of Mathematics, Nowrosjee Wadia College, 19 V. K. Jaog Path, Pune-411001, India

Email address: `svikasjadhav@gmail.com`

Contemporary Mathematics
Volume **811**, 2025
https://doi.org/10.1090/conm/811/16243

On Jacobian group and complexity of the Δ-graph

A. D. Mednykh, I. A. Mednykh, and I. N. Yudin

Dedicated to the 80-th anniversary of our friend and colleague Ravi Kulkarni

Abstract. In the present paper we investigate the set of invariants of Δ-graph $\Delta(n;k,l,m)$, namely the number of spanning trees, the number of rooted spanning forest and the Jacobian group. The Δ-graphs continue the list of families of I-, Y- and H-graphs well-known in the graph theory. In particular, graph $\Delta(n;1,1,1)$ is isomorphic to discrete torus $C_3 \times C_n$.

The main results of the paper are analytical formulas for the above mentioned invariants. In the case of graph $\Delta(n;1,1,1)$, the invariants are given through the classical Chebyshev polynomials.

1. Introduction

Let G be a finite connected graph. The notion of the complexity for a graph arises in different ways. For example, it can vary from simple parameters like the number of edges or vertices to the number of spanning trees or rooted spanning forests of a graph. All the above mentioned invariants can be found as a function of eigenvalues of the Laplacian operator of the graph. So, they are spectral invariants.

The complexity of a graph plays an important role in statistic physics, where the graphs with arbitrary large number of vertices are considered ([**14**], [**38**]). As the number of vertices of a graph grows, the degree of its Laplacian polynomial also rises and becomes quite difficult to handle. Then the most interesting invariants are given by their asymptotic formulas.

Recently, it was discovered in [**26**], that the number of spanning forests in circulant graph coincides with the order of homology group for some manifold arising as a cyclic coverings of the three dimensional sphere branched over a knot.

The notion of the Jacobian group of a graph, which is also known as the Picard group, the critical group, and the dollar or sandpile group, was independently introduced by many authors ([**2**], [**3**], [**6**], [**9**]). Jacobian group admits a natural interpretation in various areas of physics, coding theory, and financial mathematics. Also, it can be viewed as a discrete version of the Jacobian from the classical theory of Riemann surfaces. Different aspects of discrete version of Riemann surface theory were developed by Ravi Kulkarni in his paper [**4**].

2020 *Mathematics Subject Classification.* Primary 05C30, 39A12, 11C20.

Key words and phrases. Spanning tree, Jacobian group, Petersen graph, Laplacian matrix, Chebyshev polynomial.

The study was carried out within the framework of the state contract of the Sobolev Institute of Mathematics (project FWNF-2020-0005).

The Jacobian group is an important algebraic invariant of a finite graph. In particular, its order coincides with the number of spanning trees of the graph, which is known for some simple graphs such as the wheel, fan, prism, ladder, and Möebius ladder [**7**], grids [**28**], lattices [**30**], Sierpinski gaskets [**1, 32**], 3-prism and 3-anti-prism [**37**] and so on. At the same time, Jacobian group is not a spectral invariant. There are exist isospectral graphs with different Jacobian groups [**2**].

The structure of the Jacobian is known only in a several cases ([**6**], [**9**], [**20**], [**22**], [**23**], [**31**], [**33**]).

We mention that the number of spanning trees and the number of spanning rooted forests for circulant graphs are expressed in terms of the Chebyshev polynomials. See papers [**11**], [**24**], [**25**], [**34**], [**35**], and [**36**].

More generally, this result also holds for arbitrary cyclic coverings of a graph [**18**]. The family of cyclic coverings contains generalized Petersen graph $GP(n,k)$ and I-graph $I(n,j,k)$ (see, for example [**17**] and [**21**]). The later two graphs are expansions of the tree consisting of a single edge. In the same time, there is a wide family of graphs that are expansions of the tree consisting of more than one edge. The three such families of graphs were discovered by Biggs [**5**]. He calls these graphs the I-graph, Y-graph, and H-graph, because of the method by which the graphs were created. This method of construction was generalized in [**13**].

In the present paper we consider expansion of the cycle graph on 3 vertices, so called Δ-graph. We find the number of spanning trees and rooted spanning forests and investigate the structure of Jacobian. In the simplest non-trivial cases the structure of the above mentioned invariants will be find explicitly. The similar questions for I-graphs and Y-graph were considered earlier in the papers [**19**], [**21**].

A Δ-graph $\Delta(n;k,l,m)$ has $3n$ vertices $v_{x,y}$ where $x = 1,2,3$ and y in the integers modulo n. We organize the set of vertices into three groups of n vertices by first subscript x. The graph induced by the vertices $v_{x,y}$ for a fixed x, have edges which join each $v_{x,y}$ to $v_{x,y+j(x)}$, where $j(1)=k$, $j(2)=l$ and $j(3)=m$. The subscript addition is performed modulo n. On the other hand, the graph induced by the vertices $v_{x,y}$ for a fixed y is a triangle graph C_3. It has a Δ shape. This is a reason for a name of the graph family. This definition produces a connected graph if one has $\gcd(k,l,m,n)=1$, where $\gcd(k,l,m,n)$ is the greatest common divisor of k,l,m and n. In what follows, we deal with connected graphs only.

2. Basic definitions and preliminary results

Consider a connected finite graph G, allowed to have multiple edges but without loops.

A *spanning forest* F in G is an non-cyclic subgraph that contains all vertices of G. All connected components of F are trees. A spanning forest F is called *rooted* if any tree in F has a *root*, that is a labeled vertex. Connected spanning forest is a *spanning tree*.

Number of rooted spanning trees in a connected graph G is $n\,\tau(G)$, where $\tau(G)$ is the number of all spanning trees or just complexity of graph G and n is the number of vertices of G. This simple observation is not true anymore for the number of spanning forests.

To count the number of rooted spanning forests in a graph G and to count the number of all spanning forests in G are two completely different problems. Through there are about one thousand papers devoted to enumeration of spanning trees, there are just a very few papers devoted to spanning forests [8], [11], [12], [16].

We denote the vertex and edge set of G by $V(G)$ and $E(G)$, respectively. Given $u, v \in V(G)$, we set a_{uv} to be equal to the number of edges between vertices u and v. The matrix $A = A(G) = \{a_{uv}\}_{u,v \in V(G)}$ is called *the adjacency matrix* of the graph G. The degree $d(v)$ of a vertex $v \in V(G)$ is defined by $d(v) = \sum_{u \in V(G)} a_{uv}$. Let $D = D(G)$ be the diagonal matrix indexed by the elements of $V(G)$ with $d_{vv} = d(v)$. The matrix $L = L(G) = D(G) - A(G)$ is called *the Laplacian matrix*, or simply *Laplacian*, of the graph G.

Recall [20] the following useful relation between the structure of the Laplacian matrix and the Jacobian of a graph G. Consider the Laplacian $L(G)$ as a homomorphism $\mathbb{Z}^{|V|} \to \mathbb{Z}^{|V|}$, where $|V| = |V(G)|$ is the number of vertices in G. The cokernel $\operatorname{coker}(L(G)) = \mathbb{Z}^{|V|}/\operatorname{im}(L(G))$ is an abelian group. Let

$$\operatorname{coker}(L(G)) \cong \mathbb{Z}_{d_1} \oplus \mathbb{Z}_{d_2} \oplus \cdots \oplus \mathbb{Z}_{d_{|V|}}$$

be its Smith normal form satisfying the conditions $d_i \big| d_{i+1}$, $(1 \le i \le |V| - 1)$. If the graph is connected, then the groups $\mathbb{Z}_{d_1}, \mathbb{Z}_{d_2}, \ldots, \mathbb{Z}_{d_{|V|-1}}$ are finite, and $\mathbb{Z}_{d_{|V|}} = \mathbb{Z}$. In this paper, we define Jacobian group $Jac(G)$ as the torsion subgroup of $\operatorname{coker}(L(G))$. In other words,

$$Jac(G) \cong \mathbb{Z}_{d_1} \oplus \mathbb{Z}_{d_2} \oplus \cdots \oplus \mathbb{Z}_{d_{|V|-1}}.$$

Let M be an integer $n \times n$ matrix, then we can interpret M as $\mathbb{Z}$-linear operator from $\mathbb{Z}^n$ to $\mathbb{Z}^n$. In this interpretation M has a kernel $\ker M$, an image $\operatorname{im} M$, and a cokernel $\operatorname{coker} M = \mathbb{Z}^n/\operatorname{im} M$. We emphasize that $\operatorname{coker} M$ of the matrix M is completely determined by its Smith normal form. In particular, if matrices M and M' are elementary equivalent then $\operatorname{coker} M \cong \operatorname{coker} M'$.

In what follows, by I_n we denote the identity matrix of order n.

We call an $n \times n$ matrix *circulant*, and denote it by $circ(a_0, a_1, \ldots, a_{n-1})$ if it is of the form

$$circ(a_0, a_1, \ldots, a_{n-1}) = \begin{pmatrix} a_0 & a_1 & a_2 & \ldots & a_{n-1} \\ a_{n-1} & a_0 & a_1 & \ldots & a_{n-2} \\ & \vdots & & \ddots & \vdots \\ a_1 & a_2 & a_3 & \ldots & a_0 \end{pmatrix}.$$

Recall [10] that the eigenvalues of matrix $C = circ(a_0, a_1, \ldots, a_{n-1})$ are given by the following simple formulas $\lambda_j = p(\varepsilon_n^j)$, $j = 0, 1, \ldots, n-1$ where $p(x) = a_0 + a_1 x + \ldots + a_{n-1} x^{n-1}$ and ε_n is an order n primitive root of the unity. Moreover, the circulant matrix $C = p(T)$, where $T = circ(0, 1, 0, \ldots, 0)$ is the matrix representation of the shift operator $T : (x_0, x_1, \ldots, x_{n-2}, x_{n-1}) \to (x_1, x_2, \ldots, x_{n-1}, x_0)$.

Denote by $\chi_G(\lambda) = \det(\lambda I_n - L(G))$ the characteristic polynomial of the Laplacian matrix of the graph G. Its expanded form is

$$\chi_G(\lambda) = \lambda^n + c_{n-1}\lambda^{n-1} + \ldots + c_1 \lambda.$$

The theorem by Kelmans and Chelnokov [15] states that the absolute value of coefficient c_k of $\chi_G(\lambda)$ coincides with the number f_k of rooted spanning k-forests in the graph G. This theorem expand the classical Kirchhoff's Matrix Tree Theorem which states that $|c_1| = n\,\tau(n)$ is the number of spanning trees multiplied by the number of vertices of the graph G.

We note that all the roots of $\chi_G(\lambda)$ are non-negative. By Vieta's theorem the sequence c_k is alternating. So, the number of rooted spanning forests of the graph G can be found by the formula

$$\begin{aligned}(1)\qquad f(n) &= f_1 + f_2 + \ldots + f_n = |c_1 - c_2 + c_3 - \ldots + (-1)^{n-1}| \\ &= (-1)^n \chi_G(-1) = \det(I_n + L(G)).\end{aligned}$$

This result was independently obtained by many authors (P. Chebatorev and E. Shamis [**8**], O. Knill [**16**] and others).

Denote by $\lambda_0, \lambda_1, \ldots, \lambda_{n-1}$ the roots of $\chi_G(\lambda)$. If graph G is connected it has exactly one vanishing Laplacian eigenvalue, say $\lambda_0 = 0$. We can express the above mentioned results about $\tau(n)$ and $f(n)$ though λ_j, $j = 1, 2, \ldots, n-1$ as follows

$$\tau(n) = \frac{\lambda_1 \lambda_2 \ldots \lambda_{n-1}}{n} = \frac{|\chi'_G(0)|}{n}$$

and

$$f(n) = (1 + \lambda_1)(1 + \lambda_2) \ldots (1 + \lambda_{n-1}) = |\chi_G(-1)|.$$

3. Counting the number of spanning trees for the graph $\Delta(n; k, l, m)$

THEOREM 3.1. *The number of spanning trees $\tau(n)$ in the graph $\Delta(n; k, l, m)$ is given by the formula*

$$\tau(n) = \frac{n}{k^2 + l^2 + m^2} \prod_{p=1}^{s-1} |2T_n(w_p) - 2|,$$

where $s = k + l + m$, and $w_p = 1, 2, \ldots, s-1$ are all the roots different from 1 of the order s algebraic equation

$$Q(w) = -2 - A(w) - B(w) - C(w) + A(w)B(w)C(w) = 0.$$

Here $A(w) = 4 - 2T_k(w)$, $B(w) = 4 - 2T_l(w)$, $C(w) = 4 - 2T_m(w)$ and $T_j(w)$ is the order j Chebyshev polynomial of the first kind.

Proof. By the celebrated Kirchhoff theorem, the number of spanning trees $\tau(n)$ is equal to the product of non-zero eigenvalues of the Laplacian of a graph $\Delta(n; k, l, m)$ divided by the number of its vertices $3n$. To investigate the spectrum of Laplacian matrix, we denote by $T = circ(0, 1, \ldots, 0)$ the $n \times n$ shift operator. The Laplacian $L = L(\Delta(n; k, l, m))$ is given by the matrix

$$(2)\qquad L = \begin{pmatrix} 4I_n - T^{-k} - T^k & -I_n & -I_n \\ -I_n & 4I_n - T^{-l} - T^l & -I_n \\ -I_n & -I_n & 4I_n - T^{-m} - T^m \end{pmatrix}.$$

The eigenvalues of circulant matrix T are ε_n^j, $j = 0, 1, \ldots, n-1$, where $\varepsilon_n = e^{\frac{2\pi i}{n}}$. Since all of them are distinct, the matrix T is conjugate to the diagonal matrix $\mathbb{T} = diag(1, \varepsilon_n, \ldots, \varepsilon_n^{n-1})$ with diagonal entries $1, \varepsilon_n, \ldots, \varepsilon_n^{n-1}$. To find spectrum of L, without loss of generality, one can assume that $T = \mathbb{T}$. Then the $n \times n$ blocks of L are diagonal matrices. This approach helps to find eigenvalues of L. Indeed, let λ be an eigenvalue of L and $(x, y, z) = (x_1, \ldots, x_n, y_1, \ldots, y_n, z_1, \ldots, z_n)$ be the respective eigenvector. Then we have the following system of equations

$$\begin{aligned}(A-\lambda)x-y-z&=0,\\ -x+(B-\lambda)y-z&=0,\\ -x-y+(C-\lambda)z&=0,\end{aligned}$$

where $A=4I_n-T^{-k}-T^k$, $B=4I_n-T^{-l}-T^l$, and $C=4I_n-T^{-m}-T^m$.

Consequently eliminating y and z from the above equation, we obtain

$$(-2I_n-(A-\lambda)-(B-\lambda)-(C-\lambda)+(A-\lambda)(B-\lambda)(C-\lambda))x=0.$$

Recall the matrices under consideration are diagonal and the $(j+1,j+1)$-th entry of T is equal to ε_n^j. Consider the following two variable Laurent polynomial

$$P(z,\lambda)=-2-(a-\lambda)-(b-\lambda)-(c-\lambda)+(a-\lambda)(b-\lambda)(c-\lambda),$$

where $a=4-z^k-z^{-k}$, $b=4-z^l-z^{-l}$ and $c=4-z^m-z^{-m}$.

Then, for any $j=0,\dots,n-1$, matrix L has three eigenvalues, say $\lambda_{1,j},\lambda_{2,j}$, and $\lambda_{3,j}$ satisfying the order three algebraic equation $P(\varepsilon_n^j,\lambda)=0$. In particular, if $j=0$ for $\lambda_{1,0},\lambda_{2,0},\lambda_{3,0}$ we have $-\lambda(\lambda-3)^2=0$. Thus, $\lambda_{1,0}=0,\lambda_{2,0}=3$, and $\lambda_{3,0}=3$.

Let $\mathbf{e}_j=(0,\dots,\underbrace{1}_{j-th},\dots,0)$, $j=1,\dots,n$. For given $\lambda=\lambda_{i,j}$ the corresponding eigenvector is (x,y,z), where $x=(1+B-\lambda)\mathbf{e}_{j+1},y=(1+A-\lambda)\mathbf{e}_{j+1},z=((A-\lambda)(B-\lambda)-1)\mathbf{e}_{j+1}$.

By Vieta's theorem, the product $p_j=\lambda_{1,j}\lambda_{2,j}\lambda_{3,j}$ is given by the formula $p_j=P(\varepsilon_n^j)$, where $P(z)=P(z,0)$.

Now we have

$$\tau(n)=\frac{1}{3n}\lambda_{2,0}\lambda_{3,0}\prod_{j=1}^{n-1}\lambda_{1,j}\lambda_{2,j}\lambda_{3,j}=\frac{3}{n}\prod_{j=1}^{n-1}p_j=\frac{3}{n}\prod_{j=1}^{n-1}P(\varepsilon_n^j).$$

To continue the calculation of $\tau(n)$ we need Lemmas 3.2 and 3.3.

LEMMA 3.2. *The following identity holds $P(z)=Q(w)$, where $Q(w)$ is an order $k+l+m$ integer polynomial given by the formula*

$$Q(w)=-2-A(w)-B(w)-C(w)+A(w)B(w)C(w)=0,$$

where $A(w)=4-2T_k(w)$, $B(w)=4-2T_l(w)$, $C(w)=4-2T_m(w)$, $T_k(w)$ is the Chebyshev polynomial of the first kind and $w=\frac{1}{2}(z+z^{-1})$.

PROOF. Recall that $P(z)=-2-a-b-c+abc$, where $a=4-z^k-z^{-k}$, $b=4-z^l-z^{-l}$ and $c=4-z^m-z^{-m}$. Let us substitute $z=e^{i\varphi}$. It is easy to see that $w=\frac{1}{2}(z+z^{-1})=\cos\varphi$, so we have $T_k(w)=\cos(k\arccos w)=\cos(k\varphi)=\frac{1}{2}(z^k+z^{-k})$. Then the statement of the lemma follows from elementary calculations. □

Note that the leading term of Laurent polynomial $P(z)$ is $-z^{k+l+m}$. By Lemma 3.2, $P(z)=Q(w)$, where $w=\frac{1}{2}(z+z^{-1})$ and $Q(w)$ is a polynomial of degree $s=k+l+m$. We have $w-1=\frac{(z-1)^2}{2z}$. Since $Q(1)=0$ and $Q'(1)=-6(k^2+l^2+m^2)\neq 0$, polynomial $Q(w)$ has the root $w=1$ with multiplicity one. Hence, polynomial $P(z)$ has the root $z=1$ with multiplicity two. So, the roots of $P(z)$ are $1, 1, z_1, 1/z_1, \ldots, z_{s-1}, 1/z_{s-1}$, where for all $p = 1, \ldots, s-1$, $z_p \neq 1$. We get

$$P(z) = \frac{-(z-1)^2}{z^{k+l+m}} \prod_{p=1}^{s-1} (z - z_p)(z - z_p^{-1}).$$

Also, $w_p = \frac{1}{2}(z_p + z_p^{-1})$, $p = 1, \ldots, s-1$ are all the roots different from 1 of the equation $Q(w) = 0$. We set $H(z) = \prod\limits_{p=1}^{s-1} (z - z_p)(z - z_p^{-1})$. Then $P(z) = \frac{-(z-1)^2}{z^{k+l+m}} H(z)$.

The following lemma has been proved in ([**17**], Lemma 5.3).

LEMMA 3.3. *Let* $H(z) = \prod\limits_{p=1}^{s-1} (z - z_p)(z - z_p^{-1})$ *and* $H(1) \neq 0$. *Then*

$$\prod_{j=1}^{n-1} H(\varepsilon_n^j) = \prod_{p=1}^{s-1} \frac{T_n(w_p) - 1}{w_p - 1},$$

where $w_p = \frac{1}{2}(z_p + z_p^{-1})$, $p = 1, \ldots, k$ *and* $T_n(x)$ *is the Chebyshev polynomial of the first kind.*

Note that $\prod\limits_{j=1}^{n-1} (\varepsilon_n^j - 1)^2 = \lim\limits_{z \to 1} \prod\limits_{j=1}^{n-1} (z - \varepsilon_n^j)^2 = \lim\limits_{z \to 1} (\frac{z^n - 1}{z-1})^2 = n^2$ and $\prod\limits_{j=1}^{n-1} \varepsilon_n^j = (-1)^{n-1}$. As a result, by Lemma 3.3 we have

$$\begin{aligned}
\tau(n) &= \frac{3}{n} \prod_{j=1}^{n-1} P(\varepsilon_n^j) = \frac{3}{n} \prod_{j=1}^{n-1} \frac{(-1)(\varepsilon_n^j - 1)^2}{(\varepsilon_n^j)^s} H(\varepsilon_n^j) \\
&= \frac{(-1)^{(n-1)s} (-1)^{n-1} 3n^2}{n} \prod_{j=1}^{n-1} H(\varepsilon_n^j) \\
&= (-1)^{(n-1)(s+1)} 3n \prod_{p=1}^{s-1} \frac{T_n(w_p) - 1}{w_p - 1}.
\end{aligned}$$

Recall that w_p, $p = 1, \ldots, s-1$ are all the roots different from 1 of the polynomial $Q(w)$ is an order $s = k+l+m$ whose leading term is -2^{k+l+m}. We also have $Q'(1) = -6(k^2 + l^2 + m^2)$. We note that $Q(w) = -2^{k+l+m}(w-1) \prod_{p=1}^{s-1} (w - w_s)$. Hence

$$\prod_{p=1}^{s-1} (w_p - 1) = \frac{(-1)^{s-1} Q'(1)}{-2^{k+l+m}} = \frac{-6(k^2 + l^2 + m^2)}{(-2)^s}.$$

Finally, taking into account that $\tau(n)$ is a positive integer, we obtain

$$\tau(n) = \frac{(-1)^{n(s+1)} \cdot 3 \cdot 2^s n}{6(k^2 + l^2 + m^2)} \prod_{p=1}^{s-1} (T_n(w_p) - 1) = \frac{n}{k^2 + l^2 + m^2} \prod_{p=1}^{s-1} |2T_n(w_p) - 2|.$$

□

As a consequence of Theorem 3.1 we have the following example.

EXAMPLE 3.4. The number of spanning trees in the graph $\Delta(n; 1, 1, 1)$ is given by the formula

$$\tau(n) = \frac{n}{3} (2T_n(5/2) - 2)^2.$$

Indeed, we have $k^2 + l^2 + m^2 = 3$ and $Q(w) = -2(-1+w)(5-2w)^2$. Hence, $w_1 = w_2 = 5/2$. Recall that the Chebyshev polynomial $T_n(w)$ can be evaluated by the following simple formula

$$T_n(w) = \frac{(w+\sqrt{w^2-1})^n + (w-\sqrt{w^2-1})^n}{2}.$$

4. Counting the number of rooted spanning forests for the graph $\Delta(n;k,l,m)$

THEOREM 4.1. *The number of rooted spanning forests in a Δ-graph $\Delta(n;k,l,m)$ is given by the formula*

$$f(n) = \prod_{p=1}^{s} |2T_n(w_p) - 2|,$$

where w_p, $p = 1,2,\ldots,s$ are all the roots of the algebraic equation $\widetilde{Q}(w) = 0$ and

$$\widetilde{Q}(w) = \det \begin{pmatrix} K & -1 & -1 \\ -1 & L & -1 \\ -1 & -1 & M \end{pmatrix}.$$

Here $K = 5 - 2\,T_k(w)$, $L = 5 - 2\,T_l(w)$, $M = 5 - 2\,T_m(w)$ and $T_j(w)$ is an order j Chebyshev polynomial of the first kind.

Proof. Following general approach established in the papers [**8, 11, 15, 16**] we can use equation (1) for counting rooted spanning forests in a graph. In the case of Δ-graph the required number is equal to $\det(I+L)$. Since the structure of Laplacian matrix for Δ-graph is given by (2) we have to find the following determinant

$$\det(I+L) = \det \begin{pmatrix} 5I_n - T^{-k} - T^k & -I_n & -I_n \\ -I_n & 5I_n - T^{-l} - T^l & -I_n \\ -I_n & -I_n & 5I_n - T^{-m} - T^m \end{pmatrix}.$$

First all of all, we note that the eigenvalues of matrix $I+L$ are of the form $\mu_{i,j} = 1 + \lambda_{i,j}$, where $\lambda_{i,j}$, $i = 1,2,3$, $j = 0,\ldots,n-1$ are all the eigenvalues of matrix $I+L$.

Taking into account the arguments from the proof of Theorem 3.1, we make the following observation. Since for any $j = 0,\ldots,n-1$ the triple $\lambda_{1,j}, \lambda_{2,j}$ and $\lambda_{3,j}$ form all the roots of the cubic equation $P(\varepsilon_n^j, \lambda) = 0$, the numbers $\mu_{1,j}, \mu_{2,j}, \mu_{3,j}$ are roots of the equation $P(\varepsilon_n^j, \mu - 1) = 0$.

By Vieta's theorem, the product $q_j = \mu_{1,j}\mu_{2,j}\mu_{3,j}$ is given by the formula $q_j = P(\varepsilon_n^j, -1)$,

Now we have

$$f(n) = \prod_{j=0}^{n-1} \mu_{1,j}\mu_{2,j}\mu_{3,j} = \prod_{j=0}^{n-1} q_j = \prod_{j=0}^{n-1} P(\varepsilon_n^j, -1). \tag{3}$$

To continue the proof we replace the Laurent polynomial $P(z,-1)$ by $\widetilde{P}(z) = -z^s P(z,-1)$. Then $\widetilde{P}(z)$ is a monic polynomial of the degree $2s$ with the same roots as $P(z,-1)$. We note that

$$\prod_{j=0}^{n-1} \widetilde{P}(\varepsilon_n^j) = (-1)^n \varepsilon_n^{\frac{(n-1)n}{2}s} \prod_{j=0}^{n-1} P(\varepsilon_n^j, -1) = (-1)^{n+s(n-1)} \prod_{j=0}^{n-1} P(\varepsilon_n^j, -1). \tag{4}$$

One can check that Laurent polynomials $P(z,-1)$ and $\widetilde{Q}(w)$ satisfy the following relation $\widetilde{Q}(\frac{1}{2}(z+z^{-1}))=P(z,-1)$. Since $P(z,-1)=P(1/z,-1)$ all the roots of polynomial $\widetilde{P}(z)$ constitute list of the form $z_1,1/z_1,z_2,1/z_2,\ldots,z_s,1/z_s$. So, the numbers $w_j=\frac{1}{2}(z_j+z_j^{-1})$, $j=1,\ldots,s$ form the list of all the roots of $Q(w)$.

Also, we can recognize the complex numbers ε_n^j, $j=0,\ldots,n-1$ as the roots of polynomial z^n-1. By the basic properties of resultants ([**29**], Ch. 1.3) we have

$$\begin{aligned}\prod_{j=0}^{n-1}\widetilde{P}(\varepsilon_n^j)&=\mathrm{Res}(\widetilde{P}(z),z^n-1)=\mathrm{Res}(z^n-1,\widetilde{P}(z))\\ &=\prod_{z:\widetilde{P}(z)=0}(z^n-1)=\prod_{j=1}^{s}(z_j^n-1)(z_j^{-n}-1)\qquad(5)\\ &=\prod_{j=1}^{s}(2-z_j^n-z_j^{-n})=(-1)^s\prod_{j=1}^{s}(2T_n(w_j)-2).\end{aligned}$$

Combine (3), (4) and (5) we have the following formula for the number of spanning trees

$$f(n)=(-1)^{n(s+1)}\prod_{j=1}^{s}(2T_n(w_j)-2).\qquad(6)$$

Since $f(n)$ is a positive integer, from (6) we obtain

$$f(n)=\prod_{j=1}^{s}|2T_n(w_j)-2|.\qquad(7)$$

We finish the proof of the theorem. □

Theorem 3.1 is illustrated by the following example.

EXAMPLE 4.2. The number of rooted spanning forests in the graph $\Delta(n;1,1,1)$ is given by the formula

$$f(n)=(2T_n(3/2)-2)(2T_n(3)-2)^2.$$

Indeed, by direct calculations we get $s=1+1+1=3$, $\widetilde{Q}(w)=-4(-3+2w)(-3+w)^2$ and $w_1=3/2$, $w_2=w_3=3$.

5. The structure of Jacobian group for the graph $\Delta(n;k,l,m)$

We are going to investigate Jacobian group of the graph $\Delta(n;k,l,m)$ through its Laplacian matrix. Denote by $T=circ(0,1,\ldots,0)$ the $n\times n$ shift operator. Then the Laplacian $L=L(\Delta(n;k,l,m))$ can be represented in the form

$$L=\begin{pmatrix}A&-I&-I\\-I&B&-I\\-I&-I&C\end{pmatrix},$$

where $A=4-T^k-T^{-k}$, $B=4-T^l-T^{-l}$, $C=4-T^m-T^{-m}$ and $I=I_n$ is the $n\times n$ identity matrix.

Recall that $Jac(\Delta(n;k,l,m))$ is given by the torsion subgroup of the $coker(L)$, where L is considered as $\mathbb{Z}$-linear operator from $\mathbb{Z}^{3n}$ to itself. To find the structure of $coker(L)$ consider the following n-tuples of variables $x=(x_1,\ldots,x_n)$,

$y = (y_1, \dots, y_n)$, $z = (z_1, \dots, z_n)$. Then, as an Abelian group, $coker(L)$ has the following presentation

$$coker(L) = \langle x, y, z | L(x, y, z)^t = 0 \rangle.$$

Hence,

$$coker(L) = \langle x, y, z | Ax - y - z = 0,\ -x + By - z = 0,\ -x - y + Cz = 0 \rangle.$$

Taking into account that $z = Ax - y$ and eliminating z from the above presentation we obtain

$$coker(L) = \langle x, y | -(I + A)x + (I + B)y = 0,\ (-I + CA)x - (I + C)y = 0 \rangle.$$

This leads to the following result.

THEOREM 5.1. *The group $Jac(\Delta(n; k, l, m))$ is isomorphic to the torsion subgroup of the $coker(M)$, where M is the $2n \times 2n$ block matrix given by*

$$M = \begin{pmatrix} -I - A & I + B \\ -I + C\,A & -I - C \end{pmatrix},$$

where $A = 4 - T^k - T^{-k}$, $B = 4 - T^l - T^{-l}$ and $C = 4 - T^m - T^{-m}$.

The above Theorem gives a practical way to find the structure of Jacobian group. It produces the matrix whose size is smaller by one third which greatly reduce computational time. Also, it helps us to bring out the analytical form of Jacobian.

We illustrate this by the following result obtained earlier in [**27**].

EXAMPLE 5.2. Consider the following sequence

$$\nu(n) = \begin{cases} \sqrt{7}\,U_{\frac{n}{2}-1}(\frac{5}{2}) \text{ if } n \text{ is odd}, \\ U_{\frac{n}{2}-1}(\frac{5}{2}) \text{ if } n \text{ is even}. \end{cases}$$

Here $U_{m-1}(x) = \frac{\sin(m \arccos(x))}{\sin \arccos(x)}$ is the Chebyshev polynomial of the second kind. Note that $\nu(n)$ is a positive integer for all n.

Then Jacobian group $J_n = Jac(\Delta(n; 1, 1, 1))$ of the graph $\Delta(n; 1, 1, 1)$ for $n \geq 3$ has the following structure

$$J_n = \mathbb{Z}_{\frac{\gcd(n,\nu(n))}{\gcd(n,3)}} \oplus \mathbb{Z}^2_{\nu(n)} \oplus \mathbb{Z}_{\widehat{\mu}(n)\nu(n)} \oplus \mathbb{Z}_{3\mu(n)\,\mathrm{lcm}(n,\nu(n))},$$

where $\mu(n)$ is a 2-periodic sequence $\{1, 7, 1, 7, \dots\}$ and $\widehat{\mu}(n) = \gcd(n, 3)\mu(n)$.

Acknowledgments

Authors are grateful to the anonymous referee for fruitful remarks and suggestions.

References

[1] Daniele D'Angeli and Alfredo Donno, *Weighted spanning trees on some self-similar graphs*, Electron. J. Combin. **18** (2011), no. 1, Paper 16, 28, DOI 10.37236/503. MR2770121

[2] Roland Bacher, Pierre de la Harpe, and Tatiana Nagnibeda, *The lattice of integral flows and the lattice of integral cuts on a finite graph* (English, with English and French summaries), Bull. Soc. Math. France **125** (1997), no. 2, 167–198. MR1478029

[3] Matthew Baker and Serguei Norine, *Harmonic morphisms and hyperelliptic graphs*, Int. Math. Res. Not. IMRN **15** (2009), 2914–2955, DOI 10.1093/imrn/rnp037. MR2525845

[4] Hyman Bass and Ravi Kulkarni, *Uniform tree lattices*, J. Amer. Math. Soc. **3** (1990), no. 4, 843–902, DOI 10.2307/1990905. MR1065928

[5] Norman Biggs, *Three remarkable graphs*, Canadian J. Math. **25** (1973), 397–411, DOI 10.4153/CJM-1973-040-1. MR321789

[6] N. L. Biggs, *Chip-firing and the critical group of a graph*, J. Algebraic Combin. **9** (1999), no. 1, 25–45, DOI 10.1023/A:1018611014097. MR1676732

[7] F. T. Boesch and H. Prodinger, *Spanning tree formulas and Chebyshev polynomials*, Graphs Combin. **2** (1986), no. 3, 191–200, DOI 10.1007/BF01788093. MR951563

[8] P. Chebotarev and E. Shamis, *Matrix forest theorems*, Preprint, (2006), arXiv:math/0602575. DOI https://doi.org/10.48550/arXiv.math/0602575.

[9] Robert Cori and Dominique Rossin, *On the sandpile group of dual graphs*, European J. Combin. **21** (2000), no. 4, 447–459, DOI 10.1006/eujc.1999.0366. MR1756151

[10] Philip J. Davis, *Circulant matrices*, A Wiley-Interscience Publication, John Wiley & Sons, New York-Chichester-Brisbane, 1979. Pure and Applied Mathematics. MR543191

[11] Liliya A. Grunwald, Young Soo Kwon, and Ilya A. Mednykh, *Counting rooted spanning forests for circulant foliation over a graph*, Tohoku Math. J. (2) **74** (2022), no. 4, 535–548, DOI 10.2748/tmj.20210810. MR4522330

[12] Lilya A. Grunwald and Ilya Mednykh, *The number of rooted forests in circulant graphs*, Ars Math. Contemp. **22** (2022), no. 4, Paper No. 10, 12. MR4498581

[13] J. D. Horton and I. Z. Bouwer, *Symmetric Y-graphs and H-graphs*, J. Combin. Theory Ser. B **53** (1991), no. 1, 114–129, DOI 10.1016/0095-8956(91)90057-Q. MR1122300

[14] Jesper Lykke Jacobsen, Jesús Salas, and Alan D. Sokal, *Spanning forests and the q-state Potts model in the limit $q \to 0$*, J. Stat. Phys. **119** (2005), no. 5-6, 1153–1281, DOI 10.1007/s10955-005-4409-y. MR2157859

[15] A. K. Kelmans and V. M. Chelnokov, *A certain polynomial of a graph and graphs with an extremal number of trees*, J. Combinatorial Theory Ser. B **16** (1974), 197–214, DOI 10.1016/0095-8956(74)90065-3. MR345867

[16] O. Knill, *Counting rooted forests in a network*, Preprint, (2013), arXiv:1307.3810, DOI https://doi.org/10.48550/arXiv.1307.3810.

[17] Y. S. Kwon, A. D. Mednykh, and I. A. Mednykh, *On Jacobian group and complexity of the generalized Petersen graph $GP(n,k)$ through Chebyshev polynomials*, Linear Algebra Appl. **529** (2017), 355–373, DOI 10.1016/j.laa.2017.04.032. MR3659807

[18] Y. S. Kwon, A. D. Mednykh, and I. A. Mednykh, *Complexity of the circulant foliation over a graph*, J. Algebraic Combin. **53** (2021), no. 1, 115–129, DOI 10.1007/s10801-019-00921-7. MR4212987

[19] Y. S. Kwon, A. D. Mednykh, and I. A. Mednykh, *On Jacobian group and complexity of the Y-graph*, Sib. Èlektron. Mat. Izv. **19** (2022), no. 2, 662–673. MR4478155

[20] Dino Lorenzini, *Smith normal form and Laplacians*, J. Combin. Theory Ser. B **98** (2008), no. 6, 1271–1300, DOI 10.1016/j.jctb.2008.02.002. MR2462319

[21] Ilya A. Mednykh, *On Jacobian group and complexity of the I-graph $I(n,k,l)$ through Chebyshev polynomials*, Ars Math. Contemp. **15** (2018), no. 2, 467–485, DOI 10.26493/1855-3974.1355.576. MR3880002

[22] I. A. Mednykh and M. A. Zindinova, *On the structure of Picard group for Moebius ladder*, Sib. Èlektron. Mat. Izv. **8** (2011), 54–61. MR2800820

[23] A. D. Mednykh and I. A. Mednykh, *On the structure of the Jacobian group of circulant graphs* (Russian, with Russian summary), Dokl. Akad. Nauk **469** (2016), no. 5, 539–543, DOI 10.1134/s106456241604027x; English transl., Dokl. Math. **94** (2016), no. 1, 445–449. MR3561344

[24] A. D. Mednykh and I. A. Mednykh, *On the asymptotics and arithmetic properties of the complexity function of circulant graphs* (Russian, with Russian summary), Dokl. Akad. Nauk **479** (2018), no. 4, 363–367, DOI 10.1134/s1064562418020138; English transl., Dokl. Math. **97** (2018), no. 2, 147–151. MR3838534

[25] A. D. Mednykh and I. A. Mednykh, *The number of spanning trees in circulant graphs, its arithmetic properties and asymptotic*, Discrete Math. **342** (2019), no. 6, 1772–1781, DOI 10.1016/j.disc.2018.08.030. MR3922465

[26] A. D. Mednykh and I. A. Mednykh, *Plans' periodicity theorem for Jacobian of circulant graphs*, Dokl. Math. **103** (2021), no. 3, 139–142, DOI 10.1134/s1064562421030121. Translation of Dokl. Akad. Nauk **498** (2021), 51–54. MR4437982

[27] A. Mednykh, I.Mednykh, I. Yudin, *On Jacobian group of the Δ-graph.* arXiv:2211.12007 [math.CO], 2022.

[28] Stavros D. Nikolopoulos and Charis Papadopoulos, *The number of spanning trees in K_n-complements of quasi-threshold graphs*, Graphs Combin. **20** (2004), no. 3, 383–397, DOI 10.1007/s00373-004-0568-x. MR2093492

[29] Victor V. Prasolov, *Polynomials*, Algorithms and Computation in Mathematics, vol. 11, Springer-Verlag, Berlin, 2010. Translated from the 2001 Russian second edition by Dimitry Leites; Paperback edition [of MR2082772]. MR2683151

[30] Robert Shrock and F. Y. Wu, *Spanning trees on graphs and lattices in d dimensions*, J. Phys. A **33** (2000), no. 21, 3881–3902, DOI 10.1088/0305-4470/33/21/303. MR1769549

[31] Yaoping Hou, Chingwah Woo, and Pingge Chen, *On the sandpile group of the square cycle C_n^2*, Linear Algebra Appl. **418** (2006), no. 2-3, 457–467, DOI 10.1016/j.laa.2006.02.022. MR2260205

[32] Shu-Chiuan Chang, Lung-Chi Chen, and Wei-Shih Yang, *Spanning trees on the Sierpinski gasket*, J. Stat. Phys. **126** (2007), no. 3, 649–667, DOI 10.1007/s10955-006-9262-0. MR2294471

[33] Pingge Chen, Yaoping Hou, and Chingwah Woo, *On the critical group of the Möbius ladder graph*, Australas. J. Combin. **36** (2006), 133–142. MR2262614

[34] Xiebin Chen, Qiuying Lin, and Fuji Zhang, *The number of spanning trees in odd valent circulant graphs*, Discrete Math. **282** (2004), no. 1-3, 69–79, DOI 10.1016/j.disc.2003.12.006. MR2059507

[35] Yuanping Zhang, Xuerong Yong, and Mordecai J. Golin, *The number of spanning trees in circulant graphs*, Discrete Math. **223** (2000), no. 1-3, 337–350, DOI 10.1016/S0012-365X(99)00414-8. MR1782058

[36] Yuanping Zhang, Xuerong Yong, and Mordecai J. Golin, *Chebyshev polynomials and spanning tree formulas for circulant and related graphs*, Discrete Math. **298** (2005), no. 1-3, 334–364, DOI 10.1016/j.disc.2004.10.025. MR2163455

[37] Weigang Sun, Shuai Wang, and Jingyuan Zhang, *Counting spanning trees in prism and anti-prism graphs*, J. Appl. Anal. Comput. **6** (2016), no. 1, 65–75, DOI 10.11948/2016006. MR3414735

[38] F. Y. Wu, *Number of spanning trees on a lattice*, J. Phys. A **10** (1977), no. 6, L113–L115. MR489559

Sobolev Institute of Mathematics, Koptyga pr., bd. 4, Novosibirsk, 630090, Russia; and Novosibirsk State University, Pirogova st., bd. 1, Novosibirsk, 630090, Russia

Sobolev Institute of Mathematics, Koptyga pr., bd. 4, Novosibirsk, 630090, Russia; and Novosibirsk State University, Pirogova st., bd. 1, Novosibirsk, 630090, Russia

Sobolev Institute of Mathematics, Koptyga pr., bd. 4, Novosibirsk, 630090, Russia; and Gorno-Altaisk State University, Lenkina st., bd. 1, Gorno-Altaisk, 649000, Russia

Contemporary Mathematics
Volume **811**, 2025
https://doi.org/10.1090/conm/811/16244

Groups, communication and coordination

Rohit Parikh

In celebration of Ravi Kulkarni 80

ABSTRACT. We consider the dance among three notions. Communication, knowledge and action. Our actions depend on our desires and our beliefs. But it often happens that we need to coordinate with (or against) others. This coordination requires some kind of mutual knowledge about who should do what and when. Our focus here is on *the amount of coordination* that exists in a particular group, whether a political party, or members of some religion, or an orchestra. The amount of coordination can be big - as with an orchestra, or smaller, as among members of the same, large religion. Developing a measure of the amount of coordination is crucial because it will allow us to ask what a group can do and also what a group will do.

1. Introduction:

Our three cases

1. Bees
2. Chinese troops
3. BLM demonstrators

The reasons for coordination are different in the three cases.

Cooperation among animals and people

It is a fact that people and animals cooperate

FIGURE 1. A beehive.

FIGURE 2. chinese troops

FIGURE 3. Black lives matter

1. Bees cooperate because they are all sisters who have the same DNA. From a purely evolutionary point of view the survival of one is the survival of all.
2. Chinese troops cooperate because they are obeying a single commander.
3. Black lives people cooperate because they have a single goal, and they have communicated to be together at one time and place.

The need for communication. Even a master must tell his servants what he wants them to do. BLM people must rely on email or phone calls to coordinate about the time and place.

And what about bees?

Dance language of the honey bee

Social behavior in bees has a number of advantages. One of the most important of these is the ability to quickly mobilize a large number of foragers to gather floral resources that may only be available for a short period of time. The ability to communicate location with such precision is one of the most interesting behaviors of a very interesting insect.

The recruitment of foragers from a hive begins when a scout bee returns to the hive engorged with nectar from a newly found nectar source. She begins by spending 30-45 seconds regurgitating and distributing nectar to bees waiting in the hive. Once her generosity has garnered an audience, the dancing begins ...

In all cases the quality and quantity of the food source determines the liveliness of the dances. If the nectar source is of excellent quality, nearly all foragers will dance enthusiastically and at length each time they return from foraging. Food sources of lower quality will produce fewer, shorter, and less vigorous dances; recruiting fewer new foragers.

2. A measure of coordination

For the moment we will put aside communication and simply propose a measure of the amount of coordination. How much coordination is there among bees? Among Chinese troops? Among BLM activists? There should be a theory of the a) amount of information among members of the group and b) the basic similarity among agents which allows more coordination with less communication. Schelling's Focal Points make a good example. In Schelling's example, two people newly arrived in New York city have become separated. They have no prior agreement as to where to meet. Schelling suggested that they will meet at the information booth in Grand Central railway station because it is a focal point. The existence of a focal point removes some of the burdern from explicit communication.

Consider n agents $\{1, \ldots, n\}$. Each agent has a (finite) action space (or belief space) A_i. A profile of actions (or beliefs) is a tuple $(a_1, \ldots, a_n)$ where each $a_i \in A_i$. Quite often, but not always, the various A_i will be the same.

We consider the set $\mathcal{X}$ of *all possible* profiles and $\mathcal{X} = \Pi_1^n(A_i)$ We also assume a set $\mathcal{P}$ of *actually arising* profiles. If $p \in \mathcal{P}$ then it can happen that the agents act in ways a_i so that $p = (a_1, .., a_n)$ that is to say $a_i = (p)_i$

If some profile q is in $\mathcal{X}$ but not in $\mathcal{P}$ then we assume that the agents do not jointly act in such a way that the action of agent i is $(q)_i$.

2.1. Beach example: Here is an example. Suppose 1, 2, 3 are three agents and there are actions a,b for each. a stands for "take care of the baby", b for "go to the beach".

Agent 1 would prefer b but will do a if neither 2 nor 3 is taking care of the baby.

Then $\mathcal{P} = \{(a, b, b), (b, b, a), (b, a, b), (b, a, a)\}$

(b,b,b) is not in $\mathcal{P}$. Agent 1 will not do b if the other two are also doing b i.e. going to the beach.

Nor is (a,b,a). If agent 3 is taking care of the baby then agent 1 would go to the beach.

Of the eight members of $\mathcal{X}$, four are in $\mathcal{P}$. Four are not.

Here is a little program

- player 2 chooses a, player 1 chooses b
- player 2 chooses b
  -
    * player 3 chooses a, player 1 chooses b
    * player 3 chooses b, player 1 chooses a

Note that player 1 chooses a only after she knows that both 2 and 3 have chosen b (to go to the beach). So information from both 2 and 3 must pass to 1 before she makes her choice whether to take care of the baby or to go to the beach.

The degree of coordination C is defined to be $lg(\alpha)$ where $\alpha = (|\mathcal{X}|/|\mathcal{P}|)$

In this example we gave, α is two and the degree $C = 1$. If $\mathcal{P} = \mathcal{X}$ then α would be 1 and C would be zero. If 1 did not care what 2 and 3 were doing then C would be zero.

The number C indicates the extent to which the actions of the three agents are *coordinated*. The highest degree of coordination arises when $\mathcal{P}$ is a singleton.

In this preliminary draft we will not ask *why* $\mathcal{P}$ is a singleton. There could be many reasons and we will not - yet - go into the various reasons.

For any arbitrary set X, $|X|$ will be cardinalitiy of the set X. lg will be log to the base 2.

2.2. Subgroups. How much does one subgroup coordinate with another subgroup? Suppose that $G = \{1, \ldots, n\}$ is a group and there are two subgroups of this set of agents. For simplicity we will let them be $G_1 = \{1, \ldots, m\}$ and $G_2 = \{m+1, \ldots, n\}$. Given a profile P of the whole group G we let $P_1 = (p_1, .., p_m)$ and $P_2 = (p_{m+1}, \ldots, p_n)$ Abusing notation we will write $P = P_1 + P_2$.

Let $\mathcal{P}_1$ be the set of possible profiles of G_1 and $\mathcal{P}_2$ be the set of possible profiles of G_2. If $\mathcal{P} = \{P + Q | P \in \mathcal{P}_1 \text{ and } Q \in \mathcal{P}_2\}$ then we will say that G_1 and G_2 are *independent*. This will clearly happen iff $|\mathcal{P}| = |\mathcal{P}_1| \times |\mathcal{P}_2|$. Of course there could be coordination within G_1 or within G_2 and that will create coordination within G as well. But G will not have more coordination than that in G_1 and G_2. C will be just $C_1 + C_2$.

We can also define the mutual degree of coordination to be $lg(|\mathcal{P}_1| \times |\mathcal{P}_2|/|\mathcal{P}|)$. It will be 0 iff the two groups are independent.

2.3. Another example - the two kingdoms. There are two kingdoms $\mathcal{K}$ and $\mathcal{K}'$.

There are three fruits available, apples, bananas and cherries. King k likes apples and bananas but hates cherries. King k' like bananas and cherries but hates apples.

Each king may choose whatever fruit he chooses among the ones he likes but citizens are required to eat the same fruit as the king. So the profile for $\mathcal{K}$ consists of $\{(a, \ldots, a), (b, .., b)\}$. For $\mathcal{K}'$ it is $\{(b, \ldots, b), (c, \ldots)\}$.

Since the profiles of the subgroups are independent, there are four profiles for the entire group, $\mathcal{K} + \mathcal{K}'$. C is zero.

Correlation

Suppose now that king k' is more powerful and demands that the people of kingdom $\mathcal{K}$ eat the same fruit as the people of kingdom $\mathcal{K}'$ when possible.

Now kingdom $\mathcal{K}$ can still eat apples when $\mathcal{K}'$ eats cherries (since K do not have cherries) but are forbidden to eat apples when $\mathcal{K}'$ eats bananas. $\mathcal{P}_1$ and $\mathcal{P}_2$ remain the same as before and have size 2 each, but the profile $(a, \ldots, a, b, \ldots, b)$ vanishes. If king k' is eating bananas then the citizens of kingdom K are also required to eat bananas.

$\mathcal{P}$ has size 3, α is 4/3, and C rises from 0 to .41.

The demand of king k' clearly requires communication. And we conjecture that communication increases the value of C. If there is no communication, then C must of necessity be 0.

2.4. Restrictions: A restriction r on $\mathcal{P}$ is a non-empty subset $\mathcal{Q}$ of $\mathcal{P}$. We will write $\mathcal{Q} = r(\mathcal{P})$. We may also think of r not as a particular restriction but as a map from profiles to (sub)sets of profiles. The intuition is that the restriction that (say) kingdom K' imposes on kingdom K can depend on what kingdom K' itself is doing.

Suppose for instance that the speed limit on some highway is 65 mph. Cars are allowed to drive at any speed up to 65. But suppose there is another sign Construction: speed limit 45!. Then this restriction causes the allowable speed to be up to 45, and 55 is no longer available.

A system $\mathcal{S}$ of commands from king k' to kingdom $\mathcal{K}$ is a map from $\mathcal{P}'$ to the set $\mathcal{R}$ of restrictions.

The intuition is that when $\mathcal{K}'$ has a certain profile p then it sends a restriction $r(p)$ to $\mathcal{K}$ which shrinks the possible profiles for $\mathcal{K}$.

An identity command *at ease* leaves $\mathcal{P}$ as it is. Assuming that at least one of the commands is the identity, the profile $\mathcal{P}$ stays as it is. But the joint profile of $\mathcal{K}$ and $\mathcal{K}'$ together shrinks, raising the value of C.

We will say that r is at least as strict as r' if for all sets of profiles $\mathcal{P}$, $r(\mathcal{P}) \subseteq r'(\mathcal{P})$. It is more strict if if for at least some sets of profiles $\mathcal{P}$, $r(\mathcal{P}) \subset r'(\mathcal{P})$. The system of commands $\mathcal{S}$ is more strict than $\mathcal{S}'$ if all the restrictions in $\mathcal{S}$ are at least as strict as those of $\mathcal{S}'$ and at least one is more strict.

Observation

If $\mathcal{S}$ is more strict than $\mathcal{S}'$ then there is more coordination with $\mathcal{S}$ than with $\mathcal{S}'$. This is because the number of joint profiles will decrease even though the individual profiles of each group might remain the same. And note that C is anti-monotonic in the size of $\mathcal{P}$. A stronger restriction reduces the size of $\mathcal{P}$ and increases coordination.

There is obviously an algebra of restrictions. For instance there could be a restriction on the speed at which you can drive and also a restriction on the kind of car you are driving. These two restrictions commute and the algebra will be simple. On the other hand there could be a restriction on the speed at which you can drive and also a restriction on the amount of alcohol you can imbibe before driving. These two might not commute because one restriction might affect the other one. You may be required to drive at a lower speed if you have had some alcohol.

2.5. Communities defined by religion:

Suppose that there are two religions in the same community. We will call them H and M without intending any particular meaning. They also have holidays, D for H and R for M.

Most members of H observe D, but not all. Most members of M observe R but not all. Also some members of H observe R and some members of M observe D. Assume that in each community, 70% observe their own holiday and 20% observe the holiday of the other community. Also 5% in each community observe both holidays and that leaves 15% in each community who observe neither. Assuming the communities to be of equal sizes, we can calculate the degree of coordination within each community as well as the coordination within the whole society.

3. Payoffs:

Let us suppose that given a profile P it results in payoffs $u_i = u_{p,i}$ to member i of the group. We would like to suggest that ideally each group should coordinate its actions so as to maximize the welfare of the whole group. But how do we define the welfare W of *the whole group?*

We could let $W(p) = \Sigma_1^n u_{p,i}$. But we would also like the various u_i to be reasonably equal. The sum does not enforce equality.[1]

---

[1] Note that we tacitly assumed interpersonal comparison of utilities. This assumption is not widely accepted but we adopt it for convenience

Rawls might define $W(p) = n \times \mu_i(u_{p,i})$, where $u_{p,i}$ is the minimum. That does encourage fairness. But it ignores every u which is not the minimum.

From Rawls' point of view, a society of three in which the incomes are 101,101, 101 is *better* than a society with incomes of 100,200,300 since the minimum in the first society is higher.

3.1. An example of information spread and coordination: Suppose that n agents are arranged in a connected graph and are all capable of one of two actions a, b. Moreover there is a state of the world which could be A or B. The payoffs are as follows. If the state of the world is A and all agents do a then they all get a positive payoff of 1. If they act otherwise the payoff is zero. Conversely If the state of the world is B and all agents do b then they all get a positive payoff of 1. If they act otherwise the payoff is zero.

The process begins with one of the agents, say agent 1, learning the actual state of the world. She informs all agents i such that $(1, i)$ is in the graph. These in turn inform agents who are "downwind" from them. Thus i informs every j such that (i, j) is in the graph. The process terminates when all agents know the state of the world and act. (There is a time limit imposed by the diameter of the graph)

It is easy to see that it will take at least $n - 1$ messages before the steady state is reached. For every node other than 1 has to receive a message. And the amount of time it takes is of the order of d where d is the diameter of the graph. Indeed it is easy to see that it is at least $d/2$ and no more than d. for let a, b be two nodes which are distance d apart, which is the maximum possible distance. Let x be the distance from 1 to a and let y be the distance from 1 to b. The triangle inequality shows that $x + y \geq d$ and hence at least one of x and y is at least $d/2$

If the process goes as intended than only two profiles out of a possible 2^n will occur and the amount of coordination will be $lg(2^n/2) = n - 1$.

4. Knowledge in society

> The following, possibly apocryphal story is about the mathematician Norbert Wiener, well known for his absent mindedness It illustrates something even more subtle See [**Tr**].
>
> At one time the Wieners were moving and in the morning as he was going to work, Mrs. Wiener said to him, "Now don't come home to this address in the evening." And she gave him a piece of paper with the new address. However, in the evening Wiener found himself standing in front of the old address and not knowing what to do – he had already lost the slip of paper with the new address. He went to a little girl standing by and said, "Little girl, do you know where the Wieners have moved to?" The little girl replied, "Daddy, Mom knew what would happen so she sent me to fetch you."

The moral of the story, for *us*, is that communication works only if the memory of all parties involved is reliable. In this case a new profile crept in, namely Wiener not showing up at the new home despite being told where it is. Thus the amount of coordination decreased.

4.1. Wimmer and Perner on beliefs about beliefs. Wimmer and Perner are concerned primarily with the perception by children of other people's *mindsets*. The following quote from [**WP**] is a story about Maxi which they told a group of children:

Mother returns from her shopping trip. She bought chocolate for a cake. Maxi may help her put away the things. He asks her, "Where should I put the chocolate?" "In the blue cupboard," says the mother.

The story continues

Later, with Maxi gone out to play, the mother transfers the chocolate from the blue cupboard to the green cupboard. Maxi then comes back from the playground, hungry, and he wants to get some chocolate.

In Wimmer and Perner's experiment, little children who were told the Maxi story were then asked the *belief* question, "Where will Maxi look for the chocolate?"

Children at the age of three or less invariably got the answer wrong and assumed that Maxi would look for the chocolate in the *green* cupboard where *they* knew it was.

Even children aged four or five had only a one-third chance of correctly answering this question or an analgous question involving Maxi and his brother (who also wants the chocolate and whom Maxi wants to deceive).

Children aged six or more were by contrast quite successful in realizing that Maxi would think the chocolate would be in the *blue* cupboard – where he had put it and that if he wanted to deceive his brother, he would lead him towards the green cupboard.

Thus it seems that representation of other people's mindset comes fairly late in childhood, well after they have learned to deal with notions of belief and belief based action for themselves and for others who share their own view of reality. In [**St**] Chris Steinsvold investigates modal logics which are intended to represent the states of mind of young children. See also [SP].

Older children are not much better. In an experiment in my daughter's seventh grade class, I found that they were unable to deal with the muddy children puzzle beyond the first one or two levels. For further discussion see [**PW**], [**St**] and [**WP**].

4.2. The muddy children puzzle. In this by now well-known puzzle, a number of children are playing in the mud and some of them get their foreheads dirty. At this the father comes on the scene and announces, "at least one of you has got her forehead dirty."

Scenario 1: Suppose there is only one child, say Amy, who is dirty. Then she will realize that her own forehead must be dirty since she can see that the others are clean.

Scenario 2: Suppose now that there are two dirty children, Sarah and Amy, who are asked in turn, "Do you know if your forehead is dirty?" Now when Sarah is asked, she can see Amy's dirty forehead and she replies, "I don't know." However, when Amy is asked, she is able to reason, "If *my* forehead were clean, Sarah would have known that hers must be dirty since all the others are clean. But Sarah did not know. So my forehead must be dirty."

This reasoning on Amy's part requires a representation by Amy of Sarah's state of mind, and clearly Amy must be at least six for this to work. However, Sarah herself must have some reasoning ability and Amy must know that she has such abilities. It is not enough for Amy to know Sarah's view of reality, she must also represent Sarah's logical abilities in her own mind.

In particular, suppose that there are three dirty children – Jennifer, Sarah, and Amy – who are asked in turn whether they know if they are dirty, and with Amy being asked last.

If Sarah is only three, Amy would not be justified in concluding from Sarah's "I don't know" that in that case, Amy herself must be dirty.

Amy would need to know that *if* Amy were clean, Sarah would have carried out a representation in her own mind of Jennifer's state of mind and concluded from Jennifer's "I don't know" that Sarah must herself be dirty. But if Sarah is only three, Amy cannot rely on such reasoning on Sarah's part.

As the number of dirty children increases, there is a need for higher and higher levels of "I know that he knows that she knows that...." Common knowledge is at the end of this road and has been offered as the explanation of co-ordinated behaviour ([[**Lew**], [**HM**], CM, Chw]).

4.3. Levels of knowledge. Halpern and Moses in [**HM**] show that the co-ordinated attack problem requires common knowledge between the two generals, and that given the means of communication they have, such common knowledge is impossible to attain. Clark and Marshall [**CM2**] indicate similar difficulties with the referent of "the movie playing at the Roxy today."

While it is true that co-ordinated actions and, supposedly, common knowledge do happen, it may also be relevant to consider other levels of knowledge, short of the infinite, common-knowledge, level.

Such levels also arise in certain pragmatic situations, e.g. with e-mail or snail-mail or messages left on telephones as voice mail. Thus the purpose of our work is to study levels other than common knowledge and how they affect the actions of groups.[2]

4.4. Previous work. In [**PK**] we investigate the levels of knowledge which can exist in a group G of people about some fact P. The lowest level is when no one knows P. The highest is when P is common knowledge. While the notion of common knowledge is much in the news, lower levels are not well known. We showed that there are countably many levels of knowledge corresponding to regular subsets of a set of strings.

In [**PTW**] we showed how the actions in a group correspond to the levels of knowledge of some relevant facts. In [**PTW**] we bring in utilities, and also the matter of temperament. Given an action whose payoffs are not certain, a cautious agent will choose that action which has the highest safety value. Such a person will not buy a lottery ticket. An aggressive agent is someone who will choose an action whose possible payoffs are as high as possible. Such an agent will buy a lottery ticket.

In typical cooperative situations, even if a certain level of knowledge is needed, a higher level would also do. If Bob wants Ann to pick up the children at 4 PM, it is enough for him to know that she knows. Thus if he sends her e-mail at 2 PM and knows that she always reads hers at 3 PM, he can be satisfied. In such a situation Bob knows that Ann will know about the children in time, or symbolically

[2]We are not discussing here a technical discussion of levels of knowledge and their relation to joint action. The references [**PK**] and [**PTW**] contain some technical results. [**JP**] provides some background on well partial orderings which are relevant here.

$K_b(K_a(C))$ and he may feel this is enough. (Here C is "the children have to be picked up at 4 PM")

However, if he telephones her at 3 PM instead, this will create common knowledge of C, much *more* than is needed. But no harm done, since in this context, Ann and Bob have the same goals. Halpern and Zuck also state a knowledge level requirement for the sequence transmission problem, which suffices as a minimum, but since the parties are co-ordinating, a higher level does no harm.

But in other contexts one may wish for just a particular level of knowledge, no lower, and no *higher*. Suppose for instance that Bob wants Ann to know about a seminar talk he is giving, just in case she wants to come, but he does not want her to feel pressured to come – she should come only out of interest and not from politeness.

In that case he will want to arrange that $K_b(K_a(S))$ (he himself knows that she knows about the seminar), but not $K_a(K_b(K_a(S)))$ (Ann knows that Bob knows that Ann knows about the seminar), for in the latter case she *would* feel pressured.

Instead of telling her about his talk, which would create common knowledge, he may arrange for some other method, perhaps for a student to tell her, but without saying that it is a message from Bob.

4.5. The pedestrian. Suppose a pedestrian is crossing the street and sees a car approaching him. It happens in many cities, e.g., Boston, Naples, etc., that the pedestrian will pretend not to notice the car, thereby preventing $K_dK_p(C)$ with C representing the car, d being the driver and p the pedestrian. If the driver knew that the pedestrian knew, he might drive aggressively and try to bully the pedestrian into running or withdrawing. But if he does not know that the pedestrian knows, he will be more cautious.

While the social questions are fascinating and are addressed elsewhere (Cf. [**Pa**]), in this talk we shall concentrate on the technical aspects of knowledge, where it is assumed that everyone involved is logically perfect. One can still ask, what are the various levels of knowledge which can arise under various circumstances of communication and how will they affect how we may act?

4.6. Some examples. Our first example comes from the *Mahabharata*, one of the Indian epics and at 100,000 verses, believed to be the longest single work in the world. It describes the political struggle between two sets of cousins, the *Pandavas*, and the *Kauravas*. In a crucial battle between the two sets of cousins, *Krishna* is an adviser to the *Pandavas*, but out of a sense of fair play he gives his army to the *Kauravas*. At a crucial juncture *Drona*, a powerful warrior on the *Kaurava* side and also the teacher of both the *Pandavas* and *Kauravas*, turns out to be invincible in battle and the *Pandavas* are hard pressed.

However, the wily *Krishna* thinks up a strategem. *Drona's* only son is called *Ashvatthama* and so is an elephant owned by the *Pandavas*. *Krishna* suggests that the *Pandavas* kill the elephant *Ashvatthama* and then announce that the man *Ashvatthama* has been killed. After a great deal of hesitation and soul searching, self-interest prevails, the *Pandavas* do kill the elephant, and announce to *Drona* that *Ashvatthama* is dead.

Drona is a little suspicious, but knows that one of the *Pandava* brothers, *Yudhisthira* never lies. He asks *Yudhisthira*, who confirms that *Ashvatthama* is dead, muttering in an aside, "either man or elephant." *Naro va kunjaro va.* Not knowing about the elephant, *Drona* assumes it is his son who is dead, lays down his weapons, and is killed by a warrior on the *Pandava* side.

We now offer a game-knowledge theoretic analysis of this event.

Let *Drona's* two options be f for 'fight' and n for not fight. Before the announcement his preferences were $f > n$ and given his prowess, any warrior who faced him faced death.

But after the announcement that *Ashwatthama* was dead, his preferences change to $n > f$, and he can be attacked with impunity.

Ahswatthama of course was not dead and took terrible revenge on the *Pandavas* including the killing of some unborn *Pandava* children. For this he was punished by being condemned to live forever, and to wander the earth as a pariah.

4.7. The motorist and the pedestrian. Our example of the motorist and the pedestrian is a bit more complex as the deception consists of inducing a second-order false belief. Let S be the situation where a pedestrian is crossing the street and a car is coming. Let S' be the same situation without the car. In S the pedstrian has two options, g, i.e., to go, and n, i.e., to not go. The motorist also has two similar options, G and N. Here are the payoffs for the two in state S.

Motorist choices

Pedestrian choices		G	N
	g	(-100,-10)	(1,0)
	n	(0,1)	(0,0)

Figure I

Note that there are two Nash equilibria: at (g, N) and at (n, G). However, the penalty for the pedestrian (injury or loss of life) to depart from (n, G) is much greater than the penalty for the motorist (fine or loss of license) to depart from (g, N). Thus the equilibrium (g, N) is less stable than (n, G), and this fact creates the possibility for the motorist to 'bully' the pedestrian.

However, if the pedestrian is unaware of the existence of the car, then the picture is much simpler and his payoffs are 1 for g and 0 for n. g dominates n, and once this choice is made by the pedestrian, it is dominant for the motorist to choose N.

This is why the pedestrian tries to achieve the state of knowledge represented by the formulas $K_p(C), \neg K_m(K_p(C))$ indicating that the pedestrian knows the car is there but the motorist does not know that the pedestrian knows. The pedestrian chooses the action g, and knowing that the pedestrian will do this, the motorist must choose N.

However, if the motorist has a horn, he can change the knowledge situation. The existence of the car becomes common knowledge and thus the possibility for the motorist to bully the pedestrian arises again.

4.8. The election. As our final example we consider the ballot box whose function is to create certain specific states of knowledge. Suppose that five people 1, 2, 3, 4, 5 are debating whether to have lettuce or cucumbers for salad. They cast their votes into a ballot box and the final count reveals that lettuce has won by 3 votes to 2.

Let $C_i : i \leq 5$ mean that i voted for cucumber, and similarly for lettuce. Then the propositional formula which expresses that exactly two voted for lettuce and three for cucumbers is common knowledge.

This is a formula with 10 disjunctions, a typical one being $(L_1 \wedge L_2 \wedge C_3 \wedge L_4 \wedge C_5)$. Also what is common knowledge is $(\forall i, \forall j)(i \neq j \rightarrow \neg K_i(C_j))$ as well as $(\forall i, \forall j)(i \neq j \rightarrow \neg K_i(L_j))$. I.e. it is common knowledge that no one knows anyone else's vote.

I am sure the reader can see the game theoretic reasons for these two knowledge facts. Everyone must know the results of the election, and the lettuce party must not be in a position to take revenge on the two cucumbers.

We hope we have made a case that states of knowledge and belief which fall short of common knowledge and common belief do arise, are relevant, and can often be better than full common knowledge.

4.9. Future work. It should be obvious to the reader that this work is quite preliminary. Further work with more technical results will appear in due course. But I hope I have made the case that this direction is worth pursuing.

References

[CM2] H. H. Clark and C. R. Marshall, *Definite reference and mutual knowledge*, Elements of Discourse Understanding, Ed. Joshi, Webber and Sag, Cambridge U. Press, 1981.

[Dit] H. van Ditmarsch, *Knowledge games*, Bull. Econ. Res. **53** (2001), no. 4, 249–273, DOI 10.1111/1467-8586.00134. Logic and the foundations of the theory of games and decisions (Torino, 2000). MR1857519

[FHMV] R. Fagin, J. Y. Halpern, Y. Moses, and M. Y. Vardi, *Reasoning about knowledge*, MIT Press, Cambridge, MA, 1995. MR1345612

[HM] J. Y. Halpern and Y. Moses, *Knowledge and common knowledge in a distributed environment*, J. Assoc. Comput. Mach. **37** (1990), no. 3, 549–587, DOI 10.1145/79147.79161. MR1072269

[JP] D. H. J. de Jongh and R. Parikh, *Well-partial orderings and hierarchies*, Indag. Math. **39** (1977), no. 3, 195–207. Nederl. Akad. Wetensch. Proc. Ser. A **80**. MR447056

[Lew] D. Lewis, *Convention, a Philosophical Study*, Harvard U. Press, 1969.

[Lur] Lurz, Robert W. The philosophy of animal minds. Cambridge University Press, 2009.

[Pa] R. Parikh, *Social software*, Synthese **132** (2002), no. 3, 187–211, DOI 10.1023/A:1020391420768. MR1928902

[PK] R. Parikh and P. Krasucki, *Levels of knowledge in distributed systems*, Sādhanā **17** (1992), no. 1, 167–191, DOI 10.1007/BF02811342. MR1171431

[PTW] R. Parikh, Ç. Taşdemir, and A. Witzel, *The power of knowledge in games*, Int. Game Theory Rev. **15** (2013), no. 4, 1340030, 28, DOI 10.1142/S0219198913400306. MR3138702

[PW] Premack, David, and Guy Woodruff, *Does the chimpanzee have a theory of mind?*, Behavioral and brain sciences 1.4 (1978): 515-526.

[Raw] Rawls, John. *A theory of justice*, Routledge, 2004. 229-234.

[St] C. Steinsvold, *Topological models of belief logics*, ProQuest LLC, Ann Arbor, MI, 2007. Thesis (Ph.D.)–City University of New York. MR2710748

[Sch] Schelling, Thomas C. (1960). *The strategy of conflict* (First ed.). Cambridge: Harvard University Press.

[Tr] Trivedi, Shubhandu, *An anecdote to Norbert Wiener*, `https://onionesquereality.wordpress.com/2008/01/25/an-anecdote-to-norbert-wiener/`

[WP] H. Wimmer and J. Perner, *Beliefs about beliefs: representation and constraining function of wrong beliefs in young children's understanding of deception*, Cognition, **13** (1983) 103-128.

City University of New York, New York, New York 10016

Contemporary Mathematics
Volume **811**, 2025
https://doi.org/10.1090/conm/811/16245

Curvature and torsion via quadrilateral gaps

Nitin Nitsure

Abstract. An important basic figure in a Euclidean space, or more generally in a flat space, is a parallelogram. But in curved spaces, parallelograms with geodesic sides are not always possible. An attempt to draw a parallelogram with geodesic sides in an arbitrary manifold equipped with an affine connection will in general give rise to a 4-sided (quadrilateral) figure which may fail to close, as two particular vertices may not coincide, resulting in a 'quadrilateral gap' of our title.

The gap between the two vertices which fail to coincide is, to begin with, a second order infinitesimal in the linear size of the sides of the quadrilateral when the quadrilateral is infinitesimally small. It measures the torsion of the connection. If the torsion of the connection is identically zero, then the gap is a third order infinitesimal, and it measures the curvature of the connection. This article makes the above statements precise, and gives a brief account of the formulas that relate the torsion and curvature tensors to the gaps. The detailed proofs appear in [Proc. Indian Acad. Sci. Math. Sci. 131 (2021), Paper No. 4, 24].

Let M be a smooth manifold, equipped with a connection ∇ on its tangent bundle. Given a point $P \in M$ and a tangent vector u to M at P, we will denote the geodesic curve through P with tangent u at P by $\gamma_{P,u}(t)$ where t is the natural affine parameter of the geodesic. Note that $\gamma_{P,u}(t)$ is defined and unique when $|t|$ is small enough, $\gamma_{P,u}(0) = P$, $d\gamma_{P,u}/dt|_{t=0} = u$, and the tangent vector field $u(t) = d\gamma_{P,u}(t)/dt$ along $\gamma_{P,u}$ satisfies $\nabla_{u(t)}u(t) = 0$ identically.

Now let $u, v \in T_PM$ be a pair of tangent vectors to M at P. For a small enough real number s, we can construct the following figure in M (see Figure 1).

Let $\gamma_{P,u}(t)$ be the geodesic as above, and let $P_1(s) = \gamma_{P,u}(s)$. Let $v' \in T_{P_1(s)}$ be the parallel translate of v along $\gamma_{P,u}(t)$. Now consider the geodesic $\gamma_{P_1(s),v'}(t)$ with initial tangent vector v' at $P_1(s)$. Let $P_2(s) = \gamma_{P_1(s),v'}(s)$. Next, interchanging u and v in the above, let $\gamma_{P,v}(t)$ be the geodesic with initial tangent vector v at P, and let $Q_1(s) = \gamma_{P,v}(s)$. Let $u' \in T_{P_1(s)}$ be the parallel translate of u along $\gamma_{P,v}(t)$. Now consider the geodesic $\gamma_{Q_1(s),u'}(t)$ with initial tangent vector u' at $Q_1(s)$. Let $Q_2(s) = \gamma_{Q_1(s),u'}(s)$. Thus, we have constructed two points $P_2(s), Q_2(s)$ in M from the starting data P, u, v, s. Note that these two points exist and are unique whenever $|s|$ is sufficiently small.

If M is the affine space $\mathbb{R}^n$ equipped with its natural flat connection ∇, then each parameterized geodesic is a straight line with its affine parameter, and the

2020 *Mathematics Subject Classification.* Primary 53B05.

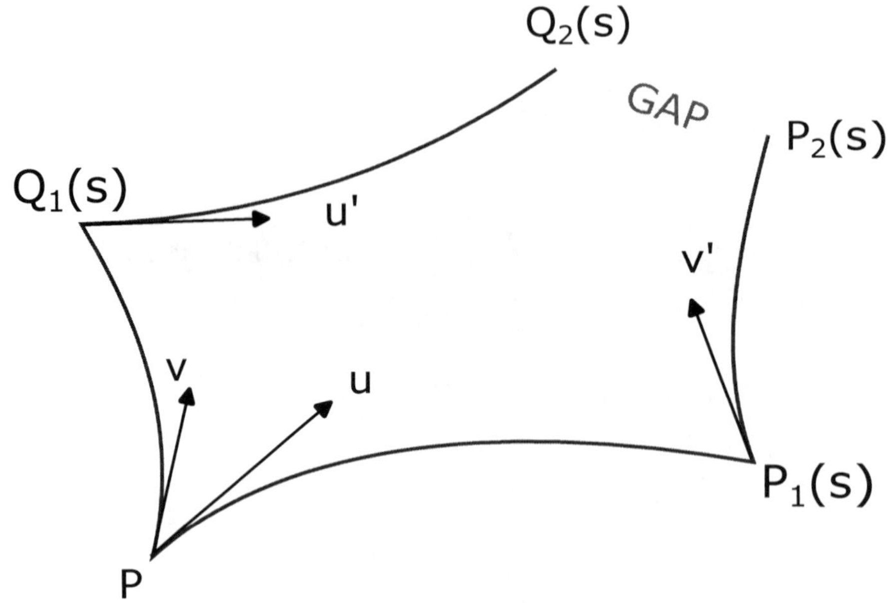

FIGURE 1

figure we get is a parallelogram, with $P_2(s) = Q_2(s)$. But in general, we will have $P_2(s) \neq Q_2(s)$. It has been intuitively known for long that the 'gap' between $P_2(s)$ and $Q_2(s)$ (which is depicted in Figure 1) comes about because the value $T(u,v) = \nabla_u v - \nabla_v u \in T_P M$ on u, v of the torsion tensor of ∇ is non-zero. For example, see [**1**] page 250 for a pictorial explanation of the gap in terms of $T(u,v)$. The word 'gap' is an informal term, so the relationship between the failure of the quadrilateral to close and the torsion and curvature tensors needs to be made precise and proved. This is done by the Theorems 1, 2 and 3 in this article, whose complete proofs appear in [**2**].

THEOREM 1. *Let M be a smooth manifold equipped with an affine connection ∇. Then for any smooth function f defined around a point P in M and any u, v $T_P M$, the limit $\lim_{s\to 0}(f(Q_2(s)) - f(P_2(s)))/s^2$ exists and equals $T(u,v)(f)$. In other words, we have*

$$\lim_{s\to 0} \frac{Q_2(s) - P_2(s)}{s^2} = T(u,v).$$

Next, suppose that the torsion tensor T of ∇ is identically zero in a neighbourhood of P. This would be the case, for example, if we are on a Riemannian or semi-Riemannian manifold (M, g) with Levi-Civita connection ∇ defined by the metric tensor g. One can ask whether the quadrilateral closes, giving a geodesic parallelogram with four vertices $P, P_1(s), Q_1(s), P_2(s) = Q_2(s)$ in this case. It can be quickly seen that even for a curved Riemannian surface, a gap exists. This author was asked by Professor Rajaram Nityananda in 2018 to quantify this gap for a curved Riemannian surface, possibly in terms its Gaussian curvature. This led to the results in [**2**], which we have reported here. As usual, we denote by $R(u,v) = \nabla_u \nabla_v - \nabla_v \nabla_u - \nabla_{[u,v]}$ the curvature operator for ∇, regarded as a tensor field on M.

THEOREM 2. *With hypothesis as in Theorem 1, suppose moreover that the torsion tensor T of ∇ is identically zero. Then for any smooth function f defined in a neighbourhood of P in M, the limit* $\lim_{s\to 0}(f(Q_2(s)) - f(P_2(s)))/s^3$ *exists, and equals* $R(u,v)(u+v)(f)$. *In other words,*

$$\lim_{s\to 0}\frac{Q_2(s)-P_2(s)}{s^3} = \frac{1}{2}R(u,v)(u+v).$$

If the connection ∇ is symmetric (means the torsion tensor T is identically zero), then is a simple exercise based on the well-known algebraic symmetries of the corresponding curvature tensor to uniquely recover the operator $R(X,Y) \in End(T_PM)$ for all $X, Y \in T_PM$ just from the values of $R(u,v)(u+v) \in T_PM$ as u and v vary over T_PM. The answer is as follows (see [**2**]). The curvature tensor satisfies $R(u,v)w = -R(v,u)w$ (skew-symmetry) and moreover if the connection ∇ is symmetric, then $R(u,v)w + R(v,w)u + R(w,u)v = 0$ (the algebraic Bianchi identity). From these, the following identities are immediate.

$$\begin{aligned} R(u,v)u &= \tfrac{1}{2}R(2u,v)(2u+v) - R(u,v)(u+v), \text{ and} \\ R(u,v)w &= \tfrac{1}{3}(R(u,v+w)(v+w) - R(u,v)v - R(u,w)w \\ &\quad + R(u+w,v)(u+w) - R(u,v)u - R(w,v)w. \end{aligned}$$

Hence the curvature operator $(u,v,w) \mapsto R(u,v)w$ can be recovered uniquely from the map $(u,v) \mapsto R(u,v)(u+v)$. Thus, if ∇ is symmetric, then the curvature gaps uniquely determine the Riemann curvature tensor.

It is possible to draw an alternative open quadrilateral, which gives rise to a parameterized curve $P_4(s)$ with $P_4(0) = P$ and $dP_4/ds|_{s=0} = 0$, which when reparametrized (as a half-curve) with parameter s^2, has tangent vector $-T(u,v)$ at P, and if the connection is symmetric, then this same curve when reparametrized with parameter s^3 has tangent vector $\frac{1}{2}R(u,v)(u+v)$ at P. This construction, which is depicted in Figure 2 on the next page, can be described as follows.

Given a point P on M, tangent vectors u and v at P and a real number s such that $|s|$ is sufficiently small, let the points $P_1(s)$ and $P_2(s)$ be plotted as in Figure 1 earlier. We denote by u_1, v_1 the tangent vectors at $P_1(s)$ obtained by parallel translating u, v along the geodesic $\gamma_{P,u}(t)$. The point $P_2(s)$ is on the geodesic $\gamma_{P_1(s),v_1}(t)$ at $t = s$. Let u_2, v_2 be tangent vectors at $P_2(s)$ obtained by parallel translating u_1, v_1 along the geodesic $\gamma_{P_1,v_1}(t)$. Next consider the geodesic $\gamma_{P_2,-u_2}(t)$ which has initial point $P_2(s)$ and initial tangent vector $-u_2$. Let $P_3(s) = \gamma_{P_2,-u_2}(s)$. Let v_3 be the tangent vector at $P_3(s)$ that is the parallel translate of v_2 along $\gamma_{P_2,-u_2}(t)$. Finally, consider the geodesic $\gamma_{P_3,-v_3}(t)$ which has initial point $P_3(s)$ and initial tangent vector $-v_3$. Let $P_4(s) = \gamma_{P_3,-v_3}(s)$. Then we have the following (see [**2**]).

THEOREM 3. *Let (M,∇) be a smooth manifold equipped with an affine connection. Then for any point P in M and any pair of tangent vectors u, v to M at P, we have*

$$\lim_{s\to 0}\frac{P_4(s)-P}{s^2} = -T(u,v).$$

If the torsion tensor T is identically zero, then we have

$$\lim_{s\to 0}\frac{P_4(s)-P}{s^3} = \frac{1}{2}R(u,v)(u+v).$$

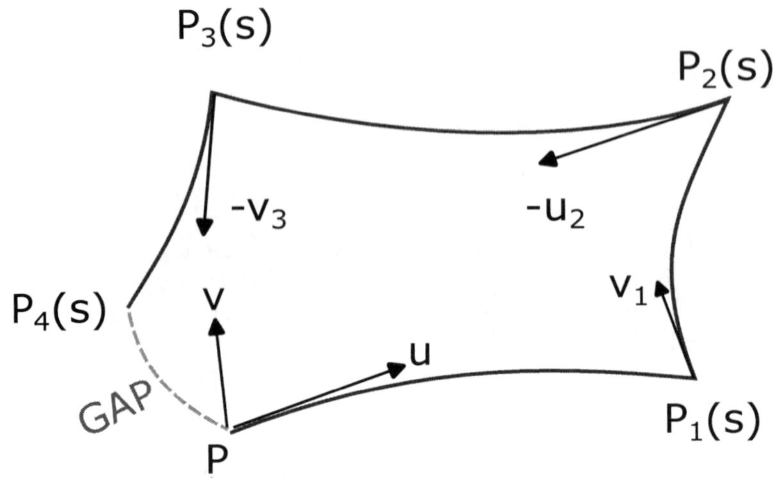

FIGURE 2

We will end with the classical case of a smooth surface M equipped with a Riemann metric g, and the resulting Levi-Civita connection ∇ which has torsion 0. Recall that if $P \in M$ and $u, v \in T_P M$ are orthonormal vectors, then the linear operator $R(u, v)$ on $T_P M$ is skewsymmetric w.r.t. the metric tensor g, and takes the matrix form $\begin{pmatrix} 0 & \kappa(P) \\ -\kappa(P) & 0 \end{pmatrix}$ in terms of the basis (u, v) of $T_P M$, where $\kappa(P)$ is the Gaussian curvature at P. Hence the tangent vector at P in terms of the new parameter s^3 to the parameterized curve $P_4(s)$ (depicted as a dashed red curve in Figure 2) that spans the gap is

$$\frac{1}{2} R(u, v)(u + v) = \frac{1}{2} \kappa(P)(u - v).$$

This vector has magnitude $\kappa(P)/\sqrt{2}$, and it is at an angle of $-\pi/4$ to u if $\kappa(P) > 0$ and at an angle of $3\pi/4$ to u if $\kappa(P) < 0$, if we define an orientation on $T_P M$ by declaring the ordered basis (u, v) to be right-handed. One can approximate the magnitude of the gap by

$$\|P_4(s) - P\| = \frac{\kappa(P)}{\sqrt{2}} s^3 + \text{higher order terms in } s.$$

Our geodesic quadrilaterals are much simpler to define than tiny 'circles' of geodesic radius r on M centered at P, but the above formula is reminiscent of the more difficult to state rigorously and to prove Bertrand and Puiseux formula that says that we can approximate the circumference $C(r)$ of the above circle for small r by

$$2\pi r - C(r) = \frac{\pi \kappa(P)}{3} r^3 + \text{higher order terms in } r.$$

Just like the formula of Bertrand and Puiseux, our formula for the magnitude of the quadrilateral gap can be regarded as an intrinsic characterization of Gaussian curvature.

We refer the reader to [**2**] for the proofs of the above theorems, and more. We will just give a brief idea of the proof here. As the first step, we take any local coordinate chart (x_i) around P. We use differential equations for the geodesics of our interest (those depicted in Figures 1) in terms of the connection coefficients Γ^i_{jk} of ∇, the resulting differential equations for parallel transports of u and v along these geodesics, and a Taylor expansion of the connection coefficients Γ^i_{jk} at P in the coordinates x_i, to get a formula for the coordinates of the points $P_2(s)$ and $Q_2(s)$ as a power series in the variable s. The coefficients of these power series are combinations of the coefficients of u and v, and the values of the connection coefficients Γ^i_{jk} and their partial derivatives of various orders at P. These appear to be messy formulas at first sight, but bunching together terms using the coordinate expressions for the components T_{jk} of the torsion tensor and the components $R^i_{jk\ell}$ of the curvature tensor in terms of the Γ^i_{jk} and their partial derivatives gives a massive simplification, resulting in the limit formulas of Theorems 1 and 2. The formulas in Theorem 3 are obtained by an argument that avoids the determination of the coordinates $P_3(s)$ and $P_4(s)$ by the above Taylor expansion method, as the power series becomes very unwieldy. Instead, a different approach is used which allows us to deduce the Theorem 3 from Theorems 1 and 2. We will also mention here that it is possible to guess the answers by heuristic reasoning and by simple examples, as detailed in [**2**]. This is what helps make the cancellations visible in a jumbled mass of terms.

Finally, I will like to express my admiration for the multifaceted talent, scholarship and activism of Professor Ravindra Kulkarni, who has been an inspiring figure for many of us younger people who know him. In addition to his mathematical research and teaching, and Sanskrit learning, he has been an organizer, as well as a generous donor of funds, for many mathematical activities and institutions over the decades. I wish him all the best on his eightieth birthday, including many more years of the vigorous intellectual life that he has always lived and enjoyed.

Acknowledgment

I thank Professor Shankar Ghosh of TIFR for producing the Figures 1 and 2 above.

References

[1] Charles W. Misner, Kip S. Thorne, and John Archibald Wheeler, *Gravitation*, W. H. Freeman and Co., San Francisco, CA, 1973. MR418833

[2] Nitin Nitsure, *Curvature, torsion and the quadrilateral gaps*, Proc. Indian Acad. Sci. Math. Sci. **131** (2021), no. 1, Paper No. 4, 24, DOI 10.1007/s12044-020-00596-2. MR4216404

RETIRED PROFESSOR, SCHOOL OF MATHEMATICS, TATA INSTITUTE OF FUNDAMENTAL RESEARCH, MUMBAI 400 005, INDIA

Email address: nitsure@gmail.com

Contemporary Mathematics
Volume **811**, 2025
https://doi.org/10.1090/conm/811/16246

Schur index and extensions of Witt-Berman's theorems

Soham Swadhin Pradhan

In honor of Professor Ravi Kulkarni

Abstract. Let G be a finite group, and F a field of characteristic 0 or prime to the order of G. In 1956, Berman proved that the number of inequivalent irreducible F-representations of G is equal to the number of F-conjugacy classes of the elements of G, where "F-conjugacy" was defined in a certain way. In this paper, we define F-conjugacy on G in a more natural way and give a proof of the Berman's theorem. In addition, we give a formula for computing a primitive central idempotent (pci) of the group algebra $F[G]$ corresponding to an irreducible F-representation of G, which can be obtained from the "F-character table" of G.

Let G contain a normal subgroup H of index p, a prime. In 1955, in case F is algebraically closed, Berman computed the pci of $F[G]$ corresponding to an irreducible F-representation of G, in terms of pci's of $F[H]$. In this paper, we give a proof of this theorem of Berman, and extend this result when F is not necessarily algebraically closed. Also, we describe decomposition of induced representation of an irreducible F-representation of H, into irreducible components.

1. Introduction

Throughout this paper, G always denotes a finite group, and F denotes a field of characteristic 0 or prime to the order of G.

Let H be a subgroup of G, and $\overline{F}$ the algebraic closure of F. In case we know the set of all F-representations of H, then the study of all F-representations of G amounts to studying how the induced representations of G induced from representations of H decompose into irreducible components. When F is not algebraically closed, there is an additional arithmetic aspect arising from how the cyclotomic polynomials split over F. The additional subtlety, in this case, is that of *Schur Index*. The account of Schur Index in the literature, I. Reiner (see [**10**]), M. Isaacs (see [**7**]), gives a "view from the top" (see Section 2). For the solution to our problem, we need to develop a view of the Schur index "from the bottom" (see Section 2).

Let u be the *exponent* of G (the lcm of order of all $g \in G$). Let ω be a primitive u-th root of unity in $\overline{F}$. Let $K = \mathrm{Gal}(F(\omega)/F)$, which is an abelian group. Then we have a homomorphism θ from the Galois group K, into the multiplicative group

2020 *Mathematics Subject Classification.* Primary 20CXX.

Key words and phrases. Group algebras, primitive central idempotents, Schur index, induced representations.

$\mathbb{Z}_u^*$, defined as follows. If $\sigma \in K$, then $\sigma(\omega) = \omega^a$, where $a \in \mathbb{Z}_u^*$, and we define $\theta(\sigma) = a$. Let $A = \theta(K) \leq \mathbb{Z}_u^*$. Following Berman [**3**], we say that two elements $x, y \in G$ are conjugate in the sense of Berman, or "*Berman-conjugate*", if there exists $g \in G$ and $j \in A$ such that $g^{-1}xg = y^j$, that is, x is conjugate to y^j. This is an equivalence relation on the set of elements of G, and so G is a disjoint union of Berman conjugacy classes. In 1952, Witt (see [**14**]) and in 1956, Berman (see [**1**], [**3**]) independently proved that *the number of equivalence classes of irreducible F-representations of G equals the number of Berman conjugacy classes of elements of G.* This theorem is known as *Witt-Berman theorem.* The result can be found in characteristic 0 as (see [**5**], p. 306, Theorem 42.8), and the proof is based on *Witt-Berman induction theorem* (see [**14**], [**3**], [**5**], p. 302, Theorem 42.3). Also the result can be found as (see [**15**], p. 641, Theorem 5.3), and the proof is based on Brauer's permutation lemma (see [**15**], p. 639, Corollary 4.3). A simplified proof in characteristic p using Brauer characters can be found in [**11**].

In this paper, we give a proof of Witt-Berman theorem when F is an arbitrary field of characteristic 0 or prime to the order of G, and the proof is different from the proof mentioned above. We define F-conjugacy as follows. We say two elements $x, y \in G$ are *F-conjugate*, denoted by $\sim_F$, if $\chi(x) = \chi(y)$ for all F-characters χ of G. Note that $\sim_F$ is an equivalence relation on the elements of G. Here, using our definition of F-conjugacy, we prove that $x \sim_F y$ if and only if x, y are Berman conjugate, and so the number of inequivalent irreducible F-representations of G is equal to the number of F-conjugacy classes of elements of G. We also show that the F-conjugacy class of an element in G of order n, is a union of certain conjugacy classes of G, and is determined by the roots of a single irreducible factor of $\Phi_n(X)$, the n-th cyclotomic polynomial, over F. Witt-Berman theorem allows us to consider *F-character table*, like the usual character table. We give an explicit formula for *primitive central F-idempotents* (pci's) of $F[G]$, and which can be read from the F-character table. All these results are known to the experts, we express them in a simple way.

Now suppose G contains a normal subgroup H of prime index p. In case F is algebraically closed, it is well known that as a consequence of *Clifford's theorem* (see [**4**], [**7**], p. 79 or [**6**], p. 72), every irreducible representation of H either induces an irreducible representation of G, or else it extends to p distinct irreducible representations of G. In 1955, Berman (see [**2**]), actually constructed primitive central idempotents of $F[G]$ in terms of the pcis of $F[H]$. In turn, this helps in actually constructing the matrices of representations of a finite solvable group over an algebraically closed field (see [**8**], p. 21, Section 4). Berman's paper is in Russian, which we couldn't read. We do not know any other reference which gives a complete proof of *Berman's theorem.* In this paper, we give a proof of Berman's theorem, and extend Berman's theorem to the case when F is not necessarily algebraically closed. This is the first main theorem of the paper (see Theorem A). In turn, this helps to construct the pci's of the semisimple group algebra of a finite solvable group.

Suppose that G contains a normal subgroup H of index p, a prime, and F is not necessarily algebraically closed. Then we can study of irreducible F-representations of G by determining the decomposition of induced representations induced from the irreducible representations of H, into irreducible components. In this paper, using classical *Schur's theory* (see [**12**], [**13**], [**10**], [**7**], Chapter 10), *Wedderburn's theory* (see [**10**]) and Theorem A, we describe decomposition of representations induced from the irreducible representations of H into irreducible components. This is the second main theorem of the paper (see Theorem B). In addition, we illustrate the splitting of idempotents and decomposition of induced representations, by several examples.

This paper is arranged as follows: After setting up the necessary notation, in Sec. 2, we give a brief introduction to Schur index in the theory of group representations. There we give two definitions of Schur index "from the top" and "from the bottom". In Sec. 3, we define F-conjugacy, give a proof of Witt-Berman theorem (Theorem 3.2), and an explicit formula for primitive central F-idempotents (Theorem 3.10). In Sec. 4, we give a proof of Berman's theorem (Theorem 4.2). In Sec. 5, we extend Berman's theorem to the case when F is not necessarily algebraically closed (Theorem A) . In Sec. 6, we describe decomposition of an representation induced from a normal subgroup of prime index, into irreducible components, when field of characteristic 0 or prime to the order of the group (Theorem B), and finally, in Sec. 7, we illustrate our results, by several examples.

NOTATION. The algebraic closure of F is denoted by $\overline{F}$. If V is a finite dimensional vector space over F, and $\rho : G \to \mathrm{GL}(V)$ is a group homomorphism, we say that ρ *is an F-representation* of G, and denote it by ρ. For an irreducible F-representation ρ, its corresponding pci in $F[G]$, we denote by e_ρ. For a positive integer n, a cyclic group of order n is denoted by C_n. For a normal subgroup H of G, and an F-representation η of H, the *conjugate representation* of η by an element $x \in G$ is denoted by η^x, and is defined by $\eta^x(h) = \eta(xhx^{-1})$, for all $h \in H$. For a prime p, and an indeterminate X, we denote $(1 + X + \cdots + X^{p-1})/p$ by e_X. The rest of our notations are standard.

2. Schur index: two definitions

2.1. Reduced dimension. Let $\rho : G \to \mathrm{GL}(V)$ be an irreducible F-representation of G. We call the F-vector space V the *representation-space* of ρ. The multiplication in G, extended by linearity, turns $F[G]$ into an F-algebra. The representation $\rho : G \to \mathrm{GL}(V)$ canonically extends to an F-algebra homomorphism $\rho^* : F[G] \to \mathrm{End}(V)$, making V an $F[G]$-module. For notational convenience, we use the same notation ρ instead of ρ^*. Let

$$D = \{A \in \mathrm{End}(V) \ : \ A \text{ commutes with } \rho(g) \text{ for all } g \in G\}.$$

By Schur's lemma, D is a finite-dimensional F-algebra which is a division ring. We call D, the *centralizer of G in* $\mathrm{End}(V)$. We can regard V as a right vector space over D^o, where D^o is D with the opposite multiplication: $x.y$ in D^o equals yx in D. Let n be the dimension of V over D^o, called the *reduced dimension* of V. Then $M_n(D^o)$ is abstractly isomorphic to the minimal 2-sided ideal of $F[G]$ which corresponds to the irreducible representation ρ. If we regard $F[G]$ as a left $F[G]$-module, then V occurs as a simple submodule of $F[G]$ with multiplicity n. In a different terminology, in the left regular F-representation of G, ρ appears with multiplicity n, the reduced dimension of V.

2.2. Schur index: view from the bottom. Now we briefly describe classical Schur's theory on group representations (see [**12**], [**13**]) and its connection with Wedderburn's theory of semisimple algebras. Let $\rho : G \to \mathrm{GL}(V)$ be an irreducible F-representation. Let $\overline{F}$ be the algebraic closure of F. Then $V \otimes_F \overline{F}$

is a $\overline{F}$-representation of G. Schur proved that $V \otimes_F \overline{F}$ decomposes into distinct irreducible $\overline{F}$-represnetations $\rho_1, \rho_2, \ldots, \rho_k$ which occur with the same multiplicity, say, m. In other words, if $\rho_i : G \to \mathrm{GL}_{\overline{F}} W_i$, then

$$\rho \otimes_F \overline{F} = m(\rho_1 \oplus \rho_2 \oplus \cdots \oplus \rho_k), \quad \text{and} \quad V \otimes_F \overline{F} \cong m(W_1 \oplus W_2 \oplus \cdots \oplus W_k).$$

Note that each mW_i is an isotypical component of $V \otimes_F \overline{F}$. The number m was later called the *Schur index of* ρ.

The main ingredients from the theory of semisimple F-algebras are the following. Let D be a finite dimensional F-algebra which is a division ring with its center Z. Then D is a (finite dimensional) Z-algebra and $[D : Z] = m^2$, for some integer m. Further, D contains maximal subfields E such that $[E : Z] = m$, and $D \otimes_Z E$ is isomorphic to $M_m(E)$. This is a *split* simple E-algebra. The *Schur index of D (or more generally, $M_n(D)$, for any n) over F* may be defined to be m. This will be used in the theory of representations of finite groups as follows.

Let $\rho : G \to \mathrm{GL}(V)$ be an irreducible F-representation. Let the minimal 2-sided ideal of $F[G]$ corresponding to ρ be abstractly isomorphic to $M_n(D)$, for suitable n, and a division ring D. Let Z be the center of $D, [Z : F] = k$, and E a maximal subfield of D. Then E contains Z. Note that $Z[G] = F[G] \otimes_F Z$ is a sum of certain number of minimal 2-sided ideals, which correspond to irreducible Z-representations of G. Now $M_n(D) \otimes_F Z$ is isomorphic to a direct summand of $Z[G]$. Since Z is the center of D, we have

$$M_n(D) \otimes_F Z \cong M_n(D \otimes_F Z) \cong M_n(\underbrace{D + \cdots + D}_{k}) \cong \underbrace{M_n(D) + \cdots + M_n(D)}_{k}.$$

Each of these summands is isomorphic to a minimal 2-sided ideal of the group algebra $Z[G]$. In terms of representations, $V \otimes_F Z$ splits into k simple $Z[G]$-summands, $V_1 \oplus \cdots \oplus V_k$. Each V_i is a Z-vector space, which by restriction of scalars may be regarded as an F-vector space. We have

$$\begin{aligned} \dim_F V &= (\dim_F Z)(\dim_Z V) = k(\dim_Z D)(\dim_D V) \\ &= km^2 n = k \dim_Z V_i \quad \text{for each } i = 1, \ldots, k. \end{aligned}$$

So, $\rho \otimes_F Z$ splits into k distinct irreducible Z-representations of G. For more precision, we shall write $M_n(D)_i$ for the i-th summand in $M_n(D) \otimes_F Z \cong M_n(D) + \cdots + M_n(D)$, which itself occurs as a summand in $Z[G]$. We have $M_n(D)_i \cong M_n(D)$, for each $i = 1, \ldots, k$, and the corresponding representation-space is V_i.

Let us now consider $E[G] = Z[G] \otimes_Z E$. Each summand of $E[G]$, isomorphic to

$$M_n(D) \otimes_Z E \cong M_n(D \otimes_Z E) \cong M_n(M_m(E)) \cong M_{nm}(E).$$

The last term is a single, split, E-algebra. Let the corresponding irreducible representation-space be denoted by an E-vector space W of dimension nm; so over Z it has dimension $(nm) \dim_Z E = nm^2$. More precisely, for each $i = 1, \ldots, k$, corresponding to the summand $M_n(D)_i$ of $E[G]$, we have an irreducible representation-space W_i.

Now consider $V_i \otimes_Z E$. It is an E-representation-space of G. As such it decomposes into irreducible E-representation-spaces of G. From the above description, its irreducible components are E-representation-spaces isomorphic to W_i. Since $\dim_Z V_i \otimes_Z E = nm^2$, it follows that $V_i \otimes_Z E$ is a sum of m copies of W_i.

THEOREM. *Let $\rho : G \to \mathrm{GL}(V)$ be an irreducible F-representation of G. Then, with D, Z, E, as defined above, we have*

$$V \otimes_F E = V \otimes_F Z \otimes_Z E = (V_1 + V_2 + \cdots + V_k) \otimes_Z E = m(W_1 + W_2 + \cdots + W_k),$$

a decomposition into distinct E-irreducible representation-spaces $W_1, \ldots, W_k$, each occurring with the same multiplicity m, where m is the Schur index of ρ, and $V_i = mW_i$, for each $i = 1, \ldots, k$.

Note that the D's occurring in the F-representation of finite groups are special. Namely, Z, the centre of D, is a *cyclotomic extension* of F, that is, it is contained in an extension of F of the form $F(\zeta_r)$, where ζ_r is a primitive r-th root of unity. This may be seen as follows. If u is the exponent of G, then by Brauer's theorem (see[**6**], Theorem17.2), $F(\zeta_u)$ is a splitting field for all F-representations of G. So, the center of $M_n(D) \otimes_F F(\zeta_u)$ is a direct sum of certain number of copies of $F(\zeta_u)$.

2.3. Schur index: view from the top. In 1961, Reiner, (see [**10**]), wrote an article on the Schur index in the theory of group representations. If we consider F as the "bottom" and $\overline{F}$ as the "top", Reiner gave a definition of the Schur index from a viewpoint of the "top". It is important to mention that this view point was also given by Schur. The fields Z, E mentioned above arise as subfields of $\overline{F}$, which depend on a specific irreducible F-representation. This needs some more terminology, for which we follow Isaacs's book (see [**7**], Chapter 10).

Let $\widetilde{\rho}$ be an irreducible representation of G over $\overline{F}$. Let $\chi = \chi_{\widetilde{\rho}}$ be the character of $\widetilde{\rho}$. If u is the exponent of G, then for each g in G, $\chi(g)$ is the sum of eigenvalues of $\widetilde{\rho}(g)$, which are in fact u-th roots of unity. So $\chi(g)$ is an element of the field $F(\zeta_u)$, where ζ_u is a primitive u-th root of unity. Let $F(\chi) = F(\chi_{\widetilde{\rho}})$ be the extension field of F obtained by adjoining all $\chi(g)$'s for g in G. It is a subfield of $F(\zeta_u)$. It is called the *character field* of $\widetilde{\rho}$ over F. Since $F(\zeta_u)$ is an abelian extension of F so is $F(\chi)$. Let $A = \mathrm{Gal}(F(\chi)/F$. It is easy to see that for each α in A, $\alpha \circ \chi$ is also a character of a representation of G over $\overline{F}$. In effect, starting with $\widetilde{\rho}$, we have obtained $|A|$ distinct representations of G, over $\overline{F}$. Any two of these representations are called *algebraically conjugate.* In this way we have obtained a class of mutually in-equivalent algebraically conjugate representations of G. These representations have different characters, but they all have the same character field.

Let Z_1 denote the field $F(\chi)$ and $e = e_{\widetilde{\rho}}$ be the pci corresponding to $\widetilde{\rho}$. Then e is in $Z_1[G]$ and $Z_1[G]e$ is a minimal 2-sided ideal of $Z_1[G]$. By the Artin-Wedderburn theorem, ring $Z_1[G]e$ (with identity e) is isomorphic to $M_n(D_1)$ (as Z_1-algebra) for some division ring D_1 such that its center is Z_1. Then $[D_1 : Z_1] = m^2$ for some integer m, and so if E_1 is any maximal subfield of D_1, then $\widetilde{\rho}$ is realizable over E_1. Then $V_1 \cong D_1^n$ is a representation space of G over Z_1. Then Z_1 is the center of D_1 and the dimension of D_1 over Z_1 is m^2 for some m. If E_1 is a maximal subfield of D_1, it can serve as a splitting field for D_1, and so the representation $\widetilde{\rho}$ is *realisable over* E_1. That is, we can choose a basis of the representation space $V_1 \otimes E_1$, w.r.t. which all the entries of the matrices $\widetilde{\rho}(g)$ for all g in G lie in E_1. Since the dimension of E_1 over Z_1 is m, and m is the least such dimension, we can take the second definition, due to Reiner (see [**10**]), of Schur index as the minimum of the dimensions of fields $\widetilde{E}$ over which the representation $\widetilde{\rho}$ is *realisable over* $\widetilde{E}$. Let V be the vector space over F obtained from V_1 by restriction of scalars from Z_1 to F. Consider it as a representation space of a representation ρ of G. It is easy to see that D_1 is the centraliser of G in $\mathrm{End}(V)$. Then the Schur index of ρ according to the first definition, equals the Schur index of the representation of $\widetilde{\rho}$ according to the second definition. These two definitions are related in the following way.

THEOREM. *Let $\rho : G \to \mathrm{GL}(V)$ be an irreducible F-representation. Let D, Z, F be as above with $[D : Z] = m^2$ and $[Z : F] = k$. Then*

(1) *Z is a cyclotomic extension of F.*
(2) *$V \otimes_F Z$ splits into distinct irreducible Z-representations $V_1, \dots, V_k$. The Galois group $\mathrm{Gal}(Z : F)$ acts simply transitively on $\{V_1, \dots, V_k\}$.*
(3) *$V \otimes_F \overline{F}$ splits into irreducible $\overline{F}$-representations $W_1, \dots, W_k$, each occurring with multiplicity m.*
(4) *Z is isomorphic to the character field of each of $W_1, \dots, W_k$.*

3. F-conjugacy, F-character table and F-idempotents

3.1. F-conjugacy. It is well known that if the characteristic of F is 0 or prime to the order of G, *character determines representation up to equivalence* in the following sense: if $\chi_{\rho_1} = \chi_{\rho_2}$ then ρ_1 is equivalent to ρ_2. We say that a character χ is irreducible if it is not a sum of two characters. Hence the number of in-equivalent irreducible F-representations of G is equal to the number of irreducible F-characters of G.

We say that $x, y \in G$ are *F-conjugate*, $x \sim_F y$ if for all finite dimensional F-representations (ρ, V) with the characters χ_ρ, we have $\chi_\rho(x) = \chi_\rho(y)$. Notice that $\sim_F$ is an equivalence relation on G, the equivalence class of an $x \in G$ is called the *F-conjugacy* class x in G, and denote it by $C_F(x)$. Let $C(x)$ denote the conjugacy class of x. In the following Lemma, we describe the factorization of n-th cyclotomic polynomial $\Phi_n(X)$ into monic irreducible factors over F.

LEMMA 3.1. *Let n be a positive integer. Let F be a field of characteristic 0 or prime to n. Let $\Phi_n(X) = f_1(X)f_2(X)\cdots f_k(X)$ be the decomposition of $\Phi_n(X)$ into irreducible monic polynomials over F. Then*

(1) *The degrees of all $f_i(X)$'s are the same.*
(2) *Let ζ be a root of one $f_i(X)$. Then all the roots of $f_i(X)$ are $\{\zeta^{r_1}, \zeta^{r_2}, \dots, \zeta^{r_s}\}$, where all r_i's are natural numbers with $r_1 = 1$, and the sequence set $\{r_1, r_2, \dots, r_s\}$ is independent of irreducible factor $f_i(x)$ of $\Phi_n(X)$ and any root of $\Phi_n(X)$.*

PROOF. The proof follows from the fact that if ζ is a primitive n-th root of $1 \in F$ then the splitting field of each $f_i(X)$ over F is $F(\zeta)$, which is also the splitting field of $\Phi_n(X)$ over F. □

THEOREM 3.2. *Let G be a finite group and F be a field of characteristic 0 or prime to the order of G. Then the number of inequivalent irreducible F-representations of G is equal to the number of F-conjugacy classes of elements of G.*

PROOF. Let u be the exponent of G and ω be a primitive u-th root of $1 \in F$. Let A be the subgroup of $\mathbb{Z}_u^*$ which gives all the F-automorphisms in $\mathrm{Gal}(F(w)/F)$. Define the following relations on G:

$g \sim h$ if g is Berman-conjugate to h, that is, g is conjugate to h^a for some $a \in A$,

$g \sim_F h$ if $\chi(g) = \chi(h)$ for all F-characters χ of G.

For any $g \in G$, let $[g]_\sim$ and $[g]_{\sim_F}$ denotes the equivalence classes of g under $\sim$ and $\sim_F$ respectively.

CLAIM. $[g]_\sim \subseteq [g]_{\sim_F}$ for all $g \in G$.

Let $h \in [g]_\sim$. Then g is conjugate to h^a for some $a \in A$, which implies $\chi(g) = \chi(h^a)$ for all F-characters χ. Note that $\sigma : w \to w^a$ is an F-automorphism of $F(\omega)$. So $\chi(g) = \chi(h^a) = \sigma(\chi(h)) = \chi(h)$ for all F-characters χ, which implies that $h \in [g]_{\sim_F}$. Therefore, $[g]_\sim \subseteq [g]_{\sim_F}$ for all $g \in G$.

Consider the sets

$$M = \{f : G \to F \ : \ f \text{ is constant on each } \sim \text{ equivalence class of } G\},$$
$$N = \{f : G \to F \ : \ f \text{ is constant on each } \sim_F \text{ equivalence class of } G\}.$$

Then M and N are F-vector spaces. Let $\phi_1, \ldots, \phi_r$ be the all distinct irreducible F-characters of G. By definition, $\text{span}\{\phi_1, \ldots, \phi_r\} \subseteq N \subseteq M$. We show that $M = \text{span}\{\phi_1, \ldots, \phi_r\}$, which will imply that $[g]_\sim = [g]_{\sim_F}$, and the proof will be complete.

Consider $\theta \in M$. Then θ will be an F-valued class function on G. Let $\chi_1, \ldots, \chi_k$ be all the irreducible $F(\omega)$-characters of G. Since $F(\omega)$ is a splitting field for G and $\{\chi_1, \ldots, \chi_k\}$ form an orthonormal basis on space of all the $F(\omega)$-valued class functions on G, we can write $\theta = \sum_{i=1}^k c_i\chi_i$ where $c_i \in F(\omega)$. Let σ be the F-automorphism $\omega \mapsto \omega^a$ of $F(\omega)$. Then for any $g \in G$, $\theta(g) = \theta(g^a)$, hence

$$\sum_{i=1}^k c_i\sigma(\chi_i(g)) = \sum_{i=1}^k c_i\chi_i(g^a) = \theta(g^a) = \theta(g) = \sum_{i=1}^k c_i\chi_i(g).$$

Therefore $c_i = c_j$ if and only if $\sigma(\chi_i) = \chi_j$ (by linear independence of χ_i's over $F(\omega)$). Therefore θ is an $F(\omega)$-linear combination of ϕ_i's, say

$$\theta = \sum_{i=1}^r d_i\phi_i.$$

CLAIM. Each d_i belongs to F.

Since $\phi_1, \ldots, \phi_r$ are linearly independent over F (see[**7**], Theorem 9.22), there exists $x_j \in G$ $(1 \le j \le r)$ such that the $r \times r$ matrix, with (i, j)-th entry $\phi_i(x_j)$, is non-singular. Therefore, d_i's are uniquely determined by the system of linear equations

$$\theta(x_j) = \sum_{i=1}^r d_i\phi_i(x_j), \quad j = 1, 2, \ldots, r.$$

It follows that $d_i = \sigma(d_i)$, for any F-automorphism σ of $F(\omega)$. Thus θ is an F-linear combination of $\phi_1, \ldots, \phi_r$. Therefore, $\text{span}\{\phi_1, \ldots, \phi_r\} = N = M$. Since $r = \dim M$ then the number of F-irreducible representations of G is equal to the number of F-conjugacy classes of elements of G. □

COROLLARY 3.3. The set of all irreducible F-characters of G forms a basis of the space of all functions $f : G \longrightarrow F$, which are constant on each F-conjugacy class of G.

REMARK 3.4. In literature, Theorem 3.2 is known as *Witt-Berman theorem* (see [**5**], p. 306, Theorem 42.8, [**15**], p. 641, Theorem 5.3).; the proof in [**5**] uses *Witt-Berman induction theorem*, and the proof in [**15**] uses *Brauer's permutation* *lemma*. Here, we have presented an elementary proof of Theorem3.2. Moreover, the notion of F-conjugacy, which we have defined in a different way than in [**5**], [**15**].

THEOREM 3.5. *Let F be a field of characteristic* 0 *or prime to the order of G and consider $x \in G$ of order n. Then $C_F(x)$ is the union of conjugacy classes $C(x^{r_1}) \cup C(x^{r_2}) \cup \cdots \cup C(x^{r_s})$, where $r_1 = 1, r_2, \ldots, r_s$ is the sequence set associated with $\Phi_n(X)$ as in the Lemma* 3.1.

PROOF. Let u be the exponent of G. Then n is a divisor of u. Let ω be a primitive u-th root and ζ be a primitive n-th root of $1 \in F$. Then $\zeta = \omega^a$, for some positive integer a. Let K and L be the Galois groups of $F(\omega)$ and $F(\zeta)$ over F respectively. Let A, B be the subgroup of $\mathbb{Z}_u^*$, $\mathbb{Z}_n^*$ respectively, which give all the F-automorphisms of $F(\omega)$, $F(\zeta)$ in K and L respectively. Therefore $B = \{\overline{r_1} = \overline{1}, \overline{r_2}, \ldots, \overline{r_s}\}$ where $\overline{r_i} \equiv= r_i \pmod{n}$. Since $C_F(x)$ is invariant under all inner automorphisms of G, it is union of conjugacy classes.

We first prove that $C(x) \cup C(x^{r_2}) \cup \cdots \cup C(x^{r_s}) \subseteq C_F(x)$. Let ρ be an arbitrary F-representation of G, and χ be its character. Then $\chi(x) = \zeta^{u_1} + \zeta^{u_2} + \cdots + \zeta^{u_t}$ for some $u_i \in \mathbb{Z}$, and it belongs to F. Since $\sigma_i : \zeta \mapsto \zeta^{r_i}$ is an element in L, then

$$\chi(x^{r_i}) = \zeta^{r_i u_1} + \zeta^{r_i u_2} + \cdots + \zeta^{r_i u_t} = \sigma_i(\zeta^{u_1} + \cdots + \zeta^{u_t}) = \sigma_i(\chi(x)) = \chi(x).$$

This implies that $C(x) \cup C(x^{r_2}) \cup \cdots \cup C(x^{r_s}) \subseteq C_F(x)$.

To prove the reverse inclusion, consider $y \in C_F(x) = [x]_{\sim_F}$. Then y is conjugate to x^a for some $a \in A$, which implies that y is conjugate to x^a, for some $a \in B$. Therefore $y \in C(x) \cup C(x^{r_2}) \cup \cdots \cup C(x^{r_s})$, and hence $C_F(x) \subseteq C(x) \cup C(x^{r_2}) \cup \cdots \cup C(x^{r_s})$. This completes the proof. □

COROLLARY 3.6. Let x be an element in G, and of order n. Then the F-conjugacy class of x is uniquely determined by the roots of just one irreducible factor of $\Phi_n(X)$ over F.

EXAMPLE 3.7. *Let n be the order of x. Let $F = \mathbb{Q}$. Then $C_{\mathbb{Q}}(x)$ is the union of $C(x^i)$, where i runs over the positive integers relatively prime to n.*

EXAMPLE 3.8. *Let n be the order of x and $F = \mathbb{R}$. Then $C_{\mathbb{R}}(x)$ is the union of $C(x)$ and $C(x^{-1})$.*

EXAMPLE 3.9. *Let n be the order of x. Let $F = \mathbb{F}_q$, $q = p^r$, p a prime and r be a positive integer. Let d be the least positive integer such that $q^d \equiv 1 \pmod{n}$. Then $C_F(x)$ is the union of $C(x^{q^i})$ for $i = 0, 1, \ldots, d-1$.*

3.2. F-character table. By Theorem 3.2, for a finite group G, the number of F-conjugacy classes of elements of G is equal to the number of F-irreducible representations of G. We can list the F-character values on F-conjugacy classes in the form of a square matrix over F, which is called the *F-character table*. The columns of F-character table are parametrized by F-conjugacy classes and rows are parametrized by irreducible F-characters. Since the number of F-conjugacy classes is at most the number of conjugacy classes, the size of the matrix representing F-character table is smaller than the usual character table.

3.3. F-idempotents. Now we give a formula for computing the pci's of $F[G]$ in terms of F-conjugacy classes. Let $R = F[G]$ and e_i $(i = 1, 2, \ldots, r)$ be the pci's of R. Then each Re_i is a minimal two-sided ideal of R. Let V denote one of these simple R-modules and e be the corresponding pci. Then by Schur's lemma, $\mathrm{End}_{F[G]}(V)$ is a division ring D, whose center Z contains F. Let $n = \dim_D V$. Then Re is isomorphic to $M_n(D^o)$, V is isomorphic to $(D^o)^n$, and Re is isomorphic to the direct sum of n copies of V. Let $\dim_F Z = \delta$ and $\dim_Z D = m^2$. Then $\dim_F V = nm^2\delta$, and so $\dim_F Re = n^2m^2\delta$.

Let $L_1, L_2, \ldots, L_r$ be the F-conjugacy classes of G and $C_1, C_2, \ldots, C_s$ be the conjugacy classes of G. Let χ be any F-character of G and $\chi(L_i)$ denote the common value of χ over L_i. If L_i be the F-conjugacy class of x, and let ${L_i}^{-1}$ denote the F-conjugacy class of x_i^{-1}. For any subset S of G, let S^* denote the formal sum of elements of S. We use the above notation in Theorem 3.10.

THEOREM 3.10. *Let F be a field of characteristic 0 or prime to the order of G. Let (ρ, V) be an irreducible F-representation of G, χ be its character and e be the corresponding pci in $F[G]$. Let n be the reduced dimension of V. Then*

$$e = \frac{n}{|G|}\sum_{i=1}^{r}\chi(L_i^{-1})L_i^*.$$

PROOF. By the classical theorem of Schur (see Section 2), $\rho \otimes_F \overline{F}$ splits into δ distinct irreducible representations over $\overline{F}$ of degree mn, and with the multiplicity m. Let

$$\rho \otimes_F \overline{F} = m(\widetilde{\rho}_1 \oplus \widetilde{\rho}_2 \oplus \cdots \oplus \widetilde{\rho}_\delta),$$

where $\widetilde{\rho}_i$'s are algebraically conjugates over F and m is the Schur index of $\widetilde{\rho}_i$ with respect to F. Let $\widetilde{\chi}_i$ be the $\overline{F}$-character of $\widetilde{\rho}_i$. The pci corresponding to the representation $\widetilde{\rho}_i$ is given by

$$\widetilde{e}_i = \frac{mn}{|G|}\sum_{k=1}^{s}\widetilde{\chi}_i(C_k^{-1})C_k^*.$$

From the Theorem 3.3.1(1) in [9], we have $e = \widetilde{e}_1 + \widetilde{e}_2 + \cdots + \widetilde{e}_\delta$. So,

$$e = \frac{mn}{|G|}\sum_{k=1}^{s}\Big\{(\widetilde{\chi}_1 + \widetilde{\chi}_2 + \cdots + \widetilde{\chi}_\delta)(C_k^{-1})\Big\}C_k^* = \frac{n}{|G|}\sum_{i=1}^{r}\chi(L_i^{-1})L_i^*.$$

This completes the proof. □

REMARK 3.11. Let $e'' = \sum_{x\in G}\chi(x^{-1})x$. Then $e'' = \{|G|/n\}e$, and $e''^2 = \{|G|/n\}^2e^2 = \{|G|/n\}^2e = \{|G|/n\}e''$. So if we know χ, then one can determine n. Therefore one can read the complete set of pci's of $F[G]$ from the F-character table.

4. Berman's theorem

Berman (see [2]) described a method for constructing the irreducible representations of a finite solvable group over algebraically closed field via the group algebra. For that, he inductively constructed all the pci's of the group algebra of a finite solvable group. In this section, we give a proof of Berman's theorem. We start with the following lemma.

LEMMA 4.1. *Let $G = \langle x : x^p = 1\rangle$ be the cyclic group of prime order p. Let F be an algebraically closed field of characteristic either 0 or coprime to p. Let ζ be a primitive p-th root of unity. Then the pci's in $F[G]$ are $e_{\zeta^i x}$, $i = 0, 1, \ldots, p-1$, where $e_X = (1 + X + \cdots + X^{p-1})/p$*

The proof of the lemma is a standard calculation and we skip it.

Let G be a finite group, H be a normal subgroup of index p, a prime, and $G/H = \langle xH \rangle$, for some $x \in G$. Let F be an algebraically closed field of characteristic either 0 or coprime to $|G|$. Let $\overline{C(x)}$ be the conjugacy class sum of x in $F[G]$. As $\overline{C(x)}$ belongs to the center of $F[G]$ and $\overline{C(x)}^p = x^p a$, for some $a \in F[H]$, then $\overline{C(x)}^p$ is a central element of $F[H]$.

Let (η, W) be an irreducible F-representation of H and e_η be the pci of η in $F[H]$. If $\eta^x \not\cong \eta$, then by Clifford's theorem (see [**4**]) $\eta \uparrow_H^G$ is an irreducible representation of G, and $\eta \uparrow_H^G \cong \eta^x \uparrow_H^G \cong \cdots \cong \eta^{x^{p-1}} \uparrow_H^G \cong \rho$, say. It is easy to see that $e_\rho = e_\eta + e_{\eta^x} + \cdots + e_{\eta^{x^{p-1}}}$.

Suppose $\eta^x \cong \eta$. Then $\overline{C(x)}^p e_\eta$ belongs to the center of $F[H]e_\eta$. By Schur's lemma, the center of $F[H]e_\eta$ is $\{\alpha e_\eta : \alpha \in F\}$. It follows that $\overline{C(x)}^p e_\eta = \lambda_x e_\eta$, where $\lambda_x \in F$. Note that λ_x depends on x. For notational convenience, we write $\lambda_x = \lambda$. We show below that $\lambda \neq 0$, for some $x \in G - H$.

Let μ be any p-th root of λ in F. Let $c = \overline{C(x)} e_\eta / \mu$. Then $c^p = e_\eta$. Let ζ be a primitive p-th root of unity in F. By the Lemma 4.1, $e_{\zeta^i c} e_\eta$ are p mutually orthogonal central idempotents in $F[G]$, and e_η is a sum of $e_{\zeta^i c} e_\eta$ in $F[G]$. Since e_η can split into at most p central idempotents in $F[G]$, then $\eta \uparrow_H^G$ splits into p distinct irreducible representations of G. In fact each of these representations are *extensions* of η, that is, the H-action on the representation space W extends to p distinct G-actions on the same vector space W.

It remains to show that x in $G-H$ may be chosen such that $\lambda \neq 0$. Suppose for all x in $G-H$ we have $\overline{C(x)}^p e_\eta = 0$. Let ρ be an extension of η, with corresponding character χ_ρ and the pci e_ρ in $F[G]$. If χ_ρ vanishes on $G - H$, then

$$e_\rho = \frac{\deg \eta}{|G|} \sum_{g \in G} \chi_\rho(g^{-1}) g = \frac{\deg \eta}{p|H|} \sum_{h \in H} \chi_\eta(h^{-1}) h = \frac{1}{p} e_\eta,$$

and this implies that e_ρ is *not* an idempotent, a contradiction. This completes the proof of Berman's theorem.

We summarise the above discussion in Theorem 4.2:

THEOREM 4.2 (Berman's theorem). *Let G be a finite group and H be a normal subgroup of index p, a prime. Let $G/H = \langle xH \rangle$, for some x in G. Let F be an algebraically closed field of characteristic either 0 or prime to the order of G. Let η be an irreducible representation of H over F. We distinguish two cases:*

(1) *If $\eta^x \not\cong \eta$, then $\rho \cong \eta \uparrow_H^G$ is irreducible, $\rho \downarrow_H^G \cong \eta \oplus \eta^x \oplus \cdots \oplus \eta^{x^{p-1}}$, and*

$$e_\rho = e_\eta + e_{\eta^x} + \cdots + e_{\eta^{x^{p-1}}}.$$

(2) *If $\eta^x \cong \eta$, then η extends to p distinct irreducible representations $\rho_0, \rho_1, \ldots, \rho_{p-1}$ of G over F, and $\eta \uparrow_H^G \cong \rho_0 \oplus \rho_1 \oplus \cdots \oplus \rho_{p-1}$. Correspondingly, $e_\eta = e_{\rho_0} + e_{\rho_1} + \cdots + e_{\rho_{p-1}}$. In the above notation, we may take $e_{\rho_i} = e_{\zeta^i c} e_\eta, i = 0, 1, \ldots, p-1$, where $e_X = (1 + X + \cdots + X^{p-1})/p$.*

As a consequence of the Berman's theorem, we get Corollary 4.3, which will be useful later.

COROLLARY 4.3. Let G be a finite group and H be a normal subgroup of index p, a prime. Let F be a field of characteristic 0 or prime to $|G|$. Let $\overline{F}$ be the algebraic closure of F. Let $G/H = \langle xH \rangle$, for some x in G. Let η be an irreducible representation of H over $\overline{F}$ with corresponding character χ_η, e_η be its corresponding pci in $\overline{F}[H]$, and $K = F(\chi_\eta)$, the character field of χ_η over F.

Suppose that $\eta^x \cong \eta$. Then $\overline{C(x)}^p e_\eta = \lambda e_\eta$, for some $\lambda \neq 0 \in \overline{F}$. Assume that there exists $z(\neq x)$ in $G - H$ such that $\overline{C(z)}^p e_\eta = \gamma e_\eta$, for some $\gamma \neq 0 \in \overline{F}$. Let μ and ν be a p-th root of λ and γ respectively, in $\overline{F}$. Then $\mu = \nu^j a$, for some $a \in K$ and $j = 1, 2, \ldots, p-1$. Moreover, the number of p-th roots of λ in K is equal to the number of p-th roots of γ in K, and the possibilities of that number are $0, 1, p$.

PROOF. As $\eta^x \cong \eta$, then from the proof of Theorem4.2 $\overline{C(x)}^p e_\eta = \lambda e_\eta$, for some $\lambda \neq 0 \in \overline{F}$. Let ζ be a primitive p-th root of unity in $\overline{F}$. Let $c = \frac{\overline{C(x)}e_\eta}{\mu}$ and $d = \frac{\overline{C(z)}e_\eta}{\nu}$. We write

$$u_i = \frac{1}{p}(e_\eta + \zeta^i c e_\eta + \zeta^{2i} c^2 e_\eta + \cdots + \zeta^{(p-1)i} c^{p-1} e_\eta) \qquad (0 \leq i \leq p-1).$$

By Theorem4.2, each u_i is a pci in $\overline{F}[G]$, and $e_\eta = u_0 + u_1 + \cdots + u_{p-1}$.
Similarly, we write

$$v_i = \frac{1}{p}(e_\eta + \zeta^i d e_\eta + \zeta^{2i} d^2 e_\eta + \cdots + \zeta^{(p-1)i} d^{p-1} e_\eta) \qquad (0 \leq i \leq p-1).$$

By Theorem4.2, each v_i is a pci in $\overline{F}[G]$, and $e_\eta = v_0 + v_1 + \cdots + v_{p-1}$.

As the decomposition of e_η into pci's in $\overline{F}[G]$ is canonical, then $\{u_0, u_1, u_2\} = \{v_0, v_1, v_2\}$. Without loss of generality, we can assume that $u_0 = v_0$. Then comparing the coefficients in both sides, we get $\mu = \nu^j a$, for some $a \in K$, and $j = 1, 2, \ldots, p-1$. As $\mu = \nu^j a$, for some $a \in K$, and $j = 1, 2, \ldots, p-1$, then it is clear that the number of p-th roots of λ in K is same as the number of p-th roots of γ in K, and the possibilities of that number are $0, 1, p$. □

COROLLARY 4.4. With the hypothesis in the Theorem 4.2, for an irreducible character χ of G, $\chi(x) \neq 0$ for some $x \in G - H$ if and only if $\chi|_H$ is an irreducible character of H.

5. Extension of Berman's theorem

Let H be a normal subgroup of prime index p in G, F be a field of characteristic 0 or prime to the order of G, and $\overline{F}$ be the algebraic closure of F. Let $G/H = \langle xH \rangle$, for some x in G. Let η be an irreducible F-representation of H, and e_η be its corresponding pci in $F[H]$. Let $\overline{C(x)}$ denote the conjugacy class sum of x in $F[G]$. Let $I(\eta) = F[H]e_\eta$. Let Z denote the center of $I(\eta) = F[H]e_\eta$. So, $\overline{C(x)}^p e_\eta = \lambda_x e_\eta$, where $\lambda_x \in Z$. Note that λ_x depends on x. For notational convenience, we take $\lambda_x = \lambda$.

We have

$$(5.1) \quad I(\eta) \otimes_F Z = F[H]e_\eta \otimes_F Z = Z[H]e_\eta \cong \underbrace{M_n(D) \oplus M_n(D) \oplus \cdots \oplus M_n(D)}_{\delta}.$$

Note that the above isomorphism is a Z-algebra isomorphism. Each summand $M_n(D)$ is a central simple algebra over Z. In terms of representations, $\eta \otimes_F Z = \eta_1 + \eta_2 + \cdots + \eta_\delta$, where η_i's are inequivalent irreducible Z-representations of G. Let e_{η_i} be the pci corresponding to η_i in $Z[H]$. So,

$$e_\eta = e_{\eta_1} + e_{\eta_2} + \cdots + e_{\eta_\delta}$$

is the canonical decomposition of e_η into pci's in $Z[H]$. By Schur's theory (see Section 2), we have

(5.2) $$\eta \otimes_F \overline{F} = (\eta \otimes_F Z) \otimes_Z \overline{F} = (\eta_1 \oplus \eta_2 \oplus \dots \eta_\delta) \otimes_Z \overline{F} = m(\widetilde{\eta_1} \oplus \widetilde{\eta_2} \oplus \dots \widetilde{\eta_\delta}),$$

where $\widetilde{\eta_i}$'s are in-equivalent irreducible $\overline{F}$-representations of H, each occurring with the same multiplicity m, the Schur index of η. Let $\psi_{\widetilde{\eta_i}}$ be the character, and $e_{\widetilde{\eta_i}}$ be the pci in $\overline{F}[H]$, corresponding to the $\overline{F}$-representation $\widetilde{\eta_i}$. Let $K_i = F(\psi_{\widetilde{\eta_i}})$. It is easy to see that $e_{\eta_i} = e_{\widetilde{\eta_i}}$. Therefore, each e_{η_i} is a pci in $\overline{F}[H]$, and

(5.3) $$e_\eta = e_{\eta_1} + e_{\eta_2} + \cdots + e_{\eta_\delta}$$

is the canonical decomposition of e_η into pci's in $\overline{F}[H]$.

If $\eta^x \ncong \eta$, then $\eta \uparrow_H^G$ is irreducible, and also,

(5.4) $$\eta \uparrow_H^G \cong \eta^x \uparrow_H^G \cong \dots \cong \eta^{x^{p-1}} \uparrow_H^G \cong \rho.$$

Moreover, $e_\rho = e_\eta + e_{\eta^x} + \cdots + e_{\eta^{x^{p-1}}}$. So, from now onward we restrict our attention to the case $\eta^x \cong \eta$.

As $\eta^x \cong \eta$, then e_η is central in $F[G]$. If e_η remains a pci in $F[G]$, we say e_η does *not split* in $F[G]$, otherwise we say e_η *splits* in $F[G]$.

Let $\Phi_p(X)$ denote the p-th cyclotomic polynomial. Let $\Phi_p(X) = f_1(X) \dots f_k(X)$ be the factorization into monic irreducible polynomials over Z. Then $X^p - 1 = f_0(X) f_1(X) \dots f_k(X)$, where $f_0(X) = X - 1$, is the factorization into monic irreducible polynomials over Z. By Proposition 3.1, for $i = 1, 2, \dots, k$, the degree of $f_i(X)$'s are the same, say, d. If ζ is a root of $f_i(X)$, then all the roots of $f_i(X)$ are $\zeta, \zeta^{r_2}, \dots, \zeta^{r_d}$, and also the sequence $\{r_1 = 1, r_2, \dots, r_d\}$ is independent of $f_i(X)$ and the roots of $f_i(X)$. Now we shall state and prove the first main theorem.

We keep the notation as above. Suppose that $\eta^x \cong \eta$. We consider two cases:

(1) $\eta_1^x \ncong \eta_1$.
(2) $\eta_1^x \cong \eta_1$.

In case (1), it is easy to see that $\overline{C(x)}^p e_\eta = 0$ for all x in $G - H$ and e_η does not split in $F[G]$.

In case (2), since $\eta_1^x \cong \eta_1$, there exists $y \in G - H$, $\overline{C(y)}^p e_\eta = \lambda e_\eta$, for some $\lambda \neq 0$. Without loss of generality, we take y to be x. So, $\overline{C(x)}^p e_\eta = \lambda e_\eta$, for some $\lambda \neq 0$. Suppose Z contains a p-th root of λ, say μ. Let $c = \overline{C(x)} e_\eta / \mu$. So c is a central element of $Z[G]$. Then $c^p = e_\eta$. Consider $e_c = (1 + c + c^2 + \cdots + c^{p-1})/p$. Then $e_c e_\eta$ is a central idempotent in $Z[G]$.

Let $\mu_1, \mu_2, \dots, \mu_p$, be the p distinct p-th roots of λ in $\overline{F}$. Let $c_i = \overline{C(x)} e_\eta / \mu_i, i = 1, 2, \dots, p$ and $e_{c_i} = (1 + c_i + c_i^2 + \cdots + c_i^{p-1})/p$. Then $e_{c_i} e_\eta, i = 1, 2, \dots, p$ are mutually orthogonal central idempotents in $\overline{F}[G]$. So, in $\overline{F}$, e_η splits into p distinct central idempotents: $e_{c_i} e_\eta, i = 1, 2, \dots, p$. Their sum is e_η.

As for Z itself, the following three mutually exclusive cases can occur:

(a) λ has no p-th roots in Z.
(b) λ has two distinct p-th roots in Z, then Z contains all p, p-th roots of λ, say, $\mu_1, \mu_2, \dots, \mu_p$. In this case, $\mu_2/\mu_1 = \zeta$, is a primitive p-th root of 1. So, ζ lies in Z. Then $\mu_1 \zeta^i, i = 0, 1, 2, \dots, p-1$, are the distinct p-th roots of λ in Z. Since Z is a field, λ has at most p distinct p-th roots in Z. So, $\{\mu_1, \mu_2, \dots, \mu_p\} = \{\mu_1 \zeta^i, i = 0, 1, 2, \dots, p-1\}$.
(c) λ has exactly one p-th root of λ, say μ, in Z.

Using all the above notations, we state and prove the first main theorem:

THEOREM A. *Let G be a finite group, and H a normal subgroup of prime index p in G. Let $G/H = \langle xH\rangle$, for some x in G. Let F be a field of characteristic 0 or prime to the order of G. Let η be an irreducible F-representation of H, and e_η be its corresponding pci in $F[H]$. Suppose that $\eta^x \cong \eta$. Let Z denote the center of $F[H]e_\eta$. Then*

(1) *If $\eta_1^x \not\cong \eta_1$, then e_η does not split in $F[G]$.*

(2) *If $\eta_1^x \cong \eta_1$, then it follows that $\overline{C(x)}^p e_\eta = \lambda e_\eta$, where $\lambda \in Z - \{0\}$. We consider three sub-cases:*

   (a) *If λ has no p-th root in Z, then e_η does not split in $F[G]$.*

   (b) *If λ has two distinct p-th roots in Z, then e_η splits into p pci's in $F[G]$. They are given by $e_{c_i}e_\eta, i = 1, 2, \ldots, p$.*

   (c) *If λ has only one p-th root in Z, then e_η splits into $1+k$ pci's in $F[G]$. Let $\zeta_{ij}, j = 1, 2, \ldots, d$ be the roots of $f_i(X)$ in $\bar{F}$. Let $c_{ij} = \frac{\overline{C(x)e_\eta}}{\mu\zeta_{ij}}$,*

$$e_{c_{ij}} = (1 + c_{ij} + c_{ij}^2 + \cdots + c_{ij}^{p-1})/p, i = 1, 2, \ldots, k; j = 1, 2, \ldots, d$$

and $e_{f_i(X)} = \sum_{j=1}^d e_{c_{ij}}e_\eta$. Then e_η splits into $e_{f_i(X)}, i = 1, 2, \ldots, k$, together with $e_c e_\eta$.

PROOF. (1) As $\eta_1^x \not\cong \eta_1$ then $\eta_i^x \not\cong \eta_i$, for all $i = 1, 2, \ldots, \delta$. It is easy to see that δ is a multiple of p. Let $\delta = lp$ where l is a positive integer. Since $\eta_i^x \not\cong \eta_i$, $\eta_i \uparrow_H^G$ is irreducible Z-representation of G, say, ρ_i. The pci e_{ρ_i} corresponding to ρ_i is:

$$e_{\rho_i} = e_{\eta_i} + e_{\eta_i^x} + \cdots + e_{\eta_i^{x^{p-1}}}.$$

It is easy to see that $e_\eta = e_{\rho_1} + e_{\rho_2} + \cdots + e_{\rho_l}$ and e_{ρ_i}'s are algebraically conjugates over F. Therefore, e_η does not split in $F[G]$. This completes the proof of (1).

(2) It is easy to show that if $\eta_1^x \cong \eta_1$, then it follows that $\overline{C(x)}^p e_\eta = \lambda e_\eta$, where $\lambda \in Z - \{0\}$.

In case (a), e_η does not split in $Z[G]$, therefore, does not split in $F[G]$.

In case (b), we see that in $Z[G]$, e_η splits into p distinct central idempotents. They are given by

$$e_{c_i}e_\eta \quad (i = 1, 2, \ldots, p).$$

But Z is contained in $F[H]$, so each $e_{c_i}e_\eta$ is defined in $F[G]$, and therefore e_η splits into p distinct central idempotents in $F[G]$. But e_η splits into at most p distinct pci's in $F[G]$. Therefore, e_η splits into p pci's $e_{c_i}e_\eta, i = 1, 2, \ldots, p$, in $F[G]$.

In case (c), Z contains exactly one p-th root μ of λ. Then the Newton's identities expressing the power symmetric functions into elementary symmetric functions, it is easy to see that

$$e_{c_{ij}}e_\eta \quad (j = 1, 2, \ldots, d)$$

add to an expression which has coefficients in Z. Since Z is in $F[H]$, these elements lie in $F[G]$. We denote this sum by $e_{f_i(X)}, i = 1, 2, \ldots, k$. Let

$$c = \frac{\overline{C(x)e_\eta}}{\mu} \text{ and } e_c = (1 + c + c^2 + \cdots + c^{p-1})/p.$$

One can easily show that $e_{f_i(X)}, i = 1, 2, \ldots, k$, together with $e_c e_\eta$, we get $1 + k$ pci's, in $F[G]$, and their sum is e_η. This completes the proof of the theorem. □

REMARK 5.1. We use the same notations as above. In Theorem A, the case (2), the sub-cases (*a*), (*b*) and (*c*) do not depend on the choice of x, which can be shown in the following way.

In case(2), since $\eta_1^x \cong \eta_1$, there exists $y \in G - H$, $\overline{C(y)}^p e_\eta = \lambda e_\eta$, for some $\lambda \in Z - \{0\}$. Without of loss of generality, we take y to be x. So, $\overline{C(x)}^p e_\eta = \lambda e_\eta$, for some $\lambda \in Z - \{0\}$. Since $\eta_1^x \cong \eta_1$, $\overline{C(x)}^p e_{\eta_i} = \lambda_i e_{\eta_i}$, $\lambda_i \neq 0$, $(1 \leq i \leq \delta)$. Therefore,

$$\overline{C(x)}^p e_\eta = \lambda e_\eta = \sum_{i=1}^{\delta} \overline{C(x)}^p e_{\eta_i} = \sum_{i=1}^{\delta} \lambda_i e_{\eta_i}.$$

It is easy to see that $\lambda_i \in K_i$, $(1 \leq i \leq \delta)$. One can show that $Z \cong K_i$ (F-isomorphism), under this F-isomorphism $\lambda \mapsto \lambda_i$.

If there exists $z(\neq x) \in G - H$ such that $\overline{C(z)}^p e_\eta = \gamma e_\eta$, for some $\gamma \in Z - \{0\}$. Then by Corollary 4.3, we easily conclude that the number of p-th roots of λ in Z is same as the number of p-th roots of γ in Z, and the possibilities of that number are $0, 1, p$.

6. Decompositon of induced representations

Let G be a finite group and H be a subgroup in G. Let F be a field of characteristic 0 or prime to the order of G. Let η be an irreducible F-representation of H, and e_η be its corresponding pci in $F[H]$. Let V be the representation space of η. Let $I(\eta)$ be the minimal 2-sided ideal of $F[H]$ corresponding to η, that is, $I(\eta) = F[H]e_\eta$. Let D be the H-centraliser of η. Then $I(\eta)$ is abstractly isomorphic to $M_n(D)$. So, $\dim_F I(\eta) = n^2 \dim_F D$. Let Z be the center of $I(\eta) = F[H]e_\eta$, which is equal to the center of D. Let $[Z : F] = \delta$. Let m be the Schur index of η. So, $\dim_Z D = m^2, \dim_F D = \delta m^2, \dim_F I(\eta) = \delta n^2 m^2$.

Let H be a normal subgroup in G. Let $\phi : F[H] \longrightarrow F[G]$ be the canonical inclusion. Let $J_1, J_2, \ldots, J_r$ be the minimal 2-sided ideals of $F[G]$, each of whose intersection with $\phi(I(\eta))$ is non-zero. So the module induced from $I(\eta)$ is $J_1 + J_2 + \cdots + J_r$, as the module induced from $F[H]$ is $F[G] \otimes_{F[H]} F[H] = F[G]$. Note that $I(\eta) = nV$ as $F[H]$-modules. So the induced module of V is contained in the induced module of $I(\eta)$, i.e., $J_1 + J_2 + \cdots + J_r$, so realizable in $F[G]$.

Now, let H be a normal subgroup of prime index p in G. Let $G/H = \langle xH \rangle$, for some $x \in G$. Let η be an irreducible F-representation of H such that $\eta^x \cong \eta$. Let $\overline{C(x)}$ denote the conjugacy class sum of x in $F[G]$. So, $\overline{C(x)}^p e_\eta = \lambda e_\eta$, where $\lambda \in Z$. Then

$$I(\eta) \otimes_F Z = F[H]e_\eta \otimes_F Z = Z[H]e_\eta \cong \underbrace{M_n(D) \oplus M_n(D) \oplus \cdots \oplus M_n(D)}_{\delta}. \tag{6.1}$$

In terms of representations, $\eta \otimes_F Z = \eta_1 + \eta_2 + \cdots + \eta_\delta$, where η_i's are inequivalent irreducible Z-representations of G. In terms of pci's, $e_\eta = e_{\eta_1} + e_{\eta_2} + \cdots + e_{\eta_\delta}$ is the splitting of e_η into pci's in $Z[G]$. So, each $Z[H]e_{\eta_i}$ is abstractly isomorphic to $M_n(D)$ as Z-algebra. Let $I(\eta_i) = Z[H]e_i$ be the minimal 2-sided ideal of $Z[H]$ corresponding to η_i. Let $I(\eta) \uparrow_H^G$ be the induced ideal in $F[G]$ from $I(\eta)$ and $I(\eta_i) \uparrow_H^G$ be the induced ideal in $Z[G]$ from $I(\eta_i)$. Therefore, $I(\eta) \uparrow_H^G = F[G]e_\eta$ and $I(\eta_i) \uparrow_H^G = Z[G]e_{\eta_i}$.

Using all the above notations and notations in the previous section, we state and prove the second main theorem:

THEOREM B. *Let G be a finite group, and H a normal subgroup of prime index p in G. Let $G/H = \langle xH \rangle$, for some x in G. Let F be a field of characteristic 0 or prime to the order of G. Let η be an irreducible F-representation of H, and e_η be its corresponding pci in $F[H]$. Let $\Phi_p(X)$ denote the p-th cyclotomic polynomial. Suppose that $\eta^x \cong \eta$. Let Z denote the center of $I(\eta) = F[H]e_\eta$. Let $\Phi_p(X) = f_1(X) \ldots f_k(X)$ be the factorization into monic irreducible polynomials over Z. Then*

(1) *If $\eta_1^x \not\cong \eta_1$, then $\eta \uparrow_H^G$ is either irreducible or equivalent to $p\rho$, where ρ is the unique extension of η to G.*
(2) *If $\eta_1^x \cong \eta_1$, then it follows that $\overline{C(x)}^p e_\eta = \lambda e_\eta$, where $\lambda \in Z - \{0\}$. Then we consider three sub-cases:*
  (a) *If λ has no p-th root in Z, then $\eta \uparrow_H^G$ is either irreducible or equivalent to $p\rho$, ρ is the unique extension of η to G.*
  (b) *If λ has two distinct p-th roots in Z, then η extends to p distinct irreducible F-representations of G, say, $\rho_0, \rho_1, \ldots, \rho_{p-1}$. Then $\eta \uparrow_H^G \cong \rho_0 \oplus \rho_1 \oplus \cdots \oplus \rho_{p-1}$.*
  (c) *If λ has only one p-th root in Z, then $\eta \uparrow_H^G$ decomposes into $1+k$ distinct irreducible F-representations of G. If $\rho_0, \rho_1, \ldots, \rho_k$ be the $1+k$ distinct irreducible F-representations of G appear in $\eta \uparrow_H^G$ then $\eta \uparrow_H^G \cong \rho_0 \oplus s(\rho_1 \oplus \cdots \oplus \rho_k)$, where ρ_0 is the unique extension of η and s divides g.c.d.(m, d), where m is the Schur index of η and d is the common degree of irreducible factors of $\Phi_p(X)$ over Z.*

PROOF. (1) As $\eta_1^x \not\cong \eta_1$, then by Theorem A, e_η does not split in $F[G]$. So, $I(\eta) \uparrow_H^G = F[G]e_\eta$ is a minimal 2-sided ideal in $F[G]$. Therefore, $\eta \uparrow_H^G$ is direct sum of $s(\rho)$, where ρ is an irreducible F-representation of G, and s is a positive integer. Now we show that s is either 1 or p. By Schur's theory (see Section 2), we have

$$\eta \otimes_F \overline{F} = (\eta \otimes_F Z) \otimes_Z \overline{F} = (\eta_1 \oplus \eta_2 \oplus \ldots \eta_\delta) \otimes_Z \overline{F} = m(\widetilde{\eta_1} \oplus \widetilde{\eta_2} \oplus \ldots \widetilde{\eta_\delta}),$$

where η_i's are in-equivalent irreducible Z-representations of H, $\widetilde{\eta_i}$'s are in-equivalent irreducible $\overline{F}$-representations of H, and m is the Schur index of η.

As $\eta_1^x \not\cong \eta_1$, then $\delta = lp$, where l is a positive integer. So, we get

$$\eta \otimes_F \overline{F} = m(\widetilde{\eta_1} \oplus \cdots \oplus \widetilde{\eta_\delta}) = m\Big\{(\widetilde{\eta_1} \oplus \widetilde{\eta_1}^x \oplus \cdots \oplus \widetilde{\eta_1}^{x^{p-1}}) \oplus \cdots \oplus (\widetilde{\eta_l} \oplus \widetilde{\eta_l}^x \oplus \cdots \oplus \widetilde{\eta_l}^{x^{p-1}})\Big\}.$$

Therefore, we get

$$(\eta) \uparrow_H^G \otimes_F \overline{F} \cong mp\Big(\widetilde{\eta_1} \uparrow_H^G \oplus \widetilde{\eta_2} \uparrow_H^G \oplus \cdots \oplus \widetilde{\eta_l} \uparrow_H^G\Big).$$

As each $\widetilde{\eta_i} \uparrow_H^G$ is irreducible and the Schur index of each $\widetilde{\eta_i} \uparrow_H^G$ with respect to F is either mp or m, then $\eta \uparrow_H^G$ is either irreducible or equivalent to $p(\rho)$, where ρ is the unique extension of η. This completes the proof of (1).

(2) It is easy to show that if $\eta_1^x \cong \eta_1$, then it follows that $\overline{C(x)}^p e_\eta = \lambda e_\eta$, where $\lambda \in Z - \{0\}$.

In case (a), if λ has no p-th root in Z, then by Theorem A, e_η does not split in $F[G]$. So, $I(\eta) \uparrow_H^G = F[G]e_\eta$ is a minimal two-sided ideal in $F[G]$. Therefore, $\eta \uparrow_H^G$ is direct sum of $s(\rho)$, where ρ is an irreducible F-representation of G, and s is a positive integer. As $\eta_1^x \cong \eta_1$ and λ has no p-th root in Z, we can easily show that s is either 1 or p. Hence, $\eta \uparrow_H^G$ is either irreducible or equivalent to $p(\rho)$, where ρ is the unique extension of η.

In case (b), λ has two distinct p-th roots in Z, then by Theorem A, e_η splits into p pci's in F[G]. So $I(\eta)\uparrow_H^G = F[G]e_\eta$ is direct sum of p minimal 2-sided ideals in $F[G]$. Consequently, $\eta\uparrow_H^G$ is direct sum of p distinct extensions of η, say, $\rho_0, \rho_1, \dots, \rho_{p-1}$ and $\eta\uparrow_H^G \cong \rho_0 \oplus \rho_1 \oplus \cdots \oplus \rho_{p-1}$.

In case (c), Z contains exactly one p-th root of λ. Since $\eta_1^x \cong \eta_1$, $\overline{C(x)}^p e_{\eta_1} = \lambda_1 e_{\eta_1}$ for some $\lambda_1 \in Z - \{0\}$ and there exist only one μ_1 in Z such that $\mu_1^p = \lambda_1$. Then

$$\begin{aligned} I(\eta_1)\otimes_{Z[H]} Z[G] &\cong M_n(D)\otimes_{Z[H]} Z[G] \\ &= M_n(D)\otimes_Z Z[G/H] \\ &= M_n(D)\otimes_Z Z[X]/(X^p-1) \\ &= M_n(D)\otimes \big(Z[X]/(X-1)\oplus Z[X]/(\Phi_p(X))\big) \\ &= M_n(D)\oplus M_n(D)\otimes_Z \big(Z[X]/(\Phi_p(X))\big) \\ &= M_n(D)\oplus M_n\big(D\otimes_Z Z[X]/(\Phi_p(X))\big). \end{aligned}$$

Also,

$$\begin{aligned} &M_n(D)\oplus M_n\big(D\otimes_Z Z[X]/(\Phi_p(X))\big) \\ &\cong M_n(D)\oplus M_n\Big(D\otimes_Z \frac{Z[X]}{(f_1(X))}\oplus D\otimes_Z \frac{Z[X]}{(f_2(X))}\oplus\cdots\oplus D\otimes_Z \frac{Z[X]}{(f_k(X))}\Big) \\ &\cong M_n(D)\oplus M_n(M_s(D_1))\oplus M_n(M_s(D_2))\oplus\cdots\oplus M_n(M_s(D_k)) \\ &\cong M_n(D)\oplus M_{ns}(D_1)\oplus M_{ns}(D_2)\oplus\cdots\oplus M_{ns}(D_k). \end{aligned}$$

Note that $\mathbb{Z}_p^*$ permutes the summands in $D\otimes_Z \frac{Z[X]}{(f_1(X))}\oplus\cdots\oplus D\otimes_Z \frac{Z[X]}{(f_k(X))}$, so all summands are abstractly isomorphic, and so all D_i's are abstractly isomorphic, and the same s works for all D_i's.

So if $V \cong D^n$ is an F-irreducible representation space for H, then its induced G-representation space splits into G-irreducible representation spaces $\cong V \oplus s(V_1 \oplus V_2 \oplus \cdots \oplus V_k)$. If $\rho_0, \rho_1, \dots, \rho_k$ are the $1+k$ distinct irreducible F-representations of G appear in $\eta\uparrow_H^G$ then $\eta\uparrow_H^G \cong \rho_0 \oplus s(\rho_1 \oplus \cdots \oplus \rho_k)$.

We have $[D:Z] = m^2$. D_1 is contained in $D\otimes_Z Z(\zeta)$. The latter is a semisimple algebra. D_1 is a division ring contained in $D\otimes_Z Z(\zeta)$. Let Z_1 be the centre of D_1. It contains $Z(\zeta)$. So $[D_1 : Z_1]$ divides $[D\otimes_Z Z(\zeta) : Z(\zeta)]$. $\dim_Z(D) = \dim_{Z(\zeta)}(D\otimes_Z(Z(\zeta)))$. Note that $Z_1 = Z(\zeta)$. Therefore, $m^2 = \dim_{Z_1}(D\otimes_Z(Z_1)) = \dim_{Z_1}(M_s(D_1)) = s^2 m_1^2$, where m_1 is the Schur index of D_1. So, we get $m^2 = s^2 m_1^2$. Therefore s^2 divides m^2. This implies that s divides m. Moreover, $m_1 = m/s$. It is clear that s divides d, and hence s divides g.c.d.(m, d). This completes the proof of the theorem. □

REMARK 6.1. Let G be a finite group, and H a normal subgroup of prime index p in G. Let $G/H = \langle xH\rangle$, for some $x \in G$. Let F be a field of characteristic 0 or prime to the order of G. Let η be an irreducible F-representation of H, and e_η be its corresponding pci in $F[H]$. If $\eta^x \ncong \eta$ then $\eta\uparrow_H^G$ is irreducible, say, ρ and also if $F[H]e_\eta \cong M_n(D)$ then $F[G]e_\rho \cong M_{np}(D)$.

REMARK 6.2. Now we make some comments on Theorem B. We use the notation as in Theorem B.

(1) $\eta_1^x \ncong \eta_1$.

In this case $\eta \uparrow_H^G$ is either irreducible or equivalent to $p\rho$, where ρ is the unique extension of η.

- In case $\eta \uparrow_H^G$ is irreducible, ρ, say, then the Schur index of ρ is mp.
- In case $\eta \uparrow_H^G$ is equivalent to $p\rho$, where ρ is the unique extension of η, then the Schur index of ρ is m.

(2) $\eta_1^x \cong \eta_1$.

(a) If λ has no p-th roots in Z, then $\eta \uparrow_H^G$ is either irreducible or equivalent to $p\rho$, where ρ is the unique extension of η.
- In case $\eta \uparrow_H^G$ is irreducible, ρ, say, then the Schur index of ρ is m.
- In case $\eta \uparrow_H^G$ is equivalent to $p\rho$, where ρ is the unique extension of η, then the Schur index of ρ is m/p.

(b) If λ has two distinct p-th roots in Z, then η extends to p distinct irreducible F-representations of G, say, $\rho_0, \rho_1, \dots, \rho_{p-1}$.
- In this case, the Schur index of each ρ_i is m.

(c) If λ has only one p-th root in Z, then $\eta \uparrow_H^G$ decomposes into $1 + k$ distinct irreducible F-representations of G; let they ber $\rho_0, \rho_1, \dots, \rho_k$. Recall that ρ_0 is an extension of η and $\rho_1, \dots, \rho_k$ appear in $\eta \uparrow_H^G$ with multiplicity s.
- The Schur index of ρ_0 is m.
- The Schur index of each $\rho_i, i = 1, 2 \dots, k$ is m/s, where s is a divisor of g.c.d(m, d).
- We always have $dk = p - 1$. So $\deg \rho_i = d/s \deg \eta$, for $i = 1, 2, \dots, k$. In the frequently occurring case, $m = 1$, so $s = 1$, and so $\deg \rho_i = d \deg \eta$.

7. Examples

In this section, we give some examples illustrating the cases in Theorem A and in Theorem B. We shall use the notations as in the previous section.

EXAMPLE 7.1. *Consider G to be Q_8 with a presentation:*

$$G = \langle x, y, z \mid x^2 = 1, y^2 = x, z^2 = y^2, z^{-1}yz = xy \rangle.$$

Take $H = C_4 = \langle x, y \mid x^2 = 1, y^2 = x \rangle$ and $F = \mathbb{Q}$. Let η be the unique faithful irreducible $\mathbb{Q}$-representation of H of degree 2, and so $\eta^z \cong \eta$. We check that $e_\eta = 1 - e_x$, where $e_x = (1 + x)/2$ and $Z \cong \mathbb{Q}(i)$, $i = \sqrt{-1}$. In this case, e_η does not split, and $\eta \uparrow_H^G$ is irreducible, say, ρ. Note that the simple components of η and ρ in their Wedderburn decompositions are $\mathbb{Q}(i)$ and $\mathbb{H}_\mathbb{Q}$ respectively, where $\mathbb{H}_\mathbb{Q}$, the quaternion algebra over $\mathbb{Q}$. So, the Schur index of ρ is 2.

Note that the irreducible $\mathbb{Q}$-representations of G are of degrees 1, 1, 1, 1 and 4, whereas the degrees of irreducible $\overline{\mathbb{Q}}$-representations of G are 1, 1, 1, 1 and 2. Let V be an irreducible representation space of G of dimension 4 over $\mathbb{Q}$. Here $V \cong \mathbb{H}_\mathbb{Q}$ (as a $\mathbb{Q}$-vector space) has reduced dimension 1, that is, it has dimension 1 regarded as a right vector space over $\mathbb{H}_\mathbb{Q}$ (as a division ring). As a ring, $\mathbb{H}_\mathbb{Q} \otimes \overline{\mathbb{Q}}$ is $M_2(\overline{\mathbb{Q}})$. Each of the two columns of $M_2(\overline{\mathbb{Q}})$ can serve as an irreducible representation space of G over $\overline{\mathbb{Q}}$. Notice that dimension of $(V \otimes_\mathbb{Q} \overline{\mathbb{Q}}) = 4$ and it is a direct sum of 2 copies of the same irreducible 2-dimensional representation space of G over $\overline{\mathbb{Q}}$.

It was the genius of Schur to realise that this indeed was a general phenomenon!

EXAMPLE 7.2. *Consider G to be $C_7 \rtimes C_3$ with a presentation:*

$$G = \langle x, y \mid x^7 = y^3 = 1, y^{-1}xy = x^2 \rangle.$$

Take $H = C_7 = \langle x \mid x^7 = 1 \rangle$ and $F = \mathbb{Q}$. Let η be the unique faithful irreducible $\mathbb{Q}$-representation of H of degree 6, and so $\eta^y \cong \eta$. We check that $e_\eta = 1 - e_x$, where $e_x = (1 + x + \cdots + x^6)/7$, and $Z \cong \mathbb{Q}(\zeta)$, where ζ is a primitive 7-th root of unity. In this case, e_η does not split in $\mathbb{Q}[G]$, and $\eta \uparrow_H^G \cong 3\rho$, where ρ is an extension of η. Note that simple components corresponding to η, ρ are $\mathbb{Q}(\zeta)$ and $M_3(\mathbb{Q}(\zeta + \zeta^2 + \zeta^4))$ respectively.

EXAMPLE 7.3. *Let p be a prime. Consider G to be $C_{p^2} = \langle x, y \mid x^p = 1, y^p = x \rangle$. Take $H = C_p = \langle x \mid x^p = 1 \rangle$, and $F = \mathbb{Q}$. Let η be the unique faithful irreducible $\mathbb{Q}$-representation of H of degree $p - 1$. We check that $e_\eta = 1 - e_x$, where $e_x = (1 + x + \cdots + x^{p-1})/p$, and $Z \cong \mathbb{Q}(\zeta_p)$, where ζ_p is a primitive p-th root of unity. In this case, e_η does not split in $\mathbb{Q}[G]$ and $\eta \uparrow_H^G$ is irreducible, say, ρ. Note that simple components corresponding to η, ρ are $\mathbb{Q}(\zeta_p)$ and $\mathbb{Q}(\zeta_{p^2})$ respectively, where ζ_{p^2} is a primitive p^2-th root of unity.*

EXAMPLE 7.4. *Consider G to be $Q_8 \times_{C_2} C_4$ (central product) with a presentation:*

$$\langle x, y, z, t | x^2 = 1, y^2 = x, z^2 = y^2, z^{-1}yz = xy, t^2 = x, t^{-1}xt = x, t^{-1}yt = y, t^{-1}zt = z \rangle.$$

Take $H = Q_8$ and $F = \mathbb{Q}$. Take η to be the unique faithful irreducible $\mathbb{Q}$-representation of H of degree 4. We check that $e_\eta = 1 - e_x$, where $e_x = (1 + x)/2$ and $Z \cong \mathbb{Q}$. In this case, e_η does not split in $\mathbb{Q}[G]$, and $\eta \uparrow_H^G \cong 2\rho$, where ρ is an extension of η. Note that simple components corresponding to η, ρ are $\mathbb{H}_{\mathbb{Q}}$ and $M_2(\mathbb{Q}(i))$ respectively.

EXAMPLE 7.5. *Let p be a prime. Consider G to be $C_p \times C_p$ with presentation*

$$\langle x, y \mid x^p = y^p = 1, xy = yx \rangle.$$

Take $H = C_p = \langle x | x^p = 1 \rangle$ and $F = \mathbb{Q}$. Let η be the degree $p - 1$ unique faithful irreducible $\mathbb{Q}$-representation of H. For an indeterminate X, let $e_X = (1 + X + \cdots + X^{p-1})/p$. One can see that $e_\eta = 1 - e_x$ and $Z \cong \mathbb{Q}(\zeta_p)$, where ζ_p is a primitive p-th root of unity. In this case, e_η splits into p pci in $\mathbb{Q}[G]$. Correspondingly, $\eta \uparrow_H^G$ decomposes into p distinct extensions of η to G, say, $\rho_0, \rho_1, \ldots, \rho_{p-1}$. Then $e_{\rho_i} = e_{x^i y}(1 - e_x), i = 0, 1, \ldots, p - 1$. Note that the simple component corresponding to each ρ_i in the Wedderburn decomposition is $\mathbb{Q}(\zeta_p)$.

EXAMPLE 7.6. *Consider G to be $\mathrm{SL}_2(3)$ with a presentation:*

$$\mathrm{SL}_2(3) = \langle x, y, z, t | x^2 = 1, y^2 = x = z^2, z^{-1}yz = xy, t^3 = 1, t^{-1}yt = z, t^{-1}zt = yz \rangle.$$

Take $H = Q_8$ and $F = \mathbb{Q}$. Let η be the degree 4 unique faithful irreducible $\mathbb{Q}$-representation of H. We can see that $e_\eta = 1 - e_x$, where $e_x = (1 + x)/2$, and $Z \cong \mathbb{Q}$. In this case, e_η splits into two pci's in $\mathbb{Q}[G]$. Correspondingly, $\eta \uparrow_H^G$ decomposes into two distinct irreducible $\mathbb{Q}$-representations of G, and of degrees 4, say, ρ_0, ρ_1. In fact, $\eta \uparrow_H^G \cong \rho_0 \oplus 2\rho_1$. Note that the simple component corresponding to η is $\mathbb{H}_{\mathbb{Q}}$, and the simple components corresponding to ρ_0, ρ_1 are $\mathbb{H}_{\mathbb{Q}}$, $\mathbb{H}_{\mathbb{Q}(\omega)} \cong M_2(\mathbb{Q}(\omega))$ respectively, where ω is a primitive cube root of unity.

Acknowledgments

I am sincerely grateful to Ravi Kulkarni for giving the guidance to write this article. I am very thankful to him for many fruitful discussions through the various states of this work. I also thank him for reading the article carefully. I would also like to thank to Rahul Dattatraya Kitture for several useful discussions.

I would also like to express my gratitude to the referee. His numerous remarks helped to improve the presentation. His remarks helped me to bring this paper to a more clear form.

References

[1] S. D. Berman, *On the theory of representations of finite groups* (Russian), Doklady Akad. Nauk SSSR (N.S.) **86** (1952), 885–888. MR52441

[2] S. D. Berman, *Group algebras of abelian extensions of finite groups* (Russian), Dokl. Akad. Nauk SSSR (N.S.) **102** (1955), 431–434. MR72139

[3] S. D. Berman, *The number of irreducible representations of a finite group over an arbitrary field* (Russian), Dokl. Akad. Nauk SSSR (N.S.) **106** (1956), 767–769. MR78360

[4] A. H. Clifford, *Representations induced in an invariant subgroup*, Ann. of Math. (2) **38** (1937), no. 3, 533–550, DOI 10.2307/1968599. MR1503352

[5] Charles W. Curtis and Irving Reiner, *Representation theory of finite groups and associative algebras*, Pure and Applied Mathematics, Vol. XI, Interscience Publishers (a division of John Wiley & Sons, Inc.), New York-London, 1962. MR144979

[6] Larry Dornhoff, *Group representation theory. Part A: Ordinary representation theory*, Pure and Applied Mathematics, vol. 7, Marcel Dekker, Inc., New York, 1971. MR347959

[7] I. Martin Isaacs, *Character theory of finite groups*, Pure and Applied Mathematics, No. 69, Academic Press [Harcourt Brace Jovanovich, Publishers], New York-London, 1976. MR460423

[8] M. Isaacs, A. Lichtman, D. Passman, S. Sehgal, N. J. A. Sloane, and H. Zassenhaus (eds.), *Representation theory, group rings, and coding theory: Papers in honor of S. D. Berman (1922–1987)*, Contemporary Mathematics, vol. 93, American Mathematical Society, Providence, RI, 1989, DOI 10.1090/conm/093. MR1003335

[9] Eric Jespers and Ángel del Río, *Group ring groups. Vol. 1. Orders and generic constructions of units*, De Gruyter Graduate, De Gruyter, Berlin, 2016. MR3618092

[10] Irving Reiner, *The Schur index in the theory of group representations*, Michigan Math. J. **8** (1961), 39–47. MR122892

[11] Irving Reiner, *On the number of irreducible modular representations of a finite group*, Proc. Amer. Math. Soc. **15** (1964), 810–812, DOI 10.2307/2034603. MR168665

[12] I. Schur, *Aritlhmetische Untersuchungen über endliche Gruppen linearer Substitutionen*, S'Ber. Akad. Wiss., (1906), pp. 164 – 184.

[13] I. Schur, *Beiträge zur Theorie der Gruppen linearer homogener Substitutionen* (German), Trans. Amer. Math. Soc. **10** (1909), no. 2, 159–175, DOI 10.2307/1988680. MR1500832

[14] Ernst Witt, *Die algebraische Struktur des Gruppenringes einer endlichen Gruppe über einem Zahlkörper* (German), J. Reine Angew. Math. **190** (1952), 231–245, DOI 10.1515/crll.1952.190.231. MR53944

[15] Gregory Karpilovsky, *Group representations. Vol. 1. Part B*, North-Holland Mathematics Studies, vol. 175, North-Holland Publishing Co., Amsterdam, 1992. Introduction to group representations and characters, DOI 10.1016/S0304-0208(08)71756-0. MR1183468

Department of Mathematics, University of Haifa, 199 Aba Khoushy Ave. Mount Carmel, Haifa, 3498838, Israel

Email address: soham.spradhan@gmail.com

Contemporary Mathematics
Volume **811**, 2025
https://doi.org/10.1090/conm/811/16247

Nonlinear retracts and the geometry of Banach spaces

M. A. Sofi

ABSTRACT. In the nonlinear geometry of Banach spaces where the objects in the category are Banach spaces as in the linear case, the morphisms in the new setting are taken to comprise of certain nonlinear maps involving say, Lipschitz maps and, in some cases, uniformly continuous or coarse(resp. coarse Lipschitz) mappings arising from the underlying metric/uniform structure attached to the norm of the Banach space. The question as to what extent the Lipschitz/uniform structure may be used to capture the full linear structure of a Banach space has been one of the most fundamental problems pertaining to the nonlinear structure of Banach spaces since this line of investigation was undertaken by Lindenstrauss in the early seventies. This line of research which is subsumed under the so- called Ribe program broadly underscores the view that metric spaces encode a much deeper and hidden structure than is apparent. It's truly surprising how the (linear) structure of a Banach space gets captured to a considerable extent by its metric space structure. This point of view has led to deep insights into Banach space theory that has paved the way for these ideas being employed in seemingly unrelated disciplines including harmonic analysis, geometric group theory, Riemannian geometry, theoretical computer science, probability theory, data structures and many other domains of mathematics.

1. Background

The notion of retracts is a generic theme in topology and analysis. In the theory of Banach spaces, the idea of a Lipschitz/uniform retract may be seen as a nonlinear counterpart onnf a complemented subspace of a Banach space which has been widely studied in the linear theory of Banach spaces. In what follows we shall see how the structure of Lipschitz retracts captures the linear structure of Banach spaces, which in most cases mimics the theory of complemented subspaces of Banach spaces. An important example of this phenomenon, to be proved in Section 3 is provided by the Lipschitz analogue of the celebrated Lindenstrauss-Tzafiriri complemented subspaces problem characterising isomorphs of Hilbert spaces. For an advanced introduction to Banach space theory, we refer to [**1**] whereas for a comprehensive treatment of Lipschitz mappings we refer to the encyclopaedic treatise [**9**] and of their role in the nonlinear geometry of Banach spaces, the best source of information is the authoritative monograph [**3**]. Another useful source is [**43**]

2020 *Mathematics Subject Classification.* Primary 6B03, 46B20; Secondary 46C15, 28B20.
Key words and phrases. Banach space, Hilbert space, Lipschitz map, extension.

providing a thorough description of the theory underlying the space $\mathrm{Lip}_0(M)$ as an algebra and its canonical predual, the free space $\mathfrak{I}(X)$ of X.

The main aim of this note is to give an introduction and overview of recent developments surrounding the nonlinear geometry of Banach spaces including the present state of art of the theory. Barring a few instances where a brief outline of the proofs has been indicated, most of the results which appeal to a much deeper and broader understanding of the technical machinery underlying the theory have been presented without proofs.

As we shall soon learn (see Section 3 and 4), an interesting feature of Lipschitz geometry of Banach spaces is the crucial role that the separability of Banach spaces plays in our considerations involving the proofs of some deep theorems. On the other side, it turns out that almost all the highly nontrivial counterexamples to certain natural questions arising in the Lipschitz classification of Banach spaces have been constructed for the nonseparable case. The construction of such counterexamples in the separable case remains a pesky issue as the main obstacle in these constructions is the huge bulk of important positive results in the theory which are valid under the assumption of separability of Banach spaces in question.

Throughout, we shall use X, Y, Z to denote Banach spaces, unless otherwise stated whereas X^* shall denote the dual of X. The closed unit ball of X shall be denoted by $B_X : B_X = \{x \in X, \|x\| \leq 1\}$. The symbol $L(X,Y)$ shall be used for the Banach space of bounded linear operators from X into Y. The following notation shall also be used in the course of our discussion:

$$\ell_p^n = (\mathbb{R}^n, \, \|\|_p), \text{ where } \|(x_i)\|_p = \left(\sum_{i=1}^{n} |x_i|^p\right)^{\frac{1}{p}}, \; 1 \leq p < \infty$$

$$\ell_\infty^n = (\|(x_i)\|_\infty = \max_{1 \leq i \leq n} |x_i|)$$

$$\ell_p = \left\{ (x_i); \|(x_i)\|_p = \left(\sum_{i=1}^{\infty} |x_i|^p\right)^{\frac{1}{p}} < \infty \right\}, \; 1 \leq p < \infty.$$

ℓ_∞, Banach space of all bounded sequences equipped with the norm

$$\|(x_i)\|_\infty = \sup_{i \geq 1} |(x_i)|.$$

c_0, (Closed) subspace of ℓ_∞ consisting of null sequences (with the induced norm).

Definition 1.1. Given a set A, B $\subset A$ and a mapping $f : A \to B$, f is said to be a *retraction* on B if $f(x) = x, \forall\, x \in B$. In this case, B is called a *retract* of A.

Continuous, uniform and Lipschitz retracts may be defined analogously, depending upon whether f is a continuous, uniformly continuous or a Lipschitz mapping acting between topological/metric spaces defined below.

Definition 1.2. A mapping $f : M \to N$ acting between metric spaces M and N is said to be a Lipschitz map if the following holds:

$$d\big(f(x), f(y)\big) \leq k\, d(x,y), \; x, y \in M, \; k \geq 1.$$

For $k = 1$, we say that f is a *nonexpansive* (NE) mapping. We begin with some examples in the finite dimensional setting.

THEOREM 1.3 (L. E. J. Brouwer). *S^n is not a (continuous) retract of B^n. Equivalently, every continuous self-map on B^n admits a fixed point.*

EXAMPLE 1.4. Not every closed, convex subset of the 3-dimensional space is an (NR). Consider the plane $P : x+y+z=4$ in ℓ_∞^3 . Then the point $D(1,1,1)$, which is distant 1 from *each* of the points $A(2,2,0), B(2,0,2)$ and $C(0,2,2)$ on P shall get mapped to some point on P which would necessarily bear a distance strictly larger than 1 *at least from one* of the points A, B and C. This also shows that the Lipschitz constant of an extension of a Lipschitz map may not be preserved.

PROPOSITION 1.5. *Closed convex subsets of ℓ_2^n are (NR).*

COROLLARY 1.6. *For X with $dim X < \infty$, every closed convex subset of X is a Lipschitz retract.*

2. Infinite dimensional setting

The following example (attributed to S. Kakutani) shows that the infinite dimensional analogue of Brouwer's fixed point theorem does not hold.

EXAMPLE 2.1. The continuous map $f : B_{\ell_2} \to B_{\ell_2}$ given by:

$$f(x) = \left(\sqrt{1 - \|x\|^2}, x_1, x_2, \dots x_n, \dots \right)$$

is a retraction of B_{ℓ_2} onto the unit sphere.

The question regarding the validity of the above phenomenon in general Banach spaces is treated in the following theorem of W.A. Kirk ([**3**, Theorem 3.3]. See also [**14**, Chapter 4]) which shows that the above phenomenon holds in all infinite dimensional Banach spaces.

THEOREM 2.2. *Every infinite dimensional Banach space admits a fixed point free continuous map on its unit ball. Furthermore, equivalently, the unit sphere of an infinite dimensional Banach space is a (continuous) retract of the unit ball.*

Theorem 2.2 may be compared with the following.

PROPOSITION 2.3. *Every closed convex subset A of a Banach space is an absolute (continuous) retract, i.e., A is a retract of every metric space containing it as a closed subset.*

The above result is a consequence of Michael's selection theorem given below:

THEOREM 2.4 ([**3**, Chapter 1]). *Let $H(X)$ denote the collection of all nonempty, closed, convex subsets of a Banach space X equipped with the Hausdorff metric:*

$$d_H(A, B) = sup(\{d(x, A); x \in B\} \cup \{d(x, B); x \in A\}).$$

Further, let $\phi : M \to H(X)$ be a (set-valued) mapping which is lower semicontinuous *(l.s.c.): $\forall\, \mathfrak{U}$, an open subset of X, the set $\{x \in M; \phi(x) \bigcap \mathfrak{U} \neq \varnothing\}$ is an open subset of M. Then ϕ admits a continuous selection $\varphi : M \to X$; $\varphi(x) \in \phi(x)$, for every $x \in M$.*

In fact, it is possible to show that there exists a continuous *selector* $\psi : H(X) \to X$, i.e., $\psi(A) \in A$, *for each* $A \in H(X)$.

DEFINITION 2.5. A subspace M of a Banach space X is said to be (linearly) complemented (continuous linear retract) if there exists a continuous linear projection $P : X \to X$ such that $P(X) = M$.

(*) Equivalently, the quotient map $\varphi : X \to \frac{X}{M}$ admits a continuous linear lifting.

However, in the Lipschitz category only one sided implication holds. Indeed, the existence of Lipschitz right inverse for the quotient map $\varphi : X \to \frac{X}{M}$ yields M as a Lipschitz retract, with the retraction r being given by given by: $r(x) = x - \psi\varphi(x) + \psi(0)$. Here, ψ is a Lipschitz right inverse of φ. The assertion that the Lipschitz analogue of the reverse implication does not hold follows from the fact that c_0 is a 2-Lipschitz retract of ℓ_∞ (but not complemented) and that the quotient map $\varphi : \ell_\infty \to \frac{\ell_\infty}{c_0}$ does not admit a Lipschitz inverse (see Theorem 3.22).

FURTHER EXAMPLES 2.6.

(i) c_0 is a 2-Lipschitz retract of ℓ_∞ and that 2 is the best constant. In fact, it can be shown that any separable subspace of ℓ_∞ containing c_0 cannot be an r-Lipschitz retract of ℓ_∞ for $1 \le r < 2$.

(ii) The space $C_u(M)$ of all bounded uniformly continuous functions on a metric space M (with the sup-norm) is a 20-absolute Lipschitz retract. (See [**3**, Theorem 1.6]). For compact metric spaces K, the constant 20 has now been improved to 2 by Kalton. See [**24**, Theorem 3.5].

(iii) For a given Banach space X, the space $UC(X)$ of sequences in X such that $\sum_{n=1}^{\infty} x_n$ is an unconditionally convergent series, is Lipschitz retract of WUC(X), the space of sequences which are weakly unconditionally Cauchy: $\sum_{n=1}^{\infty} \|f(x_n)\| < \infty$ for each $f \in X^\star$ [**26**, Theorem 5.1].

(iv) $H(X)$ is an 8 -abolute Lipschitz retract [**3**, Theorem 1.7].

Michael's theorem motivates the following question:

QUESTION 2.7. Given a Banach space X , does there exist a Lipschitz (uniformly continuous) selector $\psi : H(X) \to X$ $(\psi(A) \in A)$?

ANSWER. Yes, if X is finite dimensional.

It was conjectured for a long time that the previous question had an affirmative answer for all Banach spaces. However, the conjecture was settled in the negative by Positselskii (1971) and independently by Przeslawskii and Yost (1989).

THEOREM 2.8 ([**35**,**36**]). *Given an infinite dimensional Banach space X, there does not exist a Lipschitz (uniformly continuous) selector $\psi : H(X) \to X$ $(\psi(A) \in A)$.*

REMARK 2.9. It can also be deduced that for an infinite dimensional Banach space X, there exists no Lipschitz/uniform selector: $K(X) \to X$ where $K(X)$ is the sub-collection of $H(X)$ consisting of compact sets.

The following questions arise naturally in this circle of ideas:

a. Does there exist a Lipschitz retraction: $H(X) \to X$?

b. Does there exist a uniform retraction: $H(X) \to X$?

c. Does there exist a Lipschitz retraction: $K(X) \to X$?

d. Does there exist a uniform retraction: $K(X) \to X$?

We have the following implications:

a. Holds if and only if X is an absolute Lipschitz retract (ALR).

b. Holds if and only if X is an absolute uniform retract (AUR).
c. Open
d. $\Rightarrow X^{**}$ is injective [**36**]. Converse open.

3. Lipschitz retracts in Banach space

As in topology where the theory of retracts is an important object of study, the linear counterpart of this study in the setting of Banach spaces has led to new insights into the theory and structure of Banach spaces. It turns out that the investigation of subsets of Banach spaces which arise as Lipschitz or uniform retracts of the ambient space has far reaching implications in the study of extension of Lipschitz mappings in Banach spaces. A comprehensive treatment of these ideas has been covered in [**5**]. A parallel development involving the extension of polynomial maps between Banach spaces has been an active area of research in recent years. See [**11**] for a smorsgasbord of some deep and interesting developments surrounding the extendability of polynomial maps and their implications on the geometry of Banach spaces.

We begin with the example of an absolute retract as encountered in Proposition 2.3: A closed convex subset A of a Banach space is an absolute retract. Indeed, let M be a metric space containing A as a subset and consider the set-valued map $\phi : M \to H(X)$ where

$$\phi(x) = \begin{cases} \{x\}, x \in A \\ A, x \notin A \end{cases}$$

Clearly, ϕ is l.s.c. map taking values in $H(X)$, so Theorem 2.4 yields a continuous selection of ϕ which is the desired (continuous) retraction.In Hilbert space, the situation is far more satisfactory.

PROPOSITION 3.1. *In a Hilbert space H, every closed convex subset is a non-expansive retract (NR) of H.*

(Given a closed, convex subset C of H, consider the 'metric projection':
$P_C(x) = \{y \in C; \|x-y\| = inf(\|x-z\|; z \in C)\}$).
The following theorem of S. Reich shows that the above property of metric projections characterises Hilbert spaces.

THEOREM 3.2 ([**37**]). *X is (isometrically) a Hilbert space if (and only if) each closed, convex subset of X is an (NR).*

In the light of Theorem 2.2 and Reich's theorem quoted above, the following problems arise naturally:

PROBLEM 3.3 (Lipschitz variant). Describe the class of Banach spaces X such that

(a) Unit ball of X is a Lipschitz retract of X.
(b) The above theorem (of Reich) remains true for Lipschitz retracts in place of non-expansive retracts.

Regarding (a), it turns out that the property holds in *all* Banach spaces:
Consider the (radial) mapping:

$$r(x) = \begin{cases} x, & \|x\| \le 1 \\ \frac{x}{\|x\|}, & \|x\| \ge 1. \end{cases}$$

It follows that r maps X onto B_X which is 2-Lipschitz:

$$\|r(x) - r(y)\| \leq 2\|x - y\|, \ x, y \in X.$$

However for X, a Hilbert space, we have: $\|r(x) - r(y)\| \leq \|x - y\|, \ x, y \in X$, i.e., r is nonexpansive. Conversely, we have:

THEOREM 3.4 ([**27**]). *X is a Hilbert space if and only if r is nonexpansive.*

REMARK 3.5. Regarding the existence of the metric projection in general Banach spaces, it turns out that $P_C(x)$ may not exist or if it exists, it may be multi-valued. However, in uniformly convex Banach spaces X, $P_C(x)$ exists and is unique. In the case of C being a closed convex and bounded subset of X, it can be shown that P_C is uniformly continuous on bounded neighbourhoods of C. Combining this fact with the previous considerations, we have

THEOREM 3.6 ([**23**]). *In a super-reflexive Banach space X, each closed, convex and bounded subset C of X is a uniform retract of X.*

Indeed, assuming that C is contained in B_X and that X is (in a suitable renorming of X) uniformly convex, composing P_C with the 2-Lipschitz retraction r as described above, gives the desired conclusion.

The question regarding the Lipschitz analogue of Kirk's theorem (Theorem 2.2) is answered in the following surprising theorem of Benyamini and Sternfeld:

THEOREM 3.7 ([**4**]). *In an infinite dimensional Banach space, the unit sphere is always a Lipschitz retract of the unit ball. Equivalently, there exists a Lipchitz mapping on the unit ball of each infinite dimensional Banach space lacking fixed points.*

The previous theorem suggests that in order to ensure the possible existence of a retraction with better regularity properties acting from X onto its convex subsets, it is natural to consider a subclass of the family of all closed convex subsets and wonder if the sets in this subclass may arise as retracts of mappings which may be chosen to be, say uniformly continuous or even Lipschitz continuous. Remarkable progress along these lines was achieved by Kalton (see [**23**]) and Hajek [**16**] which considerably strengthens Proposition 2.3.

THEOREM 3.8 ([**16, 23**]). *Every compact convex subset of a Banach space X is an (absolute) uniform retract of X.*

The following theorem of Cheng et al shows that certain convex subsets of a Banach space can even be chosen to be absolute uniform retracts.

THEOREM 3.9 ([**8**]). *Every convex super-weakly compact subset of a Banach space is an absolute uniform retract.*

Here, *super-weakly compact* subset is meant in the following sense:

There exists $n > 1$ such that for each set of vectors $(x_i)_{i=1}^n \subseteq X$, there exists $1 \leq k \leq n$ such that $d(\text{co}(x_1, \ldots, x_k), co(x_{k+1}, \ldots, x_n)) = 0$, where $co(A)$ denotes the convex hull of $A \subseteq X$.

At this point, it may be noted that the somewhat technical definition of a super-weakly compact subset of a Banach space as given above falls in place when we note that the idea was chiefly motivated by the desire to provide a localised version of super reflexivity. In fact, analogous to the well-known fact that a Banach space is reflexive if and only if its closed unit ball is weakly compact, we have the following characterisation of super reflexivity in Banach spaces.

THEOREM 3.10 ([**8**]). *A Banach space X is super-reflexive if and only if B_X is super weakly compact.*

In contrast to uniform retracts in Banach spaces as indicated above, the following theorem shows that the Lipschitz retractibility of compact convex subsets places rather severe restrictions on the underlying Banach space.

THEOREM 3.11 ([**16**]). *Every compact convex subset of X is a Lipschitz retract of X if and only if X is a Hilbert space.*

As already observed in the affirmative solution to Problem 3.3(a) above which tells us that the unit ball of each Banach space X is a 2-Lipschitz retract of X, Godefroy and Ozawa [**13**] were led to ask the following question:

QUESTION 3.12. Is it true that every separable Banach space X admits a compact convex generating subset M of X which is a Lipschitz retract of X? In symbols, we ask whether every separable Banach space has GCCR?

Though the question remains open in its full generality, in a series of recent papers [**15–17**], Hajek and his collaborators have shown that the following chain of implications holds:

$$\textit{Schauder basis} \Rightarrow \textit{FDD} \Rightarrow \pi - \textit{property} \Rightarrow \textit{BAP} \Rightarrow \textit{AP} \Rightarrow \textit{CAP}.$$

Here, the $\pi-$*property* of a Banach space X refers to the existence of a uniformly bounded net of finite rank projections on X converging strongly to the identity map on X.

In the presence of GCCR involving compact convex sets which are 'small' in a certain technical sense - denoted GCCR* - we have the following string of implications [**15, 16**]:

$$\textit{FDD} \Rightarrow \text{GCCR}^* \Rightarrow \pi - \textit{property} \Rightarrow \textit{BAP}.$$

Further, in an important work [**16**] the authors show that in the special case where the set involving GCCR in X is 'small', then X has the BAP. These considerations give an indication that the answer to Question 3.12 may be negative, at least in Banach spaces lacking the bounded approximation property (BAP). However, in a recent breakthrough on a positive solution to the problem, Medina shows that the Holder analogue of the problem admits an affirmative solution.

THEOREM 3.13 ([**33**]). *In every separable Banach space X, there exists a compact convex generating subset of X that is an $\alpha-$ Holder retract of X for each given $0 < \alpha < 1$.*

In the same paper [**33**], the authors show that the following assertion also holds, thus providing further evidence to the previous statement that the Godefroy-Ozawa problem may have a negative solution.

THEOREM 3.14. *In every separable Banach space X lacking BAP and for each $0 < \alpha < 1$, there exists a compact convex generating subset of X that is an $\alpha-$ Holder retract but not a Lipschitz retract of X.*

Note: At the time of resubmitting the paper after revision, an important breakthrough indicating a final push towards a negative solution to the Godefroy-Ozawa problem has been announced in a recent work by Medina [**34**]. In the said work, the author is able to construct a separable Banach space X and some $\lambda > 1$ such that no compact convex generating subset M of X admits a λ-Lipschitz retraction onto M.

Going back to Problem 3.3(b), the next result by this author provides a complete answer which is the isomorphic analogue of Reich's theorem involving Lipschitz retracts.

THEOREM 3.15 ([**40**]). *A Banach space X has the property that each closed and convex subset of X is a Lipschitz retract of X if and only if X is isomorphic to a Hilbert space.*

The above statement follows from the next theorem where for a given metric space M the symbol $\mathrm{Lip}_0(M)$ shall be used for the space of all *Lipschitz* functions $f: M \to \mathbb{R}$, vanishing at a distinguished point, say $\theta \in M$. It is easily checked that under pointwise operations, $\mathrm{Lip}_0(M)$ is a Banach space when equipped with the norm:

$$\|f\|_{\mathrm{Lip}} = \sup_{x \neq y} \frac{|f(x) - f(y)|}{d(x,y)}.$$

Also, a subspace Y of a Banach space X is said to be *locally complemented* if there exists $c > 0$ such that for each finite dimensional subspace M of X, there exists a continuous linear map $f : M \to Y$ with $\|f\| \leq c$ and $f(x) = x$ for all $x \in M \cap Y$. In this case, we shall say that Y is *c-locally complemented.*

REMARK 3.16. It can be shown that Y is locally complemented in X if and only if there exists a bounded linear 'extension' map $\psi : Y^* \to X^*$, $\psi(g)|_Y = g$ for each $g \in Y^*$. ([**21**], see also [**18**, Proposition 5.5]).

In the theorem below, $Lip_0(X)$ shall denote the of Lipschitz functions that vanish at the origin.

THEOREM 3.17. *Let X be a Banach space and Y a subspace of X such that there exists a bounded linear 'extension' map $L : Lip_0(Y) \to Lip_0(X)$, i.e., $L(g)|_Y = g$ for each $g \in \mathrm{Lip}_0(Y)$. Then L restricted to Y^* gives rise to a bounded linear 'extension' map $G : Y^* \to X^*$; $G(g)|_Y = g$ for each $g \in Y^*$. Equivalently, Y is locally complemented. In particular, a Lipschitz retract is always locally complemented.*

Proof (Brief sketch): The idea of the proof of the above theorem as given in [**40**] (see also [**39**]) is based on the existence of Banach limits on an abelian group which is used to produce a new bounded linear 'extension' map $G : \mathrm{Lip}_0(Y) \to \mathrm{Lip}_0(X)$ from L and which takes values in X^*. In other words, the map ψ restricted to Y^* yields a bounded linear extension map $L : Y^* \to X^*$. Recall that a Banach limit- hereinafter to be denoted by $\int .dx-$ is a continuous linear functional on $\ell_\infty(X)$ satisfying the following conditions:

(a) $\left\|\int ..dx\right\| = 1.$
(b) $\int 1.dx = 1.$
(c) $\int f\left(x + x'\right) dx = \int f(x)\, dx, \ \forall f \in \ell_\infty(X) \text{ and } \forall x' \in X.$

The map G is now defined for $f \in Lip_0(Y)$, $z \in X$ by the formula

$$G(f)(z) = \int \left\{ \int [(\mathrm{L\ f})(x+y+z) - (\mathrm{L\ f})(x+y)]\, \mathrm{dy} \right\} \mathrm{dx}.$$

The linearity and boundedness of L together with (a) yield that $G : \mathrm{Lip}_0(Y) \to \mathrm{Lip}_0(X)$ is a well-defined bounded linear map. Using translation invariance of the Banach limit as indicated in (c) and an appropriate manipulation of the Banach limit, it follows that for $f \in Lip_0(Y)$ and $z_1, z_2 \in X$ we have,

$$G(f)(z_1 + z_2) = G(f)(z_1) + G(f)(z_2).$$

and that $G(g)(z) = g(z)$ for $g \in Y^*$ and $z \in Y$. In other words, $G : Y^* \to X^*$ is an 'extension' operator. Together with Fakhoury's theorem [**10**], this completes the proof.

CONSEQUENCES. The above theorem may be looked upon as a Lipschitz analogue of the celebrated complemented subspaces theorem of Lindenstrauss and Tzafiriri characterising Hilbert spaces isomorphically in terms of complementedness of its closed subspaces.

REMARK 3.18. An inspection of the proof of Theorem 3.17 shows that the adjoint map $G^* : X^{**} \to Y^{**}$ is such that for $y \in Y \subseteq X \subseteq X^{**}$, $G^*(y) = y$ where y is identified with the evaluation map $\hat{y} \in X^{**}$. Indeed, for $y^* \in Y^*$, we have: $\langle y^*, G^*(\hat{y}) \rangle = \langle G(y^*), \hat{y} \rangle = \langle y, G(y^*) \rangle = \langle y, y^* \rangle = \langle y^*, \hat{y} \rangle$. Thus, G^* is the identity map on Y. This yields an alternative proof of the following well known theorem of Lindenstrauss [**32**].

COROLLARY 3.19. *If Y is a subspace of X which is a c -Lipschitz retract and is complemented in its bidual, then Y is c -complemented in X.*

Proof: Let $P : Y^{**} \to Y$ be a bounded linear projection and let G: $Lip_0(Y) \to \mathrm{Lip}_0(X)$ be an 'extension' operator induced by a Lipschitz retraction of X onto Y. By Theorem 3.17, G may be considered as an extension operator $G : Y^* \to X^*$. Then $Q = Po\ \ G^*|_X$ defines a projection $Q : X \to Y$: indeed, since $Q(x) \in Y$, the previous remark gives that

$$\begin{aligned} Q^2(x) &= (Po\ \ G^*|_X (Q(x)) \\ &= P(Q(x)) = P^2(\ G^*|_X)(x) \\ &= Po\ \ G^*|_X(x) = Q(x). \end{aligned}$$

EXAMPLE 3.20. An application of the previous corollary to $Y = P^2(X)$, the space of 2-homogeneous polynomials on the Banach space X yields, since Y is a dual space (as the dual of a symmetric tensor product), that Y is a complemented subspace of $\mathrm{Lip}_0(X)$ as soon as Y is a Lipschitz retract of $\mathrm{Lip}_0(X)$. However, in certain special cases that include Hilbert spaces, it was shown by Hajek and Russo [**20**] that $P^2(X)$ is not complemented in- and is therefore, not a Lipschitz retract of - $\mathrm{Lip}_0(X)$. This is in sharp contrast with an old result of Lindenstrauss asserting that the space $P^1(X)(= X^*)$ of 1-homogeneous polynomials is always 1-complemented in $\mathrm{Lip}_0(X)$ for every Banach space X.

REMARK 3.21. A Banach space X is said to be $\lambda-$ *injective* ($\lambda \geq 1$) if every X - valued continuous linear mapping f on any subspace of a given Banach space can be extended to a continuous linear map g on the whole space such that $\|g\| \leq$

$\lambda \|f\|$. It can be shown that a $\lambda-$ injective Banach space is a λ- absolute Lipschitz retract (λ-ALR) i.e., it is a λ-Lipschitz retract of every metric space containing it and that an (ALR) has the property that it's an L_1-predual, i.e., its bidual is $\lambda-$ injective. One way of seeing why the latter statement is true is by noting that the canonical (isometric) inclusion of X into $C(B_{X^*},\ w^*)$ - situating X as a Lipschitz retract of $C(B_{X^*},\ w^*)$ - produces a continuous linear extension operator $\text{Lip}_0(X) \to \text{Lip}_0(C(B_{X^*},\ w^*))$. Theorem 3.17 yields a continuous linear extension map $\psi:\ X^* \to C(B_{X^*},\ w^*)^*$ with $\|\psi\| \leq \lambda$. Taking conjugates produces a continuous linear extension map $\psi^*:\ C(B_{X^*},\ w^*)^{**} \to X^{**}$ which when restricted to X^{**} is the identity map on X^{**}. This yields X^{**} as a λ-Lipschitz retract of $C(B_{X^*},\ w^*)^{**}$. Finally, since the latter space is 1-injective (see [**1**, Proposition 4.3.8]) it follows that X^{**} is λ- injective. However, a Banach space X with an injective bidual may not be an ALR. A proof of this fact depends upon the following important theorem of Kalton.

THEOREM 3.22 ([**25**]). *There does not exist a Lipschitz right inverse for the quotient map* $\varphi : \ell_\infty \to \frac{\ell_\infty}{c_0}$.

Interestingly, the existence of a Lipschitz right inverse for quotient maps leads to the following useful characterisation of Hilber space among infinite dimensional Banach spaces.

THEOREM 3.23 ([**40**]). *Given a Banach space X such that for each closed subspace M of X and the quotient map $\varphi : X \to \frac{X}{M}$, there exists a Lipschitz map $\psi : \frac{X}{M} \to X$ with $\varphi\psi =$ identity map on $\frac{X}{M}$, then X is (isomorphically) a Hilbert space.*

This follows because under the conditions of the theorem, the map: $r(x) = x - \psi\varphi(x) + \psi(0)$ defines a Lipschitz retraction onto M, and so we are in the situation of Theorem 3.15 to arrive at the desired conclusion.

A local theoretic analogue of the above theorem also holds as stated below.

THEOREM 3.24. *Assume that there exists $c > 0$ such that for each finite dimensional subspace M of X the quotient map $\varphi : X \to \frac{X}{M}$, admits a Lipschitz right inverse Lipschitz map $\psi : \frac{X}{M} \to X$: $\varphi\psi =$ identity map on $\frac{X}{M}$ with $\|\|\psi\|\|_{\text{Lip}} \leq c$. Then X is (isomorphically) a Hilbert space.*

Indeed, as was noted in the argument following Definition 2.5, the stated condition gives that each finite dimensional subspace of X is a c -Lipschitz retract. Combining this observation with Corollary 3.19 yields that finite dimensional subspaces of X are c -complemented. An application of the local version of Lindenstrauss-Tzafiriri theorem (see [**1**, Theorem 12.4.4]) asserting that a Banach space X is (isomorphically) a Hilbert space if (and only if) its finite dimensional subspaces are uniformly complemented completes the argument.

In view of its relevance to the theme of an earlier work by the author [**39**] devoted to certain aspects of nonlinear analysis in Banach spaces, Theorem 3.17 (and some of its corollaries) also appear in the said volume with proofs of some of these results having since yielded to slight modifications/improvements as included in the present paper.

Our main theorem (Theorem 3.17) motivates the following question:

PROBLEM 3.25. Does a locally complemented subspace Y of a Banach space X imply that Y is a Lipschitz retract of X.

It turns out (see [**22**, Proposition 3.2]) that Problem 3.25 is equivalent to one of the most important open problems in Banach space theory asking if every Banach space is a Lipschitz retract of its bidual.

4. Extension maps involving subsets of a Banach space

It's natural to ask if an appropriate analogue of the Theorem 3.15 holds in case Y is chosen to be an arbitrary subset of the metric space (M, d). In other words, the question involves the existence of a bounded linear 'extension' operator $\psi : \operatorname{Lip}(A) \to \operatorname{Lip}(M)$ where A is now chosen to be an arbitrary subset of M. If such is the case, we say that the metric space M has the *'simultaneous Lipschitz extension property'* (SLEP). An answer to this question turns out to be highly nontrivial, especially in the case of X being a non-Banach metric space. We note that the existence of an 'extension' operator entails (i) extendability of Lipschitz mappings from arbitrary subsets to the entire space (ii) a suitable choice of the extended map which may be effected in a bounded linear manner. Whereas the validity of (i) is guaranteed by an on old result of McShane (see [**38**]), in the case of Banach spaces, the existence of a suitable choice of the extended map as required in (ii) forces the space to be finite dimensional as we shall see shortly. Further, the extension procedure involving (i) applied to the mappings consisting of contractions and taking values inside ℓ_2 results in X being a Hilbert space [**38**, Theorem 5.8]. This is a well-known converse to Kirszbraun's theorem [**28**].

Beyond the class of Banach spaces, this question has been a subject of intensive research spanning different areas of mathematics including, in particular, geometry and group theory. There are interesting examples of situations where such extensions always exist which include, for example, $\mathbb{R}^n$ (with respect to any norm), the Heisenberg group, doubling metric spaces and metric trees (See [**5**]). On the other hand, making use of some deep but well known facts from the local theory of Banach spaces, it is possible to show that the extension procedure for arbitrary subsets breaks down for all infinite dimensional Banach spaces, including of course, infinite dimensional Hilbert spaces where the situation was seen to be extremely pleasant from the viewpoint of the extension procedure involving linear functionals and Lipschitz mappings acting on linear subspaces. We provide below a sketch of the proof of this assertion as given in [**40**] which is a simplification of the proof as provided in [**6**].

THEOREM 4.1. *A Banach space X has (SLEP) if and only if it is finite dimensional.*

Proof (Sketch): We recall the following notation to be used in the proof:

Given a metric space M and $A \subset M$, denote by $Ext(A, M)$ the set of *extension operators*:

$$T : Lip(A) \to Lip(M), \ i.e, \ T(g)|_Y = g, \ \text{for each } g \in Lip_0(A),$$
$$\lambda\,(A, M) = inf\{\, \|T\|\,; T \in \operatorname{Ext}(A,\ M)\,\}$$

and put

$$\lambda\,(M) = sup\,\{\lambda\,(A, M)\,, A \subset M\}\,.$$

Claim: $\lambda\,(X) = \infty$ for each infinite dimensional Banach space X.

In particular, this would imply that each infinite dimensional Banach space admits a subset A without the 'simultaneous extension property'. We shall need the following tools to prove our statement.

(a) Dvoretzky's spherical sections theorem: Each infinite dimensional Banach space X contains for each n, an n-dimensional subspace $X_n$which is 2-isometric with ℓ_2^n.

(b) There exists $c > 0$ (independent of n) such that $\lambda\left(\ell_2^n\right) \geq cn^{\frac{1}{8}}$.

(c) If $r : B \to A$ is a c -Lipschitz retraction of B onto A, then $\lambda(A) \leq c\lambda(B)$.

Whereas (a) is already well known as a highly nontrivial fact from the local theory of Banach spaces (see [**3**, Theorem 12.10 1]), the proof of (b) hinges on another deep result involving an alternative description of $\lambda(M)$ given by the following formula (See [**6**]):

$$\lambda(M) = \sup\left\{\nu(M, Z);\ Z \in FD\right\}.$$

Here FD denotes the class of all finite dimensional Banach spaces and $\nu(M, Z)$ is given by

$$\nu(M, Z) = sup\left\{\nu(A, M, Z); A \subset M\right\}$$

where $\nu(A, M, Z)$ is defined to be the infimum of all constants $c > 0$ such that each $g \in Lip(A, Z)$ admits an extension to $f \in Lip(M, Z)$ such that

$$\|f\|_{\mathrm{Lip}(M,Z)} \leq c\,\|g\|_{\mathrm{Lip}(A,Z)}.$$

Finally, considering that the proof of (c) does not seem to be explicitly available in the existing literature, we include a proof of it as follows. Let $S \subset A$ and choose T$\subset B$ with $r(T) = S$. Now given a Lipschitz map $f : S \to \mathbb{R}$, we see that $\circ|T$ is a Lipschitz function on T. Fix $\epsilon > 0$ and choose an extension operator $E : Lip(T) \to Lip(B)$ with $\|E\| \leq \lambda(T, B) + \epsilon$. The assignment $E_1(f) = E(\circ_{|A})$ then defines a continuous linear map $E_1 :$ Lip$(S) \to$ Lip(A). From the inequality: $\left\|\circ_{|T}\right\|_{Lip(T)} \leq \|r\|_{Lip(B,A)} \|f\|_{Lip(S)}$,we conclude that

$$\|E_1\| \leq \|r\|_{Lip(B,A)} \|E\|.$$

This gives:

$$\lambda(S, A) \leq \|E_1\| \leq \|r\|_{\mathrm{Lip}_0(B,A)} \|E\| \leq \|r\|_{\mathrm{Lip}_0(B,A)} (\lambda(T, B) + \epsilon).$$

Taking supremum over S and T and letting $\epsilon \to 0$ yields the desired inequality.

Now to prove the desired assertion, for each n> 0, choose $X_n \subseteq X$ according to (a). Thus there exists a linear isomorphism $\varphi_n : \ell_2^n \to X_n$ such that $\|\varphi_n\| \left\|\varphi_n{}^{-1}\right\| < 2$. Fix $\epsilon > 0$ and let $A \subset \ell_2^n$. There exists $E \in Ext(\varphi_n(A),\ X_n)$ such that $\|E\| < \lambda(\varphi_n(A),\ X_n) + \epsilon$. Consider the mappings $\Psi_n : Lip(X_n) \to Lip(\ell_2^n)$ and $L_n : Lip(A) \to Lip(\varphi_n(A))$ given by

$$\Psi_n(f) = fo\varphi_n \text{ and } L_n(g) = go\varphi_n{}^{-1},\ f \in Lip(X_n),\ g \in Lip(A).$$

Then F given by $F = \Psi_n$oEoL_n defines a linear map $F : Lip(A) \to Lip(\ell_2^n)$ such that

$$\|F\| \leq \|\Psi_n\| \|E\| \|L_n\| \leq \|\varphi_n\| \|E\| \left\|\varphi_n{}^{-1}\right\| \leq 2\|E\|.$$

Using the above estimates, we get

$$\lambda(A, \ell_2^n) \leq \|F\| \leq 2\|E\| < 2\ \left(\lambda(\varphi_n(A),\ X_n) + \epsilon\right) \leq 2(\lambda(X_n) + \epsilon).$$

Letting $\epsilon \to 0$ gives $\lambda(A, \ell_2^n) \leq 2(\lambda(X_n)$.

Applying (c) to (the 2-Lipschitz retract) X_n of X, it follows that $\lambda(X_n) \leq 2\lambda(X)$. Combined with the above estimate, this yields

$$\lambda(X) \geq \frac{1}{2}\lambda(X_n) \geq \frac{1}{4}\lambda(A, \ell_2^n).$$

Taking supremum over $A \subset \ell_2^n$ and using (b), we get

$$\lambda(X) \geq \frac{1}{4}\lambda(\ell_2^n) \geq \frac{c}{4}n^{\frac{1}{8}}.$$

Since n is arbitrary, it follows that $\lambda(X) = \infty$. On the other hand, the assertion that $\lambda(X) < \infty$ for finite dimensional Banach spaces is a well known result due to Whitney (see [**41**, Chapter (6.2.3)].

REMARKS 4.2. In contrast to Theorems 3.15 and 4.1 involving the existence/ nonexistence of extension operators from spaces of Lipschitz functions on (closed) subspaces and arbitrary subsets of Banach spaces respectively, Kopecka [**29**] shows that in the latter case involving arbitrary subsets S of a Hilbert space H, there exist 'extension' operators $\psi: \; Lip_b(S, H) \to Lip_b(H, H)$ preserving the Lipschitz norm and acting between spaces of vector-valued *bounded* Lipschitz functions such that ψ is continuous but *not* linear. Here the spaces of bounded Lipschitz functions are equipped with the sup-norm.

(ii) In the light of Theorem 3.15 and the corollaries following from it, a natural question arises regarding the extent of the validity of these conclusions under the assumption that for a subspace Z of X, the extension operator $F: \; \mathrm{Lip}_0(Z) \to \mathrm{Lip}_0(X)$ is assumed to be Lipschitz rather than linear. To the best of our knowledge, the question does not seem to have been addressed in the literature so far.

Accordingly, we pose it as an open problem:

QUESTION 4.3. Let X be a Banach space. Does the existence of a Lipschitz extension operator $F: \; \mathrm{Lip}_0(Z) \to \mathrm{Lip}_0(X)$ for each subspace Z of X yield that X is a Hilbert space?

We note that for subspaces Z of X for which $\mathrm{Lip}_0(Z)$ is separable, the answer is in the affirmative. In particular, this holds if X is assumed to be separable. The conclusion of the theorem follows from an important theorem of Godefroy and Kalton [**12**] asserting that given Banach spaces X and Y with X separable and a bounded linear map $T: Y \to X$, then the existence of a Lipschitz right inverse $S: X \to Y$ of T yields the existence of a bounded linear right inverse $L: X \to Y$ of T: $ToL = \mathrm{Id}_X$.

For a subspace Z of X, consider the bounded linear map $\Psi: \; \mathrm{Lip}_0(X) \to \mathrm{Lip}_0(Z)$ given by: $\Psi(f) = f_{|Z}$. Now the presumed existence of $F: \; \mathrm{Lip}_0(Z) \to \mathrm{Lip}_0(X)$ as a Lipschitz extension operator gives that F is a Lipschitz right inverse of Ψ. By the Godefroy-Kalton theorem just stated, we get a bounded linear right inverse G of Ψ, *i.e.*, $G: \mathrm{Lip}_0(Z) \to \mathrm{Lip}_0(X)$ is a bounded linear map such that $\Psi \odot G = \mathrm{Id}_{\mathrm{Lip}_0(Z)}$. In other words, G is a bounded linear extension operator and so we are in the situation of Theorem 3.17 to arrive at the desired conclusion.

We conclude with the following problems which remain open.

PROBLEM 4.4. Describe sufficient conditions on a closed, convex subset C of a (separable) Banach space X such that C is a Lipschitz retract of X.

We observe that by combining the fact that $\lambda(X) = \infty$ for each infinite dimensional Banach space X (Theorem 4.1) with the radial mapping $r : X \to B_X$ being a 2-Lipschitz retraction, it follows that $\lambda(B_X) = 2\lambda(X) = \infty$.

As noted in the discussion of proof of Theorem 4.1(b), an important ingredient of the proof was the following equality of Brudnyi:

$$\lambda(M) = \sup\{\nu(M, Z);\ Z \in FD\} \ (*)$$

In a recent work, Basso [**2**] shows that FD in the above equality can be replaced by the class of all dual Banach spaces. This is achieved by showing the following inequality:

$$\lambda(M, N) \le v(M, N, Z) \text{ for every dual Banach space} \ldots (\star\star)$$

This motivates the following question.

PROBLEM 4.5. Is it true that the above equality $(\star)$ holds for the class of all Banach spaces? In particular, is it possible to replace FD by the class of all Banach spaces which are complemented in their bidual?

In the same work [**2**], Basso shows how $\lambda(M)$ is related to $ae(M)$, the absolute Lipschitz constant of the metric space M:

$$\mathrm{ae}(M) = \nu(M, Z) = sup\{\nu(M, N, Z); M \subset N\}.$$

Recalling the notation

$$\lambda(M, N) = inf\{\|T\|; T \in \mathrm{Ext}(M,\ N)\}$$

we see that (**) yields:

THEOREM 4.6 ([**2**]). *For a metric space M , we have*

$$\mathrm{ae}(M) \ge \sup_{N \supset M} \lambda(M, N),$$

where the equality holds for finite metric spaces M and finite metric spaces N (containing M).

COROLLARY 4.7 ([**2**]). (a) $\mathrm{ae}(3) = \dfrac{3}{4}$.

(b) $\dfrac{5 + 4\sqrt{2}}{7} \le ae(4) \le \dfrac{3 + 6\sqrt{2}}{7}$.

Here the symbol $ae(n)$ denotes $ae(M)$ where M is an n-point metric space.

We conclude with a brief mention of a recent breakthrough on a nonlinear version of Argyros's discovery of (hereditarily) indecom-posable Banach spaces: there exist Banach spaces X which do not admit (proper) infinite dimensional complemented subspaces. In other words, X does not admit (linear) continuous projections onto closed infinite dimensional subspaces of X.

The following theorem of Hajek and Quilis provides a nonlinear analogue of the previous statement.

THEOREM 4.8 ([**19**]). *There exists a complete metric space M with $|M| = c$ such that no non-singleton closed separable subspace of M is a Lipschitz retract of M.*

Some more examples of retracts in analysis

In the linear theory of Banach spaces, it's well known, and easy to prove, that $K(\ell_2)$, the space of compact operators is not complemented in $L(\ell_2)$. It's natural to ask if $K(\ell_2)$ may be a Lipschitz retract of $L(\ell_2)$. In a recent paper, R. Tanaka [**42**] shows that, more generally, every 2-sided ideal in a Von Neumann algebra is a Lipschitz retract in the ambient space. In the more general case of $p \geq 1$ it's possible to show that $K(\ell_p)$ is not complemented in $L(\ell_p)$ whereas the two classes of operators $K(\ell_p, \ell_q)$ and $L(\ell_p, \ell_q)$ coincide (by Pitt's theorem) if $\infty > p > q \geq 1$. Surprisingly, regardless of the range of p and q, the Lipschitz analogue of this phenomenon always holds, as the following theorem shows.

THEOREM 4.9 ([**7**]). *For $1 \leq p, q < \infty$, the space $K(\ell_p, \ell_q)$ of compact operators is a 8-Lipschitz retract of $L(\ell_p, \ell_q)$.*

The previous result may be seen as a far reaching strengthening of Example 2.6(iii).

Symmetric products (Kovalev theory)

By definition, an injective Banach space is complemented in each Banach space containing it as a subspace. The metric version of this property is provided by the notion of hyperconvexity which is characterised by the property that a hyperconvex metric space is an absolute Lipschitz retract (ALR) - a Lipschitz retract of each larger metric space. The Hahn Banach theorem yields that $\mathbb{R}$ is injective as a Banach space. As a metric space, $\mathbb{R}$ is also hyperconvex (ALR).

In an important work, Kovalev [**30**] shows that the symmetric product $\mathbb{R}^{(n)}$ of $\mathbb{R}$ is also an ALR for all $n \geq 1$. Here, $\mathbb{R}^{(n)}$ is defined to be the set of nonempty subsets of $\mathbb{R}$ with cardinality at most n equipped with the Hausdorff metric . The same also holds for higher dimensional Euclidean space $\mathbb{R}^d$. This follows from his main theorem in a subsequent work [**31**], asserting that for all $1 \leq k \leq n$ there exists a Lipschitz retraction from $H^{(n)}$ onto $H^{(k)}$ where H is a Hilbert space, finite or infinite dimensional. Here the proof is based on the existence of gradient flow trajectories in a finite dimensional subspace which is guaranteed by classical theory of ordinary differential equations. Another delightful consequence of the latter assertion is that $H^{(n)}$ embeds in a biLipschitz manner into an Euclidean space. In fact, a metric space X admits a bi-Lipschitz embedding into an Euclidean space if and only if its symmetric product $X^{(n)}$ does. From this, it can be deduced that $\mathbb{R}^{(n)}$ is an ALR. However, the author also remarks that it remains unknown if the property of being an ALR passes from X to its symmetric product $X^{(n)}$. The question whether for a Banach space X, $X^{(k)}$ is a Lipschitz retract of $X^{(n)} (1 \leq k \leq n)$ remains wide open.

Acknowledgments

The author would like to thank the referee for a thorough reading of the paper and for making suggestions that have helped improve the presentation of the article.

References

[1] Fernando Albiac and Nigel J. Kalton, *Topics in Banach space theory*, 2nd ed., Graduate Texts in Mathematics, vol. 233, Springer, [Cham], 2016. With a foreword by Gilles Godefory, DOI 10.1007/978-3-319-31557-7. MR3526021

[2] Giuliano Basso, *Absolute Lipschitz extendability and linear projection constants*, Studia Math. **264** (2022), no. 3, 335–359, DOI 10.4064/sm210708-21-9. MR4416024

[3] Yoav Benyamini and Joram Lindenstrauss, *Geometric nonlinear functional analysis. Vol. 1*, American Mathematical Society Colloquium Publications, vol. 48, American Mathematical Society, Providence, RI, 2000, DOI 10.1090/coll/048. MR1727673

[4] Y. Benyamini and Y. Sternfeld, *Spheres in infinite-dimensional normed spaces are Lipschitz contractible*, Proc. Amer. Math. Soc. **88** (1983), no. 3, 439–445, DOI 10.2307/2044990. MR699410

[5] Alexander Brudnyi and Yuri Brudnyi, *Metric spaces with linear extensions preserving Lipschitz condition*, Amer. J. Math. **129** (2007), no. 1, 217–314, DOI 10.1353/ajm.2007.0000. MR2288741

[6] A. Brudnyi and Y. Brudnyi, *Linear and nonlinear extensions of Lipschitz functions from subsets of metric spaces*, Algebra i Analiz **19** (2007), no. 3, 106–118, DOI 10.1090/S1061-0022-08-01003-0; English transl., St. Petersburg Math. J. **19** (2008), no. 3, 397–406. MR2340707

[7] Lixin Cheng, Wuyi He, and Sijie Luo, *$K(\ell_p, \ell_q)$ is a Lipschitz retract of $B(\ell_p, \ell_q)$*, Proc. Amer. Math. Soc. **150** (2022), no. 9, 3915–3925, DOI 10.1090/proc/15970. MR4446240

[8] Li Xin Cheng, Qing Jin Cheng, and Jian Jian Wang, *On absolute uniform retracts, uniform approximation property and super weakly compact sets of Banach spaces*, Acta Math. Sin. (Engl. Ser.) **37** (2021), no. 5, 731–739, DOI 10.1007/s10114-021-0273-1. MR4261244

[9] Ştefan Cobzaş, Radu Miculescu, and Adriana Nicolae, *Lipschitz functions*, Lecture Notes in Mathematics, vol. 2241, Springer, Cham, 2019, DOI 10.1007/978-3-030-16489-8. MR3931701

[10] Hicham Fakhoury, *Sélections linéaires associées au théorème de Hahn-Banach* (French), J. Functional Analysis **11** (1972), 436–452, DOI 10.1016/0022-1236(72)90065-1. MR348457

[11] Maite Fernández-Unzueta and Ángeles Prieto, *Extension of polynomials defined on subspaces*, Math. Proc. Cambridge Philos. Soc. **148** (2010), no. 3, 505–518, DOI 10.1017/S0305004110000022. MR2609305

[12] G. Godefroy and N. J. Kalton, *Lipschitz-free Banach spaces*, Studia Math. **159** (2003), no. 1, 121–141, DOI 10.4064/sm159-1-6. Dedicated to Professor Aleksander Pełczyński on the occasion of his 70th birthday. MR2030906

[13] Gilles Godefroy and Narutaka Ozawa, *Free Banach spaces and the approximation properties*, Proc. Amer. Math. Soc. **142** (2014), no. 5, 1681–1687, DOI 10.1090/S0002-9939-2014-11933-2. MR3168474

[14] Kazimierz Goebel and W. A. Kirk, *Topics in metric fixed point theory*, Cambridge Studies in Advanced Mathematics, vol. 28, Cambridge University Press, Cambridge, 1990, DOI 10.1017/CBO9780511526152. MR1074005

[15] P. Hajek and R. Medina, *Compact retractions and the π- property of Banach spaces*, arXiv:2106.00331v1, 2021.

[16] Petr Hájek and Rubén Medina, *Compact retractions and Schauder decompositions in Banach spaces*, Trans. Amer. Math. Soc. **376** (2023), no. 2, 1343–1372, DOI 10.1090/tran/8807. MR4531677

[17] Petr Hájek and Rubén Medina, *Retractions and the bounded approximation property in Banach spaces*, Mediterr. J. Math. **20** (2023), no. 2, Paper No. 75, 13, DOI 10.1007/s00009-023-02270-z. MR4537521

[18] Petr Hájek and Andrés Quilis, *Lipschitz retractions and complementation properties of Banach spaces*, J. Funct. Anal. **283** (2022), no. 1, Paper No. 109494, 35, DOI 10.1016/j.jfa.2022.109494. MR4404554

[19] Petr Hájek and Andrés Quilis, *A complete metric space without non-trivial separable Lipschitz retracts*, J. Funct. Anal. **285** (2023), no. 2, Paper No. 109941, 41, DOI 10.1016/j.jfa.2023.109941. MR4575682

[20] Petr Hájek and Tommaso Russo, *Projecting Lipschitz functions onto spaces of polynomials*, Mediterr. J. Math. **19** (2022), no. 4, Paper No. 190, 22, DOI 10.1007/s00009-022-02075-6. MR4452916

[21] N. J. Kalton, *Locally complemented subspaces and $\mathcal{L}_p$-spaces for $0 < p < 1$*, Math. Nachr. **115** (1984), 71–97, DOI 10.1002/mana.19841150107. MR755269

[22] Nigel J. Kalton, *The nonlinear geometry of Banach spaces*, Rev. Mat. Complut. **21** (2008), no. 1, 7–60, DOI 10.5209/rev_REMA.2008.v21.n1.16426. MR2408035

[23] N. J. Kalton, *Spaces of Lipschitz and Hölder functions and their applications*, Collect. Math. **55** (2004), no. 2, 171–217. MR2068975

[24] N. J. Kalton, *Extending Lipschitz maps into $\mathcal{C}(K)$-spaces*, Israel J. Math. **162** (2007), 275–315, DOI 10.1007/s11856-007-0099-2. MR2365864

[25] N. J. Kalton, *Lipschitz and uniform embeddings into ℓ_∞*, Fund. Math. **212** (2011), no. 1, 53–69, DOI 10.4064/fm212-1-4. MR2771588

[26] N. J. Kalton, *The uniform structure of Banach spaces*, Math. Ann. **354** (2012), no. 4, 1247–1288, DOI 10.1007/s00208-011-0743-3. MR2992997

[27] L. A. Karlovitz, *The construction and application of contractive retractions in 2-dimensional normed linear spaces*, Indiana Univ. Math. J. **22** (1972/73), 473–481, DOI 10.1512/iumj.1972.22.22041. MR308755

[28] M. D. Kirszbraun, Uber die Zusammenziehenden und Lipschitzchen Transformationen, Fund. Math. 22(1934), 77-108.

[29] Eva Kopecká, *Bootstrapping Kirszbraun's extension theorem*, Fund. Math. **217** (2012), no. 1, 13–19, DOI 10.4064/fm217-1-2. MR2914919

[30] Leonid V. Kovalev, *Symmetric products of the line: embeddings and retractions*, Proc. Amer. Math. Soc. **143** (2015), no. 2, 801–809, DOI 10.1090/S0002-9939-2014-12280-5. MR3283666

[31] Leonid V. Kovalev, *Lipschitz retraction of finite subsets of Hilbert spaces*, Bull. Aust. Math. Soc. **93** (2016), no. 1, 146–151, DOI 10.1017/S0004972715000672. MR3436023

[32] Joram Lindenstrauss, *On nonlinear projections in Banach spaces*, Michigan Math. J. **11** (1964), 263–287. MR167821

[33] Rubén Medina, *Compact Hölder retractions and nearest point maps*, Adv. Math. **428** (2023), Paper No. 109140, 13, DOI 10.1016/j.aim.2023.109140. MR4600059

[34] Rubén Medina, *On the constant of Lipschitz approximability*, Trans. Amer. Math. Soc. **377** (2024), no. 4, 2925–2945, DOI 10.1090/tran/9110. MR4744823

[35] E. Positselski, *Lipschitz maps in the space of convex bodies*, Optimisation, 4(21), (1971), 83-90.

[36] Krzysztof Przesławski and David Yost, *Continuity properties of selectors and Michael's theorem*, Michigan Math. J. **36** (1989), no. 1, 113–134, DOI 10.1307/mmj/1029003885. MR989940

[37] Simeon Reich, *Extension problems for accretive sets in Banach spaces*, J. Functional Analysis **26** (1977), no. 4, 378–395, DOI 10.1016/0022-1236(77)90022-2. MR477893

[38] M. A. Sofi, *Some problems in functional analysis inspired by Hahn-Banach type theorems*, Ann. Funct. Anal. **5** (2014), no. 2, 1–29. MR3192006

[39] M. A. Sofi, *Nonlinear aspects of certain linear phenomena in Banach spaces*, Applied analysis in biological and physical sciences, Springer Proc. Math. Stat., vol. 186, Springer, New Delhi, 2016, pp. 407–425, DOI 10.1007/978-81-322-3640-5_26. MR3613752

[40] M. A. Sofi, Extension operators and nonlinear structure of Banach spaces, J. Nonlinear Func. Anal, to appear.

[41] Elias M. Stein, *Singular integrals and differentiability properties of functions*, Princeton Mathematical Series, No. 30, Princeton University Press, Princeton, NJ, 1970. MR290095

[42] Ryotaro Tanaka, *Constructing a Lipschitz retraction from $\mathcal{B}(\mathcal{H})$ onto $\mathcal{K}(\mathcal{H})$*, Proc. Amer. Math. Soc. **147** (2019), no. 9, 3919–3926, DOI 10.1090/proc/14536. MR3993784

[43] Nik Weaver, *Lipschitz algebras*, World Scientific Publishing Co. Pte. Ltd., Hackensack, NJ, 2018. Second edition of [MR1832645]. MR3792558

Department of Mathematics, Kashmir University, Srinagar-190006, India
Email address: aminsofi@gmail.com

Contemporary Mathematics
Volume **811**, 2025
https://doi.org/10.1090/conm/811/16248

Computations for invariant bilinear forms under an invertible linear transformation: an expository article

Jagmohan Tanti

Abstract. Let $\mathbb{F}$ be a field of characteristic different from 2, V a vector space of finite dimension over $\mathbb{F}$ and $T : V \to V$ an invertible linear map. Then the set of T-invariant bilinear forms on V is a subspace of space of bilinear forms. In this article we give an algorithm to compute the dimension of this invariant subspace and also discuss about the existence of a non-degenerate invariant form.

1. Introduction

Let $\mathbb{F}$ be a field of characteristic $\neq 2$, for $n \in \mathbb{N}$, V a vector space of dimension n over $\mathbb{F}$ and T an invertible linear map from V onto V. A bilinear form B on V is called T-invariant if $B(Tu, Tv) = B(u, v)$ for all $u, v \in V$. If we fix an ordered basis say $\underline{e} = \{e_1, \cdots, e_n\}$ for V, then from linear algebra we know that with respect to this basis T and B are expressible as square matrices $[T]_{\underline{e}}$ and $[B]_{\underline{e}}$ respectively of order n with $[T]_{\underline{e}}^t[B]_{\underline{e}}[T]_{\underline{e}} = [B]_{\underline{e}}$. It can be checked easily that the set of all T-invariant forms is a vector space. For more information about bilinear forms one can refer to [**7**]. In the study of physical sciences it is a fundamental question to find a non-degenerate T-invariant bilinear form and also to compute the basis and dimension of the space of invariant bilinear forms.

The problem in concern has been theoretically dealt in many different ways in the literature. Gongopadhyay and Kulkarni [**4**] discussed about the existence of T -invariant non-degenerate symmetric (resp. skew-symmetric) bilinear forms. Kulkarni and Tanti [**8**] studied about the space of T-invariant (resp. infinitesimal invariant) forms and also discussed about the existence of a non-degenerate invariant form over a field of characteristic $\neq 2$. Frobenius [**3**] proved that every endomorphism of a finite dimensional vector space V is self-adjoint for at least one non-degenerate symmetric bilinear form on V. Stenzel [**10**] studied about when an endomorphism could be skew- self adjoint for a non-degenerate quadratic form, or self-adjoint or skew-self adjoint for a symplectic form on complex vector spaces. Pazzis [**9**] studied the case of automorphisms on a finite dimensional vector space which are orthogonal (resp. symplectic) for at least one non-degenerate quadratic form (resp. symplectic

2020 *Mathematics Subject Classification.* Primary 15A03, 15A63; Secondary 15A04, 15A20 13E15.

Key words and phrases. Vector space, linear transformation, bilinear forms, finitely generated modules, Jordan forms.

form) over an arbitrary field of characteristics 2. Gongopadhyay, Mazumder and Sardar [**6**] investigated for an invertible linear map $T : V \to V$, when does the vector space V over $\mathbb{F}$ admit a T-invariant non-degenerate c-hermitian form.

In the present article we discuss a methodology and propose an algorithm to compute the space of T-invariant bilinear forms in a systematic way. Also we discuss the criteria for existence of a non-degenerate form. Basically its a compilation of author's two earlier articles [**1, 8**] with a computational point of view. Explicit computations have been exhibited for dimension 3 as an example.

2. Preliminaries

If T is a linear transformation from V to V with $\chi_T(x)$ the characteristic polynomial for T, then naturally T induces an $\mathbb{F}[x]$-module structure on V. Thus V is a finitely generated module over $\mathbb{F}[x]$ (a principal ideal domain). By Structure theorem for finite dimensional vector space [**1**], we have

$$V = \oplus_{i=1}^{l} C_i \tag{1}$$

where for each $1 \leq i \leq l$, C_i is a cyclic $\mathbb{F}[x]$-submodule of V and $C_i \simeq \mathbb{F}[x]/ < f_i(x) >$ for some $f_i(x)$, a divisor of $\chi_T(x)$ with $C_i = \ker f_i(T)$ and $f_{i+1}|f_i \ \forall\, 1 \leq i \leq l-1$. Thus we have $\chi_T(x) = f_1(x) \cdot f_2(x) \cdots f_l(x)$ with decreasing order of degrees of the polynomials. Here $f_1(x)$ is the minimal polynomial for T. Now if $\chi_T(x) = p_1(x)^{r_1} \cdots p_k(x)^{r_k}$ is the factorization into irreducible factors, then for every $1 \leq i \leq l$, we have $f_i(x) = p_1(x)^{r_{i1}} \cdots p_k(x)^{r_{ik}}$ for some non-negative integers r_{i_j} with $r_{(i+1)_j} \leq r_{i_j}$ $\forall 1 \leq j \leq k$. Also by Chinese remainder theorem we have for each $1 \leq i \leq l$, $C_i \simeq \mathbb{F}[x]/ < p_1(x)^{r_{i1}} > \oplus \cdots \oplus \mathbb{F}[x]/ < p_k(x)^{r_{ik}} >$. Suppose $C_{ij} = \mathbb{F}[x]/ < p_j(x)^{r_{ij}} > (= \ker p_j(T)^{r_{ij}})$, then we get $C_i = \oplus_{j=1}^{k} C_{ij}$ $\forall 1 \leq i \leq l$. Hence $V \simeq \oplus_{i=1}^{l} \left(\oplus_{j=1}^{k} C_{ij}\right)$ with $d_i(say) = \dim C_i$. Let $T_i = T|_{C_i}$ and $T_{ij} = T_i|_{C_{ij}}$ then T_i is cyclic and T_{ij} is indecomposable.

In view of invariance under field extension [**8**] (Proposition 2.5) without loss of generality we may assume that $\mathbb{F}$ is algebraically closed. Then for every $1 \leq j \leq k$ we have $p_j(x) = x - \lambda_j$ for some $\lambda_j \in \mathbb{F}$. Then $d_i = r_{i_1} + \cdots + r_{i_k}$ and $\dim C_{ij} = r_{i_j}$. If $J(T)$ (resp. $J(T_i)$) denotes the Jordan canonical form for T (resp. T_i), then we have $J(T) = diag(J(T_1) \cdots J(T_l))_{n \times n}$ and for each $1 \leq i \leq l$, $J(T_i) = diag(J(T_{i1}), \cdots, J(T_{ik}))_{d_i \times d_i}$, with

$$J_{ij}(say) = J(T_{ij}) = \begin{pmatrix} \lambda_j & 1 & & \\ & \ddots & 1 & \\ & & \lambda_j \end{pmatrix}_{r_{i_j} \times r_{i_j}}, \quad \forall 1 \leq j \leq k$$

are the indecomposable Jordan blocks.

For T invertible, we have to compute the T-invariant bilinear forms on V. With respect to the same basis $\underline{e}$, which has been utilised to get the matrix representation $J(T)$ of T, a bilinear form B on V can be expressed as a $n \times n$ matrix $X = (x_{ij})$. Thus B is T-invariant if $J(T)^t X J(T) = X$, so we have to characterise all such matrices.

3. Outline of the method

Consider any two Jordan blocks J_{ij} & J_{uv}, $1 \leq i, u \leq l$, $1 \leq j, v \leq k$ of $J(T)$, then we have following cases.

Case I: If $\lambda_j \cdot \lambda_v \neq 1$, then the corresponding invariant blocks $X_{r_{i_j} \times r_{u_v}}$ and $X_{r_{u_v} \times r_{i_j}}$ of sizes $r_{i_j} \times r_{u_v}$ and $r_{u_v} \times r_{i_j}$ respectively in X are 0 by Lemma 4.1 in [**8**].

Case II: If for some $1 \leq j \leq k$, $\lambda_j \neq \pm 1$, then the corresponding invariant blocks $X_{r_{i_j} \times r_{i_j}}, X_{r_{i_j} \times r_{u_j}}$ & $X_{r_{u_j} \times r_{i_j}}$ $\forall$ $1 \leq i, u \leq l$ in X are 0 by Lemmas 3.4 and 4.7 in [**8**].

Case III: If $\pm 1 \neq \lambda_j$, λ_v & $\lambda_j \cdot \lambda_v = 1$, then each of the corresponding invariant blocks $X_{r_{i_j} \times r_{u_v}}$ & $X_{r_{u_v} \times r_{i_j}}$ $\forall 1 \leq i, u \leq l$ in X is of rank=min$\{r_{i_j}, r_{u_v}\}$ with possible non-degeneracy if $r_{i_j} = r_{u_v}$ by Lemma 4.2 in [**8**].

Subcase (i): For $r_{i_j} = r_{u_v}$, if X is symmetric, we have $X_{r_{i_j} \times r_{u_v}} = X_{r_{u_v} \times r_{i_j}}$ and thus they have a contribution of min$\{r_{i_j}, r_{u_v}\}$ to the total rank.

Subcase (ii): For $r_{i_j} = r_{u_v}$, if X is skew symmetric, we have $X_{r_{i_j} \times r_{u_v}} = -X_{r_{u_v} \times r_{i_j}}$ and thus they have a contribution of min$\{r_{i_j}, r_{u_v}\}$ to the total rank.

Case IV: If for some $1 \leq j \leq k$, $\lambda_j = \pm 1$ & $1 \leq i, u \leq l$, then by Lemma 4.4 in [**8**] each of the corresponding invariant blocks $X_{r_{i_j} \times r_{u_j}}$ & $X_{r_{u_j} \times r_{i_j}}$ in X is of rank min$\{r_{i_j}, r_{u_j}\}$ with admitting non-degeneracy when $r_{i_j} = r_{u_j}$ and the diagonal blocks $X_{r_{i_j} \times r_{i_j}}$ & $X_{r_{u_j} \times r_{u_j}}$ are of ranks r_{i_j} & r_{u_j} respectively with admitting non-degeneracy. Thus in this case for every pair (i, u), $1 \leq i, u \leq l$ their contribution is $r_{i_j} + r_{u_j} + 2 \cdot \min\{r_{i_j}, r_{u_j}\}$ to the total rank of X.

4. Algorithm

Given that $\mathbb{F}$ is a field of characteristic different from 2, for n a natural number, V a vector space over $\mathbb{F}$ of dimension n and $T : V \to V$ an invertible linear map. Thus in view of the above outline we may incorporate following steps to compute a T-invariant bilinear form X on V.

Step 1. Represent T by a matrix $[T]_{\underline{e}}$ with respect to an ordered basis $\underline{e}$ of V.
Step 2. Find the Jordan canonical form $J(T)$ for $[T]_{\underline{e}}$ over $\overline{\mathbb{F}}$: algebraic closure of $\mathbb{F}$ as described in the preliminary section.
Step 3. For any pair of indecomposable Jordan blocks J_{ij} & J_{uv}, $1 \leq i, u \leq l$, $1 \leq j, v \leq k$ of $J(T)$ if $\lambda_j \cdot \lambda_v \neq 1$ then place a zero matrix for each of the blocks $X_{r_{i_j} \times r_{u_v}}$ and $X_{r_{u_v} \times r_{i_j}}$ for X of respective order.
Step 4. . For any $1 \leq j \leq k$ if $\lambda_j \neq \pm 1$, then place a zero matrix for each of the blocks $X_{r_{i_j} \times r_{i_j}}, X_{r_{i_j} \times r_{u_j}}$ & $X_{r_{u_j} \times r_{i_j}}$ $\forall$ $1 \leq i, u \leq l$ in X of respective order.
Step 5. For any pair $1 \leq j, v \leq k$, if $\pm 1 \neq \lambda_j$, λ_v & $\lambda_j \cdot \lambda_v = 1$, then each of the blocks $X_{r_{i_j} \times r_{u_v}}$ & $X_{r_{u_v} \times r_{i_j}}$ $\forall 1 \leq i, u \leq l$ in X is of rank=min$\{r_{i_j}, r_{u_v}\}$. Compute their bases in terms of sets of matrices of respective orders.
Step 6. For any $1 \leq j \leq k$ if $\lambda_j = \pm 1$, then $\forall$ $1 \leq i, u \leq l$ each of the blocks $X_{r_{i_j} \times r_{u_j}}$ & $X_{r_{u_j} \times r_{i_j}}$ in X is of rank min$\{r_{i_j}, r_{u_j}\}$ and the diagonal blocks $X_{r_{i_j} \times r_{i_j}}$ & $X_{r_{u_j} \times r_{u_j}}$ are of ranks r_{i_j} & r_{u_j} respectively. Compute their bases in terms of sets of matrices of respective orders.
Step 7. Return the final symbolic form of X and get a basis for space of invariant bilinear forms.
Step 8. If $\det X \neq 0$ then X is a non-degenerate form.

5. Example

Let V be a vector space of dimension 3 over $\mathbb{C}$, $T : V \to V$ an invertible linear map with $\chi_T(x)$ as it's characteristic polynomial.

5.1. Case 1. If $\chi_T(x) = (x-\alpha)^3$ for some $\alpha \in \mathbb{C}$, then the corresponding Jordan forms for T are (i) $\begin{pmatrix} \alpha & 1 & 0 \\ 0 & \alpha & 1 \\ 0 & 0 & \alpha \end{pmatrix}$ with minimal polynomial $(x-\alpha)^3$, (ii) $\begin{pmatrix} \alpha & 1 & 0 \\ 0 & \alpha & 0 \\ 0 & 0 & \alpha \end{pmatrix}$ with minimal polynomial $(x-\alpha)^2$ and (iii) $\begin{pmatrix} \alpha & 0 & 0 \\ 0 & \alpha & 0 \\ 0 & 0 & \alpha \end{pmatrix}$ with minimal polynomial $(x-\alpha)$.

Subcase (i) For $J(T) = \begin{pmatrix} \alpha & 1 & 0 \\ 0 & \alpha & 1 \\ 0 & 0 & \alpha \end{pmatrix}$ with minimal polynomial $(x-\alpha)^3$ if $X \in M_3(\mathbb{C})$ satisfies $J(T)^t X J(T) = X$ then we have

$$X = \begin{cases} \begin{pmatrix} 0 & 0 & x_{13} \\ 0 & -x_{13} & x_{23} \\ x_{13} & \alpha^{-1}x_{13} - x_{23} & x_{33} \end{pmatrix} & \text{if } \alpha = \pm 1; \\ 0 & \text{otherwise.} \end{cases}$$

Here basis elements for $\alpha = \pm 1$ are given as $e_{13} - e_{22} + e_{31} - e_{32}$, $e_{23} + e_{32}$ and e_{33}, with e_{ij} the 3×3 matrix with all entries 0 except 1 at the ij^{th} place. Also for $x_{13} \in \mathbb{C} - 0$ the invariant form given by X is non-degenerate.

Subcase (ii) For $J(T) = \begin{pmatrix} \alpha & 1 & 0 \\ 0 & \alpha & 0 \\ 0 & 0 & \alpha \end{pmatrix}$ with minimal polynomial $(x-\alpha)^2$ then the invariant forms can be seen as

$$X = \begin{cases} \begin{pmatrix} 0 & x_{12} & 0 \\ -x_{12} & x_{22} & x_{23} \\ 0 & x_{32} & x_{33} \end{pmatrix} & \text{if } \alpha = \pm 1; \\ 0 & \text{otherwise.} \end{cases}$$

Here for $\alpha = \pm 1$ the basis elements are given as $e_{12} - e_{21}$, e_{22}, e_{23}, e_{32}, e_{33}. Also if x_{12} and x_{33} are non zero then the form X is non-degenerate.

Subcase (iii) For $J(T) = \begin{pmatrix} \alpha & 0 & 0 \\ 0 & \alpha & 0 \\ 0 & 0 & \alpha \end{pmatrix}$ with minimal polynomial $(x-\alpha)$, similarly we get the invariant forms given by

$$X = \begin{cases} (x_{ij}) & \text{if } \alpha = \pm 1; \\ 0 & \text{otherwise.} \end{cases}$$

Also for $\alpha = \pm 1$ the basis elements here are e_{ij} $\forall 1 \le i, j \le 3$. Plenty of non-degenerate invariant forms are there.

5.2. Case 2. If $\chi_T(x) = (x-\alpha)^2(x-\beta)$ for some $\alpha \neq \beta \in \mathbb{C}$, then the Jordan forms for T are (i) $\begin{pmatrix} \alpha & 1 & 0 \\ 0 & \alpha & 0 \\ 0 & 0 & \beta \end{pmatrix}$ with minimal polynomial $(x-\alpha)^2(x-\beta)$ and (ii) $\begin{pmatrix} \alpha & 0 & 0 \\ 0 & \alpha & 0 \\ 0 & 0 & \beta \end{pmatrix}$ with minimal polynomial $(x-\alpha)(x-\beta)$.

5.2.1. *Subcase (i).* For $J(T) = \begin{pmatrix} \alpha & 1 & 0 \\ 0 & \alpha & 0 \\ 0 & 0 & \beta \end{pmatrix}$ with minimal polynomial $(x-\alpha)^2(x-\beta)$, the invariant forms are given by

$$X = \begin{cases} 0 & \text{if } \beta,\, \alpha \neq \pm 1 \ \&\ \alpha\beta \neq 1; \\ x_{23}e_{23} + x_{32}e_{32} & \text{if } \beta,\, \alpha \neq \pm 1 \ \&\ \alpha\beta = 1; \\ x_{33}e_{33} & \text{if } \beta = \pm 1 \neq \alpha \ \&\ \alpha\beta \neq 1; \\ x_{12}(e_{12} - e_{21}) + x_{22}e_{22} & \text{if } \beta \neq \pm 1 = \alpha \ \&\ \alpha\beta \neq 1; \\ x_{12}(e_{12} - e_{21}) + x_{22}e_{22} + x_{33}e_{33} & \text{if } \beta = \mp 1,\, \pm 1 = \alpha. \end{cases}$$

Here for each of the different conditions involving α and β one can get a basis for the space of invariant forms like in the last condition it is $\{e_{12} - e_{21},\, e_{22},\, e_{33}\}$. Also this is to note that among all the conditions, only last one admits a non-degenerate form.

5.2.2. *Subcase (ii).* For $J(T) = \begin{pmatrix} \alpha & 0 & 0 \\ 0 & \alpha & 0 \\ 0 & 0 & \beta \end{pmatrix}$ with minimal polynomial $(x-\alpha)(x-\beta)$, the invariant forms are given by

$$X = \begin{cases} 0 & \text{if } \beta,\, \alpha \neq \pm 1 \ \&\ \alpha\beta \neq 1; \\ x_{13}e_{13} + x_{23}e_{23} + x_{31}e_{31} + x_{32}e_{32} & \text{if } \beta,\, \alpha \neq \pm 1 \ \&\ \alpha\beta = 1; \\ x_{33}e_{33} & \text{if } \beta = \pm 1 \neq \alpha \ \&\ \alpha\beta \neq 1; \\ x_{11}e_{11} + x_{12}e_{12} + x_{21}e_{21} + x_{22}e_{22} & \text{if } \beta \neq \pm 1 = \alpha \ \&\ \alpha\beta \neq 1; \\ x_{11}e_{11} + x_{12}e_{12} + x_{21}e_{21} + x_{22}e_{22} + x_{33}e_{33} & \text{if } \beta = \mp 1,\, \pm 1 = \alpha. \end{cases}$$

Here also for each of the different conditions one can get a basis for the corresponding space of invariant forms like in the last condition it is $\{e_{11}, e_{12}, e_{21}, e_{22}, e_{33}\}$. Further only in the last condition a non-degenerate invariant form exists.

5.3. Case 3. If $\chi_T(x) = (x-\alpha)(x-\beta)(x-\gamma)$ for some $\alpha,\, \beta,\, \gamma \in \mathbb{C}$ distinct elements, then the Jordan form for T is given by $J(T) = \begin{pmatrix} \alpha & 0 & 0 \\ 0 & \beta & 0 \\ 0 & 0 & \gamma \end{pmatrix}$ with minimal polynomial $(x-\alpha)(x-\beta)(x-\gamma)$. The invariant forms in this case can be obtained in the following manners.

5.3.1. *Subcase (i).* If $\alpha, \beta, \gamma \notin \{1, -1\} \& \alpha\beta, \beta\gamma, \gamma\alpha \neq 1$ we have $X = 0$ as the only invariant form.

5.3.2. *Subcase (ii).* If $\alpha \neq \beta \in \{1, -1\} \,\&\, \gamma \neq \pm 1$, then none of the pair product is equal to 1 and the corresponding invariant form is

$$X = \begin{pmatrix} x_{11} & 0 & 0 \\ 0 & x_{22} & 0 \\ 0 & 0 & 0 \end{pmatrix}$$

with e_{11}, e_{22} as bases elements. Similarly if $\alpha \neq \gamma \in \{1, -1\}$ & $\beta \neq \pm 1$, then

$$X = \begin{pmatrix} x_{11} & 0 & 0 \\ 0 & 0 & 0 \\ 0 & 0 & x_{33} \end{pmatrix}$$

with e_{11}, e_{33} as bases elements and if $\beta \neq \gamma \in \{1, -1\}$ & $\alpha \neq \pm 1$, then

$$X = \begin{pmatrix} 0 & 0 & 0 \\ 0 & x_{22} & 0 \\ 0 & 0 & x_{33} \end{pmatrix}$$

with e_{22}, e_{33} as bases elements.

5.3.3. *Subcase (iii).* If α, β, $\gamma \notin \{1, -1\}$ and exactly one of the pairs has product equals to 1, then

$$X = \begin{cases} \begin{pmatrix} 0 & x_{12} & 0 \\ x_{21} & 0 & 0 \\ 0 & 0 & 0 \end{pmatrix} & \text{if } \alpha\beta = 1; \\ \begin{pmatrix} 0 & 0 & 0 \\ 0 & 0 & x_{23} \\ 0 & x_{32} & 0 \end{pmatrix} & \text{if } \beta\gamma = 1; \\ \begin{pmatrix} 0 & 0 & x_{13} \\ 0 & 0 & 0 \\ x_{31} & 0 & 0 \end{pmatrix} & \text{if } \alpha\gamma = 1. \end{cases}$$

Here by inspection one can easily find the bases elements for the spaces of invariant forms.

5.3.4. *Subcase (iv).* If exactly one of α, β, γ is equals to ± 1, then the invariant forms are given by

$$X = \begin{cases} \begin{pmatrix} x_{11} & 0 & 0 \\ 0 & 0 & x_{23} \\ 0 & x_{32} & 0 \end{pmatrix} & \text{if } \pm 1 = \alpha \text{ and } \beta\gamma = 1; \\ \begin{pmatrix} x_{11} & 0 & 0 \\ 0 & 0 & 0 \\ 0 & 0 & 0 \end{pmatrix} & \text{if } \pm 1 = \alpha \text{ and } \beta\gamma \neq 1; \\ \begin{pmatrix} 0 & 0 & x_{13} \\ 0 & x_{22} & 0 \\ x_{31} & 0 & 0 \end{pmatrix} & \text{if } \pm 1 = \beta \text{ and } \alpha\gamma = 1; \\ \begin{pmatrix} 0 & 0 & 0 \\ 0 & x_{22} & 0 \\ 0 & 0 & 0 \end{pmatrix} & \text{if } \pm 1 = \beta \text{ and } \alpha\gamma \neq 1; \\ \begin{pmatrix} 0 & x_{12} & 0 \\ x_{21} & 0 & 0 \\ 0 & 0 & x_{33} \end{pmatrix} & \text{if } \pm 1 = \gamma \text{ and } \alpha\beta = 1; \\ \begin{pmatrix} 0 & 0 & 0 \\ 0 & 0 & 0 \\ 0 & 0 & x_{33} \end{pmatrix} & \text{if } \pm 1 = \gamma \text{ and } \alpha\beta \neq 1. \end{cases}$$

By inspection one can see that in this case the non-degenerate invariant forms can be obtained only in Subcase (iv).

Thus in this article we have discussed a methodology and accordingly proposed an algorithm to compute the invariant bilinear forms under an invertible linear map on a finite dimensional vector space in a systematic way. We have also discussed about the existence of non-degenerate invariant forms. The study may be extended to a group of linear operators on a vector space in future.

Funding

Not applicable.

Conflicts of interest/Competing interests

Not applicable.

Availability of data and material

The manuscript has no associated data.

Code availability

Not applicable.

Acknowledgment

The author is thankful to the Babasaheb Bhimrao Ambedkar university for providing an excellent environment to prepare this article.

References

[1] Puja Bharti and Jagmohan Tanti, *Similarity classes of linear transformations*, J. Indian Math. Soc. (N.S.) **87** (2020), no. 3-4, 148–164, DOI 10.18311/jims/2020/25448. MR4123469

[2] David S. Dummit and Richard M. Foote, *Abstract algebra*, 3rd ed., John Wiley & Sons, Inc., Hoboken, NJ, 2004. MR2286236

[3] G. Frobenius, *Uber die mit einer Matrix vertauschbaren matrizen*, Sitzungsber. Preuss. Akad. Wiss. (1910) 3 – 15.

[4] Krishnendu Gongopadhyay and Ravi S. Kulkarni, *On the existence of an invariant non-degenerate bilinear form under a linear map*, Linear Algebra Appl. **434** (2011), no. 1, 89–103, DOI 10.1016/j.laa.2010.08.009. MR2737234

[5] R. Gow and T. J. Laffey, *Pairs of alternating forms and products of two skew-symmetric matrices*, Linear Algebra Appl. **63** (1984), 119–132, DOI 10.1016/0024-3795(84)90139-3. MR766504

[6] Krishnendu Gongopadhyay, Sudip Mazumder, and Sujit Kumar Sardar, *Conjugate real classes in general linear groups*, J. Algebra Appl. **18** (2019), no. 3, 1950054, 12, DOI 10.1142/S0219498819500543. MR3924832

[7] Kenneth Hoffman and Ray Kunze, *Linear algebra*, 2nd ed., Prentice-Hall, Inc., Englewood Cliffs, NJ, 1971. MR276251

[8] Ravi S. Kulkarni and Jagmohan Tanti, *Space of invariant bilinear forms*, Proc. Indian Acad. Sci. Math. Sci. **128** (2018), no. 4, Paper No. 47, 22, DOI 10.1007/s12044-018-0426-z. MR3834388

[9] Clément de Seguins Pazzis, *When does a linear map belong to at least one orthogonal or symplectic group?*, Linear Algebra Appl. **436** (2012), no. 5, 1385–1405, DOI 10.1016/j.laa.2011.08.035. MR2890926

[10] H. Stenzel, *Über die Darstellbarkeit einer Matrix als Produkt von zwei symmetrischen Matrizen, als Produkt von zwei alternierenden Matrizen und als Produkt von einer symmetrischen und einer alternierenden Matrix* (German), Math. Z. **15** (1922), no. 1, 1–25, DOI 10.1007/BF01494381. MR1544555

[11] V. V. Sergeĭchuk, *Classification problems for systems of forms and linear mappings* (Russian), Izv. Akad. Nauk SSSR Ser. Mat. **51** (1987), no. 6, 1170–1190, 1358, DOI 10.1070/IM1988v031n03ABEH001086; English transl., Math. USSR-Izv. **31** (1988), no. 3, 481–501. MR933960

Babasaheb Bhimrao Amedkar University, Lucknow-226025, U. P. India
Email address: jagmohan.t@gmail.com

Contemporary Mathematics
Volume **811**, 2025
https://doi.org/10.1090/conm/811/16249

A comparative overview of Riemannian and Finsler geometry

Bankteshwar Tiwari

Abstract. The aim of this article is to present a comparative overview of Riemannian and Finsler geometry, starting from some historical developments to the various directions of current research. This includes the discussion on classification of Finsler spaces of constant flag curvature and constant Ricci curvature, cut locus, conjugate locus, Comparison theorems, Gauss-Bonnet-Chern Theorem and sphere theorem in Finsler geometry. The topological, differential and metric structures on Riemannian manifolds in the presence of convex functions have been active fields of research in the second half of the last century. We discuss some of these results on Riemannian manifolds with convex functions and their recently extended analogues on Finsler manifolds.

Contents

1. Introduction
2. Definitions and preliminaries
3. Examples of Finsler metric
4. The Finsler metrics of constant/scalar flag/Ricci curvature
5. Global Riemann-Finsler geometry
6. Busemann functions and convex functions
References

1. Introduction

Roughly speaking, Finsler geometry is related to Riemannian geometry in the same way that Banach space is related to Hilbert space. A Finsler metric on a smooth manifold is a smoothly varying family of Minkowski norms on the tangent spaces (see Definition 2.1), rather than a family of inner products, as in the case of a Riemannian metric. It turns out that every Finsler metric induces an inner product, one in each direction of a tangent space at each point of the manifold. Thus, a Finsler metric associates to the manifold a family of inner products parametrized by the tangent spaces of the manifold instead of being parametrized by the manifold, as in the case of a Riemannian metric.

This work was presented in the International Conference in honor of Ravi S. Kulkarni's 80th Birthday on Geometry, Groups and Mathematical Philosophy.

The fundamental idea of Finsler geometry may be traced back to Riemann's 1854 Habilitation address "Uber die Hypothesen, welche der Geometrie zu grunde liegen" (On the Hypotheses, which lie at the Foundations of Geometry) (see [**72**]). In this memoir, Riemann discussed various possibilities through which an n-dimensional manifold might be endowed with a metric, however, he focused particular attention on a metric defined by the positive square root of a positive definite quadratic differential form. Thus, the geometry with such a metric is well recognized as Riemannian geometry. In the same memoir, it was also suggested by Riemann that the positive fourth root of a fourth order differential form might serve as a metric function. The first systematic study of manifolds endowed with such a metric was started after more than 60 years by Paul Finsler in 1918 ([**29**]), after whom such spaces were eventually named, under the supervision of Carathéodory, who introduced the techniques of calculus of variations to develop the new geometrical background. In fact, the close association between differential geometry and the calculus of variations had been noticed even prior to the publication of Finsler's thesis, in particular by Bliss, Landsberg and Blasche (See [**73**] and references therein for more historical developments). However, the geometrical theories of Bliss and Landsberg could not fulfil the true objectives of the generalisation of Riemann's proposal as they were developed against Euclidean background and therefore, Finsler's thesis must be regarded as the first step in this direction. Finsler did not make use of the tensor calculus, being guided in principle by the notions of the calculus of variations. A few years later, in 1925, the methods of the tensor calculus were applied to the theory independently but almost simultaneously by Synge, Taylor and Berwald (see [**11**, **88**–**91**]). It was found that the second derivatives of the square of the fundamental function [see Definition 2.3] along vertical directions served as components of a metric tensor in analogy with Riemannian geometry, and from the differential equations of geodesics, connection coefficients could be derived by means of which a generalisation of Levi-Civita's parallel displacement could be defined. While the corresponding covariant derivatives as introduced by Synge and Taylor coincide, the theory of Berwald showed a marked difference in the sense that, in his geometry, the lemma of Ricci (which in Riemannian geometry implies the vanishing of the covariant derivative of the metric tensor) is no longer valid. Nevertheless, Berwald continued to develop his theory, with particular reference to the theory of curvature. The significance of this work was enhanced by the advent of the general geometry of paths (a generalisation of the so-called non-Riemannian geometry) due to Douglas and Knebelman, for the initial approach of Berwald was such as to establish a close affinity between these branches of metric and non-metric differential geometry. Again, the theory took a new and unexpected turn in 1934, when E. Cartan published his tract on Finsler spaces. He showed that it was indeed possible to define connection coefficients and hence a covariant derivative, such that the preservation of Ricci's lemma was ensured. On this basis, Cartan developed a theory of curvature, and practically all subsequent investigations concerning the geometry of Finsler spaces were dominated by this approach. Several mathematicians expressed the opinion that the theory had thus attained its final form. To a certain extent, this was correct, but not altogether so, as very little progress was made towards global differential geometry.

One of the basic differences between Riemannian and Finsler geometry is the possible asymmetry of distance functions. In Riemannian geometry, the Levi-Civita connection is the canonical connection. It is torsion-free and metrical. There is no such connection in Finsler geometry, which is torsion-free and metrical. There are different connections that have their own importance. The Chern connection is important from two points of view: firstly, when the Finsler metric induces a Riemannian metric, it reduces to the Levi-Civita connection, and secondly, it solves the problem of equivalence in Finsler geometry (see **[23]**). The Chern connection is torsion-free but not metrical.

The flag curvature in Finsler geometry is the analogue of the sectional curvature in Riemannian geometry. By the Beltrami theorem, it is well known that a complete Riemannian manifold with constant sectional curvature is locally isometric to one of the three spaces: the Euclidean space, the hyperbolic space or the sphere with the corresponding standard Riemannian metric. However, in Finsler geometry, the classification of constant flag curvature spaces is far from reach. Since the start of the current century, some progress has been made in this direction for special Finsler spaces. We will sketch a brief review of this progress, in particular, the classification of Randers spaces with constant flag curvature and its relation to Zermelo navigation will be presented.

It is interesting to note that *convex sets and convex functions* are valid notions on Finsler manifolds in spite of the lack of symmetry of the Finsler distance function. In this article, we discuss several topics on Finsler manifolds with non-symmetric distance functions, such as the cut locus, the conjugate locus, convex sets, convex functions, and Busemann functions. We also discuss some recent developments on the analogues of comparison theorems, the Gauss-Bonnet-Chern theorem and the sphere theorem on the Finsler manifold. To make the article concise and to include various directions of current research in Finsler geometry, we prefer not to write proofs of the results; however, references are being provided, wherever it seems necessary.

The article is organized as follows: Definitions and notation are set up in §2. Some examples are given and explored in §3. Finsler metrics of constant, scalar flag/Ricci curvature are discussed in connection with the Zermelo navigation problem in §4. The forward cut locus and the forward conjugate locus and their properties, some comparison theorems, the Gauss-Bonnet-Chern theorem, the sphere theorem on the Finsler manifold are discussed in §5. In §6, we discuss the properties of Busemann functions and convex functions on complete and non-compact Riemannian and Finsler manifolds. Finally, we summarize the Riemannian and Finsler results on convex functions. For the basic tools in Riemannian and Finsler geometry, we refer to **[17]**, **[47]**, **[18]**, **[12]**, **[75]**, **[24]**,**[23]**,**[6]**, **[59]**, **[79]**.

2. Definitions and preliminaries

As well known, Riemannian metric on a smooth manifold is a smoothly varying family of inner products, whereas Finsler metric on a smooth manifold is a smoothly varying family of norms, what is called Minkowski norm. After giving the definition of Minkowski norm on a vector space, we give the definitions of Riemannian and Finsler metrics on a smooth manifold and discuss an important relation between them.

DEFINITION 2.1 (see section 1.1, **[25]**). Let V be an n-dimensional real vector space. A non negative function F defined on V is said to be a Minkowski norm, if the following conditions are satisfied:

- $F(u) = 0$ if and only if $u = 0$;
- $F(cu) = cF(u)$ for any $u \in V$ and $c > 0$;
- F is C^∞ on $V \setminus \{0\}$ such that for any $u \in V \setminus \{0\}$, the bilinear symmetric functional g_u on V, defined as $g_u(v, w) := \frac{1}{2}\frac{\partial^2}{\partial s \partial t}[F^2(u + sv + tw)]_{s=t=0}$, is an inner product, i.e., for a fix basis $\{b_i\}_{i=1}^n$ of V and $u = u^i b_i \neq 0$ [1] the matrix $(g_{ij}(u))$ defined by $g_{ij}(u) := g_u(b_i, b_j) = \frac{1}{2}\frac{\partial^2}{\partial u^i \partial u^j}[F^2](u)$ should be positive definite.

EXAMPLE 2.1. Let $V = R^n$ and for $v = (v^i) \in \mathbb{R}^n$, $|v| := \sqrt{\sum_{i=1}^n (v^i)^2}$ be standard Euclidean norm. If $b \in \mathbb{R}$ with $0 < b < 1$, then the standard Euclidean norm $F(v) = |v|$, as well as $F(v) = |v| + bv^1$ are Minkowski norms on $\mathbb{R}^n$.

REMARK 1. The second condition for the function F in above definition is known as 'F is positively homogeneous of degree 1'. More generally, a real valued function $f : \mathbb{R}^n \to \mathbb{R}$ is said to be positively homogeneous of degree k if $f(cx) = c^k f(x)$, for all $c > 0$. In addition, if f is differentiable, possibly away from the origin of $\mathbb{R}^n$, then by Euler's theorem the function f is positively homogeneous of degree k iff $\sum_{i=1}^n x^i \frac{\partial f(x)}{\partial x^i} = cf(x)$ (see section 1.2 A of [**6**]).

2.1. Riemannian and Finsler metrics. Let M be a smooth manifold of dimension $n \geq 2$ and T_xM be the tangent space to the manifold M at the point x. The tangent bundle $TM := \cup_{x \in M} T_xM$ over M is a smooth $2n$-manifold, with projection $\tau : TM \to M$ and $T_xM = \tau^{-1}x$ for $x \in M$. The projection $\tau : TM \to M$ induces the double tangent bundle $d\tau : TTM \to TM$; for any $X \in T_xM$, the space $T_X(TM)$ is called the double tangent space at $X \in TM$ and has dimension $2n$. The vertical tangent space, denoted by $T_X^v(TM) = T_X(T_xM)$ and defined as the kernel of the map $d\tau$, viz., $T_X^v(TM) = ker(d\tau : T_X(TM) \to T_xM)$. Collectively the vertical subspaces form the verical subbundle $VTM \subset TTM$. In local coordinate system (x^i, y^i) in TM, where (x^i) is a local coordinate system in M and y^i are coefficients of $y = y^i \left.\frac{\partial}{\partial x^i}\right|_x$. Then $\{\frac{\partial}{\partial x^i}, \frac{\partial}{\partial y^i}\}$ be the natural frame for TTM and the vertical tangent bundle $VTM = \text{span}\{\frac{\partial}{\partial y^i}\}$. To write TTM as direct sum of vertical tangent bundle and another subbundle(called horizontal subbundle HTM), we need an extra structure on the manifold, what is called, non-linear connection, which we discuss in subsection 2.3.

DEFINITION 2.2 (Riemannian metric). A Riemannian metric g on M is a smoothly varying family of inner products on M in the sense that for smooth vector fields X, Y defined in a neighborhood U of x in M, the function $g(X, Y) : U \to \mathbb{R}$ defined as $x \mapsto g_x(X(x), Y(x))$ is a smooth function. The pair (M, g) is called a Riemannian manifold.

DEFINITION 2.3 (Finsler metric; see Section 1.1, [**6**]). A Finsler metric F on a smooth manifold M is a continuous function $F : TM \to \mathbb{R}$ satisfying the following:

(1) F is smooth on $TM \setminus \{0\}$ (regularity);
(2) $F(x, cu) = cF(x, u)$ for all $c > 0$ and for all $(x, u) \in TM$ (positive homogeneity);

[1] Here Einstein summation convention has been applied and without explicitly mentioning we will frequently use it in this article, for instance, if in a term $A^i B_i$ where same suffix is appearing once in the lower position and then again in the upper position, we mean $A^i B_i = \sum_{i=1}^n A^i B_i$.

(3) The $n \times n$ Hessian matrix $(g_{ij}(x,u)) := (\frac{1}{2}\frac{\partial^2 F^2(x,u)}{\partial u^i \partial u^j})$ is positive definite for all $(x,u) \in TM \setminus \{0\}$ (strong convexity).

The pair (M, F) is called a Finsler manifold and the function F is called fundamental function of the Finsler manifold.

The positive homogeneity and the strong convexity of F leads to the following facts:

LEMMA 2.1 (see Section 1.2 B, [**6**]). *Let (M, F) be a Finsler manifold and $x \in M$. If $v, w \in T_xM$, then we have*

- *F is positive on $T_xM \setminus \{0\}$.*
- *F satisfies triangle inequality, i.e., $F(v+w) \le F(v) + F(w)$ for all $v, w \in T_xM$.*

REMARK 2. In view of Lemma 2.1, the restriction of the function F to each tangent space to the manifold M is a non-symmetric norm, in the sense of functional analysis. Moreover, a Finsler manifold can be thought of as a smooth manifold with a smoothly varying family of Minkowski norms in each tangent space.

2.2. Fundamental and Cartan tensors. Let (M, F) be a Finsler manifold. Choose a local coordinate system $(x^i)_{i=1}^n$ on an open set $U \subset M$, and let $x \in U$. For $u \in T_xM \setminus \{0\}$, we define the components of the fundamental tensor $g_{(x,u)}$ by

$$g_{ij}(x,u) := \frac{1}{2}\frac{\partial^2 F^2(x,u)}{\partial u^i \partial u^j}, \quad i, j = 1, 2, \dots, n.$$

Recall that this symmetric matrix is positive-definite by the strong convexity. Therefore, for fix u, $(g_{ij}(x,u))_{i,j=1}^n$ defines an inner product on T_xM, denoted by g_u in the following sense.

DEFINITION 2.4 (Riemannian metric $g_{\mathfrak{U}}$; see Definition 3.1, [**59**]). Let $\mathfrak{U}$ be a nowhere vanishing vector field on an open set $U \subset M$ such that $\mathfrak{U}(x) = u \in T_xM \setminus \{0\}$. We define a Riemannian metric $g_{\mathfrak{U}}$ on U such that for vectors $a = a^i \frac{\partial}{\partial x^i}, b = b^j \frac{\partial}{\partial x^j} \in T_xM$, we have

$$g_{\mathfrak{U}(x)}(a^i \frac{\partial}{\partial x^i}, b^j \frac{\partial}{\partial x^j}) = a^i b^j g_{ij}(u)$$

It is interesting to note that the expression in the above definition is independent of the choice of the coordinates, which ensures the well-definedness of the Riemannian metric $g_{\mathfrak{U}}$. In particular, if F is induced from a Riemannian metric g (by $F(u) = \sqrt{g(u,u)}$), then g_u coincides with the original metric g for all $u \in T_xM \setminus \{0\}$ and $x \in \mathfrak{U}$. The Riemannian metric $g_{\mathfrak{U}}$ will play a significant role in the Riemannian geometric approach to Finsler geometry. Since $F^2|_{T_xM}$ is positively 2-homogeneous and its derivative $(\frac{\partial [F^2]}{\partial u^i})|_{T_xM}$ is positively 1-homogeneous (see Remark 1), we have

$$g_u(u,u) = \frac{1}{2}\frac{\partial^2 [F^2]}{\partial u^i \partial u^j} u^i u^j = \frac{\partial [F^2]}{\partial u^i} u^i = F^2(u)$$

for all $u \in T_xM \setminus \{0\}$. By this equality and the definition of g_u, the unit sphere of the inner product g_u is tangent to that of $F|_{T_xM}$ at $u/F(u)$ up to second order. In this sense, g_u can be regarded as the best Riemannian approximation of F in the direction u. Thereby the metric g_u is useful when we are interested in the behavior of some geometric or analytic quantity in the direction u.

Let νU be the space of smooth vector fields on the open subset U and $\nu U^+ \subset \nu U$ be the subset of nowhere vanishing vector fields.

DEFINITION 2.5 (Cartan tensor; see Section 3, [**69**]). For $\mathcal{U} \in \nu U^+$ and for all $X, Y, Z \in \nu U$ define a trilinear form $\langle \cdot\,,\cdot\,,\cdot\rangle_{\mathcal{U}}$ by $\langle X, Y, Z\rangle_{\mathcal{U}} = \frac{1}{4}\frac{\partial^3}{\partial r \partial s \partial t} F^2(\mathcal{U} + rX + sY + tZ)|_{r=s=t=0}$, which is a symmetric $(0,3)$ tensor, called the Cartan tensor. The components of the Cartan tensor can be written as $C_{ijk}(u) := \frac{1}{4}\frac{\partial^3 F^2}{\partial u^i \partial u^j \partial u^k}(u) = \frac{1}{2}\frac{\partial g_{ij}}{\partial u^k}(u)$, $u \in T_xM \setminus \{0\}$. It measures the variation of g_u in the vertical direction. In some other literature Cartan tensor is also denoted by A_{ijk} and defined by $A_{ijk}(u) := F(u)A_{ijk}(u)$ (see Section 1.4, [**6**]). Since $g_{ij}(u)$ is positively 0-homogeneous in u, we deduce that $C_{ijk}(u)u^i = C_{ijk}(u)u^j = C_{ijk}(u)u^k = 0$, for all $u \in T_xM \setminus \{0\}$.

On a Riemannian manifold (M, g), the Cartan tensor vanishes on whole $TM \setminus \{0\}$, since $g_u = g$ for all $u \in T_xM \setminus \{0\}$. Conversely, if $C_{ijk} = 0$ holds on $T_xM \setminus \{0\}$ for all $i, j, k = 1, 2, \ldots, n$, then g_u is independent of $u \in T_xM \setminus \{0\}$ and $F|_{T_xM}$ coincides with the norm induced from g_v. Therefore, the Cartan tensor is a genuinely non-Riemannian quantity appearing only in the Finsler world. Thus we have

PROPOSITION 2.1. *A Finsler metric reduces to a Riemannian metric if and only if its Cartan tensor vanishes.*

2.3. Intrinsic distance and geodesics. Let (M, F) be a Finsler manifold of dimension ≥ 2. For a smooth curve $\gamma : [a, b] \to (M, F)$ the length $L(\gamma)$ is given by

$$L(\gamma) := \int_a^b F(\gamma(t), \dot\gamma(t))\, dt, \quad \gamma'(t) = \frac{d\gamma}{dt}.$$

The reversed curve of γ, viz. $t \mapsto \gamma(a + b - t)$, $t \in [a, b]$, is denoted by γ^{-1}. The length of γ^{-1} is in general different from that of γ:

$$L(\gamma^{-1}) = \int_a^b F(\gamma^{-1}(t), \dot\gamma^{-1}(t))\, dt.$$

The intrinsic distance $d(x, y)$ from a point $x \in M$ to a point $y \in M$ is defined by

$$d(x, y) := \inf\{L(\gamma) \,|\, \gamma \text{ is a smooth curve from } x \text{ to } y\}.$$

We note that in general $d(x, y) \neq d(y, x)$.

The indicatrix $\Sigma_x \subset T_xM$ at a point x is the set of all unit vectors with respect to F:

$$\Sigma_x := \{u \in T_xM \,|\, F(x, u) = 1\}.$$

The *reversibility constant* $\lambda(C)$ of a compact set $C \subset M$ is defined by

$$\lambda(C) := \sup\left\{\frac{F(x, -u)}{F(x, u)} \,|\, x \in C, \quad u \in T_xM \setminus \{0\}\right\}. \tag{2.1}$$

The *reversibility constant* plays a very important role in the non-symmetric Finsler manifolds. It actually controls the distance in reverse direction between two points (see Lemma 6.2.1, [**6**]), viz.,

$$\lambda(C)^{-1} d(x, y) \leq d(y, x) \leq \lambda(C) d(x, y), \text{ for all } x, y \in C.$$

A Finsler manifold (M, F) satisfying $F(x, u) = F(x, -u)$, for all $x \in M$ and $u \in T_xM \setminus \{0\}$ is said to be reversible Finsler manifold otherwise non-reversible. For a reversible Finsler manifold the reversibility constant $\lambda(M) = 1$.

In Riemannian geometry, the Levi-Civita connection is the unique connection which is metric compatible and torsion-free. In the Finsler case, however, one cannot always find a connection which satisfies both of these properties. Therefore, we need to modify (at least) one of the requirements, and so there arise a number of connection, for instance, Cartan connection, Berwald connection, Chern connection etc. An important fact is that, despite the non-uniqueness of connections, Riemann curvatures can be uniquely determined. In this article we discuss only on the Chern connection. Let $U \subset (M, F)$ be an open set. A linear connection ∇ on U is a map $\nabla : (X, Y) \in \nu U \times \nu U \to \nabla_X Y \in \nu U$, linear in Y (not necessarily linear in X) and satisfying the following conditions $\nabla_X(fY) = f\nabla_X Y + X(f)Y$ and $\nabla_{fX} Y = f\nabla_X Y$ for all $f \in C^\infty(U)$ and $X, Y \in \nu U$.

THEOREM 2.1 (see Theorem 3.10, [**69**]). *Let (M, F) be a Finsler manifold, $U \subset M$ an open set and $V \in \nu U^+$, then there is a unique linear connection ∇^V associated with V, called the Chern connection, satisfying the following conditions:*

(1) *∇^V is torsion free, that is, $\nabla^V_X Y - \nabla^V_Y X = [X, Y]$ for all $X, Y \in \nu U$.*
(2) *∇^V is almost metrical, that is,*

$$Xg_V(Y, Z) = g_V(\nabla^V_X Y, Z) + g_V(Y, \nabla^V_X Z) + 2\langle \nabla^V_X V, Y, Z\rangle_V \text{ for all } X, Y, Z \in \nu U.$$

Moreover, such connection ∇ can be explicitly written as follows, what is known as generalized Koszul formula:

$$\begin{aligned}2g_V(\nabla^V_X Y, Z) = X(g_V(Y, Z) + Y(g_V(Z, X)) - Z(g_V(X, Y)) + g_V([X, Y], Z)\\ -g_V([Y, Z], X) + g_V([Z, X], Y) - 2(\nabla^V_X V, Y, Z)_V\\ -2(\nabla^V_Y V, Z, X)_V + 2(\nabla^V_Z V, X, Y)_V\end{aligned}$$

In local coordinate system the components $\Gamma^k_{ij}(x, y)$ of the Chern connection are given by (see [**69**], [**6**])

$$\Gamma^k_{ij} = \gamma^k_{ij} - g^{im}(C_{mjr}N^r_k - C_{jkr}N^r_m + C_{kmr}N^r_j)$$

where $\gamma^h_{ij} = \frac{1}{2}g^{hk}(\frac{\partial g_{jk}}{\partial x^i} + \frac{\partial g_{ki}}{\partial x^j} - \frac{\partial g_{ij}}{\partial x^k})$ and $N^i_j = \gamma^i_{jk}y^k - C^i_{jk}\gamma^k_{rs}y^r y^s$. The quantities N^i_j are called components of non-linear connection. These connections play an important role in the construction of the horizontal subbundle HTM of the tangent bundle TTM. In fact, $HTM := \text{span}\{\frac{\delta}{\delta x}\}$, where $\frac{\delta}{\delta x^i} = \frac{\partial}{\partial x^i} - N^j_i \frac{\partial}{\partial y^j}$. Thus the double tangent buble TTM can be written as $TTM = HTM \oplus VTM$.

Using the connection ∇^V, we introduce the covariant derivative $\frac{\nabla^V}{dt}$ along a smooth curve $\gamma : [0, l] \to M$. For a vector field X along the curve γ with tangent vector field $\dot\gamma$, define $\frac{\nabla^V}{dt}X(t) = \nabla^V_{\dot\gamma}X(t)$. If the vector fields V and $\dot\gamma$ along γ coincide, we also write $\frac{\nabla^V}{dt}X(t) = \frac{\nabla}{dt}X(t)$.

A smooth curve $\gamma : [0, l] \to (M, F)$ on a Finsler manifold (M, F) is said to be a forward geodesic if $\frac{\nabla}{dt}\dot\gamma(t) = 0$, for all $t \in [0, 1]$. In the local coordinates, if $\dot\gamma(t) = \frac{dx^i}{dt}\frac{\partial}{\partial x^i}$, then the forward geodesics are the solutions of the second order non-linear differential equations, which can also be obtained by the standard variational argument

$$\frac{d^2x^k}{dt^2} + \Gamma^k_{ij}(x, y)\frac{dx^i}{dt}\frac{dx^j}{dt} = 0. \tag{2.2}$$

For a smooth curve $\gamma : [0, l] \to M$, the *energy* is defined as

$$E(\gamma) = \frac{1}{2}\int_0^1 F^2(\gamma'(t))dt.$$

A smooth variation of the curve $\gamma : [0, 1] \to M$ is a smooth map $\sigma : [0, l] \times (-\epsilon, \epsilon) \to M$ such that $\sigma(t, 0) = \gamma(t)$ for all $t \in [0, l]$. For each $s \in (-\epsilon, \epsilon)$ the curves $\gamma_s : [0, l] \to M$ defined by $\gamma_s(t) = \sigma(t, s)$ are called t-curve and for each $t \in [0, l]$ the curves $\gamma_t(s) : (-\epsilon, \epsilon) \to M$ defined by $\gamma_t(s) = \sigma(t, s)$ are called s-curve. The velocity vector $V(t)$ of s-curve $\gamma_t(s)$ at $s = 0$ is called variation vector field, i.e., $V(t) = \dot{\gamma}_t(0) = (\partial/\partial s)|_{s=0}(\sigma(t, s))$.

First variation formula. The first variation formula is closely related to the geodesics in the Finsler manifold while the second variation formula (see (2.5)) is related to the flag curvature. If $\sigma : [0, l] \times (-\epsilon, \epsilon) \to M$ is a smooth variation of the curve γ with variation vector field $V(t) = (\partial/\partial s)|_{s=0}\sigma(t, s)$, then the first variation formula is given by

$$\left.\frac{d}{ds}\right|_{s=0} E(\gamma_s) = g_{\dot{\gamma}(1)}(\dot{\gamma}(1), V(1)) - g_{\dot{\gamma}(0)}(\dot{\gamma}(0), V(0)) - \int_0^1 g_{\dot{\gamma}}(\frac{\nabla}{dt}\dot{\gamma}, V)dt.$$

PROPOSITION 2.2. *If $\gamma : [0, l] \to M$ is a piecewise smooth curve such that no other piecewise smooth curve joining $p = \gamma(0)$ and $q = \gamma(l)$ is shorter, then γ is a geodesic, that is, γ is smooth and $\frac{\nabla}{dt}\dot{\gamma} = 0$.*

A vector field $V \in \nu U$ is said to be a geodesic vector field if $\nabla_V^V V = 0$, that is, if all the flow lines of V are geodesics. A unit speed forward geodesic $\gamma : [0, \infty) \to (M, F)$ is by definition a *forward ray* if every subarc $\gamma|_{[a,b]}$, $0 \le a < b < \infty$ of γ is minimizing. We say that a geodesic $\gamma : (-\infty, \infty) \to (M, F)$ is a straight line if it satisfies that $d(\gamma(s), \gamma(t)) = t - s$ for all $s, t \in (-\infty, \infty)$ with $s < t$.

PROPOSITION 2.3 (see Rademacher [**69**]). *Let V be a nowhere-vanishing geodesic vector field defined on an open subset $U \subset M$. Denote by $\overline{\nabla}$, the Levi-Civita connection of the Riemannian manifold (U, g_V) then $\nabla_X^V V = \overline{\nabla}_X V$, for all vector fields X, in particular, the vector field V is also a geodesic vector field for the Riemannian manifold (U, g_V).*

2.4. Exponential map and geodesic completeness. A forward geodesic $\gamma_u : [0, h) \to (M, F)$ with the initial conditions $\gamma_u(0) := x$ and $\dot{\gamma}_u(0) := u$ is the solution of a non-linear second order differential equation with smooth coefficients. Let the set $\Omega_x \subset T_x M$ be defined as follows:

$$\Omega_x = \{u \in T_x M \mid \gamma_u(1) \text{ is defined}\}. \tag{2.3}$$

In view of the homogeneity property $\gamma_{\lambda u}(1) = \gamma_u(\lambda)$ of the forward geodesic, it can be observed that the set Ω_x is convex with respect to origin of $T_x M$. We then define the exponential map $\exp_x : \Omega_x \to (M, F)$ at x by

$$\exp_x u := \gamma_u(1), \quad u \in \Omega_x.$$

It is an easy consequence of the definition of the exponential map that it satisfies the following homogeneity property

$$\exp_x \lambda u = \gamma_u(\lambda), \quad \lambda > 0 \quad \text{and} \quad u, \lambda u \in \Omega_x.$$

The exponential map exp_p is of class C^1 at the origin, where as C^∞ away from the origin. Its derivative at the zero section is the identity map, just like in Riemannian geometry. However, due to a theorem by Akbar-Zadeh's (see [**6**]), the exponential map is C^2 at the zero section iff the Finsler structure is Berwald.

We say that (M, F) is *forward geodesically complete* if $\Omega_x = T_xM$ at some point $x \in M$. It then follows that $\Omega_y = T_yM$ for all $y \in M$. A forward (resp. backward) Cauchy sequence $\{q_j\}_{j=1}^{\infty}$ is defined by the condition that for every $\varepsilon > 0$ there exists an integer N_ε such that

$$d(q_j, q_k) < \varepsilon \,(\text{resp. } d(q_k, q_j) < \varepsilon), \text{ for all } N_\varepsilon < j < k.$$

We say that (M, F) is *forward complete* (resp. *backward complete*) if every forward (resp. backward) Cauchy sequence converges.

REMARK 3. If $F(p, u) = F(p, -u)$ holds for all $(p, u) \in TM$, then all the completeness conditions as above are equivalent to each other.

The classical Hopf-Rinow theorem in Finsler geometry can be stated as follows:

THEOREM 2.2 (Hopf–Rinow theorem, [**59**],[**6**]). *The following statements are equivalent:*

1. *(M, F) is forward complete.*
2. *We have $\Omega_x = T_xM$ for some $x \in M$ and (M, F) is forward geodesically complete.*
3. *Any closed, forward bounded set is compact.*

Furthermore, if (M, F) is forward complete, then for any points $x, y \in M$, there exists a minimal geodesic from x to y.

2.5. Curvature. By analyzing the behavior of geodesics, we can introduce the notion of Jacobi fields and through Jacobi fields, one can introduce the flag curvature, which corresponds to the sectional curvature in Riemannian geometry. In this strategy, one naturally encounters an important observation that the flag curvature coincides with the sectional curvature with respect to a certain Riemannian metric induced from the original Finsler metric (see Definition 2.4).

In Riemannian geometry, curvature is intimately linked to the behavior of geodesics. Such a connection can be seen from the Jacobi equation

$$D_{\dot\gamma} D_{\dot\gamma} J + R(J, \dot\gamma)\dot\gamma = 0, \tag{2.4}$$

where $\gamma : [0, l] \to M$ is a geodesic, $\sigma : [0, l] \times (-\epsilon, \epsilon) \to M$ is a C^∞-map such that $\sigma(t, 0) = \gamma(t)$ and for each s the t-curve $c_s(t) = \sigma(t, s)$ is a geodesic, $J(t) := \partial_t\sigma(t, 0)$ is the variational vector field along γ, and R is the curvature operator. A vector field along a geodesic satisfying (2.4) is called a Jacobi field and, conversely, every Jacobi field arises in this way as the variational vector field of some geodesic variation. Having the Finsler versions of geodesics and covariant derivatives in hand, we follow the samc line in the Finsler setting.

DEFINITION 2.6 (Jacobi fields and curvature tensor; see Definition 5.1 [**59**]). A C^2-vector field J along a geodesic $\gamma : [0, l] \to (M, F)$ is called a Jacobi field if it satisfies the Jacobi equation

$$D_{\dot\gamma}^{\dot\gamma} D_{\dot\gamma}^{\dot\gamma} J + R_{\dot\gamma}(J) = 0$$

on $(0, l)$. Here the curvature tensor $R : T_xM \ni v \to R_v \in T_x^*M \otimes T_xM$ is defined by

$$R_v(w) := R^i_j(v) w^j \left.\frac{\partial}{\partial xi}\right|_x$$

for $v, w \in T_xM$, where

$$R^i_k = 2\,\partial_k G^i - y^j\,\partial_j\,\dot{\partial}_k G^i + 2G^j\,\dot{\partial}_k\,\dot{\partial}_j G^i - \dot{\partial}_j G^i\,\dot{\partial}_k G^j$$

and G^i are spray coefficients of the Finsler metric F, given by

$$G^i(x,y) := \frac{1}{4} g^{ir}\left(y^k F^2_{x^k y^r} - F^2_{x^r}\right).$$

Index form For a smooth curve $\gamma : [0,1] \to (M,F)$, denote by W_γ, the set of all smooth vector fields along γ. The index form of a geodesic $\gamma : [0,1] \to M$ is the symmetric bilinear form $I_\gamma : W_\gamma \times W_\gamma \to \mathbb{R}$ defined by

$$I_\gamma(X,Y) = \int_0^1 (g_{\dot\gamma}(\frac{\nabla}{dt}X, \frac{\nabla}{dt}Y) - g_{\dot\gamma}(R_{\dot\gamma}(X),Y))dt$$

for $X, Y \in W_\gamma$. It is an interesting fact that $I_\gamma(X,Y) = I_\gamma(X,Y)$.

Second variation formula Let $\sigma : [0,1] \times (-\epsilon,\epsilon) \to (M,F)$ be a smooth variation of the geodesic γ with fixed end points, i.e., $\gamma_s(0) = \gamma(0)$, and $\gamma_s(1) = \gamma(1)$. Let the variation vector field be $V(t) = \frac{\partial}{\partial s}\big|_{s=0}\sigma(t,s)$. Then

$$\frac{d^2}{ds^2}\bigg|_{s=0} E(\gamma_s) = I_\gamma(V,V) \tag{2.5}$$

2.6. Flag and Ricci curvatures and their characterizations. We define the flag curvature, which generalizes the sectional curvature in Riemannian geometry, by using the curvature tensor introduced in Definition 2.6.

DEFINITION 2.7 (Flag curvature). For linearly independent vectors $y, V \in T_xM$, we define the flag curvature of (y,V) by

$$K(x,y;V) := \frac{g_y(R_y(V),V)}{g_y(y,y)g_y(V,V) - g_y(y,V)^2}.$$

Notice that $K(av,bw) = K(v,w)$ for all $a,b>0$ by the positive homogeneity. Moreover, $K(v,w)$ is independent of the choice of w in the 2-plane $v\wedge w \subset T_xM$ spanned by v and w. The 2-plane $v \wedge w$ is called a flag, and v is called a flagpole in it. Thus it is more precise to denote the flag curvature by $K(v; v\wedge w)$, whereas we write $K(v,w)$ for convenience. In the Riemannian case, the flag curvature coincides with the sectional curvature of the flag (regardless of the choice of a flagpole). In the Finsler case, however, the flag curvature depends not only on the flag, but also on the choice of a flagpole in it. For instance, $K(w,v) \neq K(v,w)$ may happen.

If $K(x,y,P) = K(x,y)$, then the Finsler metric F is said to be of scalar flag curvature. If $K(x,y,P) = K(x)$ then the Finsler metric is said to be of isotropic flag curvature. If $K(x,y,P) = \mathit{constant}$ then the Finsler metric is said to be of constant flag curvature.

DEFINITION 2.8 (Ricci curvature). For $v \in T_xM$, we define the Ricci curvature $Ric(v)$ of v as the trace of the endomorphism $R_v : T_xM \to T_xM$. If $v \neq 0$, then $Ric(v)$ can be written as

$$Ric(v) = \Sigma_{i=1}^{n-1} g_v(R_v(e_i),e_i) = F^2(v)\Sigma_{i=1}^{n-1}K(v,e_i)$$

where $\{\frac{v}{F(v)}\} \cup \{e_i\}_{i=1}^{n-1} \subset T_xM$ is an orthonormal basis with respect to g_v.

It is naturally expected from the construction of the curvature tensor via Jacobi fields that these curvatures have close connections with the behavior of geodesics. we have the following important and useful characterizations of the Finsler curvatures in terms of the Riemannian curvatures.

THEOREM 2.3 (Riemannian characterizations of curvatures). *Given a nonzero vector $v \in T_xM\backslash\{0\}$, take a nowhere vanishing C^∞-vector field V on a neighborhood U of x such that $V(x) = v$ and every integral curve of V is a geodesic. Then, for any $w \in T_xM$ linearly independent of v, the flag curvature $K(v,w)$ coincides with the sectional curvature of the 2-plane $v \wedge w$ with respect to the Riemannian metric g_V. Similarly, the Finsler Ricci curvature $Ric(v)$ coincides with the Riemannian Ricci curvature of v with respect to g_V.*

2.7. Volume form and S-curvature. It is well known that there is no canonical volume form on the Finsler manifolds, like in Riemannian case (see section 9.1 [**59**]). Some well known volume forms on Finsler manifolds are the Busemann-Hausdorff volume form, the Holmes-Thompson volume form, the maximum volume form and the minimum volume form. A Finsler space with a volume form $d\mu$ is called a Finsler m-space and is denoted by $(M, F, d\mu)$.

The canonical volume form in a Riemannian manifold $(M^n, \alpha), \alpha = \sqrt{a_{ij}dx^i dx^j}$, is given by

$$dV = \sqrt{\det(a_{ij})}\, dx.$$

The *Busemann-Hausdorff* volume form is defined as: $dV_{BH} = \sigma_{BH}(x)dx$, where

$$\sigma_{BH}(x) = \frac{vol(B^n(1))}{vol\,\{(v^i) \in T_xM : F(x,v) < 1\}}. \tag{2.6}$$

The *Holmes-Thompson* volume form is defined as $dV_{HT} = \sigma_{HT}(x)dx$, where

$$\sigma_{HT}(x) = \frac{1}{vol(B^n(1))} \int_{F(x,v)<1} \det(g_{ij}(x,v))dv. \tag{2.7}$$

Here, $B^n(1)$ is the Euclidean unit ball in $\mathbb{R}^n$ and vol is the *Euclidean volume.* Until unless specifically mentioned in this article, by a volume form, we mean Busemann-Hausdorff volume form.

DEFINITION 2.9 (S-curvature). Suppose $\gamma(t)$ be the geodesic with $\gamma(0) = x$ and $\gamma'(0) = y$ on M. Then the S-curvature $S(x,y)$ is defined as

$$S(x,y) = \frac{d}{dt}[\tau(\gamma(t), \dot{\gamma}(t))]_{t=0},$$

where $\tau(x,y) := \frac{\sqrt{\det(g_{ij}(y))}}{\sigma(x)}$ is the distortion of F and $\sigma(x)$ is a volume form. The Finsler metric F is said to have isotropic S-curvature if there exists a function $c = c(x)$ on M so that $S = (n+1)cF$ and is of constant S-curvature if c is a constant on M.

3. Examples of Finsler metric

3.1. Minkowski normed space. Let F be a Minkowski norm on $\mathbb{R}^n$ (see Definition 2.1), then the pair $(\mathbb{R}^n, F)$ is a Minkowski normed space. A Minkowski normed space $(\mathbb{R}^n, F)$ is a Finsler manifold by canonically identifying each of its tangent spaces with $(\mathbb{R}^n, F)$.

3.2. Berwald metric. Berwald spaces were introduced by Ludwig Berwald himself, where he called them as "affinely connected Finsler spaces"; (see [**10**]). These spaces can be defined as follows:

DEFINITION 3.1 ([**6**]). A Finsler manifold (M, F) is called Berwald if the connection coefficients Γ^i_{jk} are constant on the (slit) tangent space $T_xM \setminus \{0\}$, for every $x \in M$.

Riemannian spaces and Minkowski normed spaces are examples of Berwald space. There are several Berwald spaces which are neither Riemannian nor Minkowski. For instance, here is one such example:

On the manifold $M = \mathbb{S}^2 \times \mathbb{S}^1$ define a Finsler metric F, as follows: $F(x,y) = \sqrt{(\sin^2 x^2)(y^1)^2 + (y^2)^2 + (y^3)^2} + \epsilon y^3$. Here $(x,y) = (x^1, x^2, x^3, y^1, y^2, y^3)$ are coordinates on the tangent bundle TM, with (x^1, x^2) as spherical coordinates on $\mathbb{S}^2$, whereas (x^3) is the coordinate on $\mathbb{S}^1$ and $\epsilon < 1$ is a real constant. It can be verified that this Finsler metric is actually Berwald (For details see section 11.6 B, [**6**]). In fact any Randers metric $F := \sqrt{g} + \beta$ (see Definition 3.3) is Berwald iff the one form β is parallel with respect to the Riemmanian metric g (see Theorem 6.17, [**59**]) and the metric in above example is clearly Randers as well as the one form $\beta = \epsilon dx^3$ is parallel with respect to the Riemmanian metric g.

The Berwald space has interesting property that all tangent spaces are linearly isometric to each other. Further the exponential map is C^∞ in the neighbourhood of origin for every point of the manifold, while it is only C^1 for general Finsler metric (see [**6**]).

3.3. Randers metric. ([**21**]) Let (M, g) be a Riemannian manifold and β be a 1-form on M with $|\beta(v)| < \sqrt{g(v,v)}$ for all $v \in \mathbb{T}M \setminus \{0\}$. Then $F(v) := \sqrt{g(v,v)} + \beta(v)$ is a Finsler structure of M. A Finsler manifold constructed in this way is called a Randers space (named after the physicist Gunnar Randers) and the metric F is called a Randers metric. Randers himself originally considered this metric in connection with general relativity (see [**70**]). Notice that F is reversible if and only if $\beta = 0$ (and then F is Riemannian). One could regard that the 1-form β represents the effect of the wind blown on the Riemannian manifold (M, g). This interpretation leads to a resolution of Zermelo's navigation problem by means of Randers metrics (see subsection 4.1).

3.4. Funk metric. ([**30**]) The Funk metric was introduced as supporting example to solve the *Hilbert fourth problem*: *Find all metrics for which line segments are geodesics.* The Funk metric on unit disc can also be interpreted as Randers metric on the unit disc obtained by the deformation of the well known Klein metric on the disc by a closed 1-form. Let us first introduce Funk distance in a strongly convex domain $\Omega \subset \mathbb{R}^n$. Let $\partial\Omega = \bar{\Omega} \setminus \Omega$ denote the boundary of Ω. For any two points $x_1, x_2; x_1 \neq x_2$ in Ω, let $a_\Omega(x_1, x_2)$ denotes the ray starting at x_1 and passing through x_2.

DEFINITION 3.2 (Funk distance, [**30, 64**]). The *Funk* distance on a strongly convex domain Ω is denoted by $d_{F,\Omega}$ and is defined as: For any x_1 and x_2 in Ω and $a = a_\Omega(x_1, x_2) \cap \partial\Omega$

$$d_{F,\Omega}(x_1, x_2) = \log\left(\frac{|x_1 - a|}{|x_2 - a|}\right),$$

where $|p-q|$ is the Euclidean distance between the points p and q in $\mathbb{R}^n$.

It can be shown that the Funk distance $d_{F,\Omega}$ is realized by the smooth Finsler structures F_F on $T\Omega$, provided the boundary $\partial\Omega$ of Ω is smooth ([**60**], [**78**] §2.3). The Finsler structure F_F on $T\Omega$ is given by,

$$F_F(x,v) = \frac{|v|}{|x-a|}, \tag{3.1}$$

where $x \in \Omega$, $v \in T_x\Omega$. It is interesting to note that the unit ball with respect to Finsler structure F_F at a point $x \in \Omega$ is the domain Ω itself in the tangent space $T_x\Omega \cong \mathbb{R}^n$, but with the point x as centre. If the convex set Ω be the unit ball $\mathbb{B}^n$ then

$$a = \left(x + \frac{v}{F_F(x,v)}\right) \in \partial\mathbb{B}^n \text{ equivalently } |x + \frac{v}{F_F(x,v)}|^2 = 1.$$

Rewriting this condition as

$$F_F^2(1-|x|^2) - 2F_F\langle x,v\rangle - |v|^2 = 0$$

where $\langle x,v\rangle$ denote Euclidean inner product of x and v. The non-negative root of the above quadratic equation is

$$F_F(x,v) = \alpha_F(x,v) + \beta_F(x,v), \tag{3.2}$$

where $\alpha_F(x,v) = \frac{\sqrt{(1-|x|^2)|v|^2+\langle x,v\rangle^2}}{1-|x|^2}$ is the well known Klein metric on the unit ball and $\beta_F(x,v) = \frac{\langle x,v\rangle}{1-|x|^2}$ is a one form on the ball. Thus the metric F_F, given by (3.2) is the Funk metric on $\mathbb{B}^n$, which is actually a Finsler metric of Randers type.

There is another viewpoint of Funk metric on unit $\mathbb{B}^n$, recently observed in [**49**], and explained below. It is well known that the pullback of Lorenzian metric on the upper sheet of the hyperboloid of two sheets is the realization of Klein metric on the unit disc. In [**49**] the authors have constructed a non-positive definite Randers metric in upper half space such that that its pullback on the upper sheet of the hyperboloid of two sheets is the realization of the Funk metric on the unit disk.

Let $(\mathbb{R}^3_+, \alpha_L)$ denote the upper half space $\mathbb{R}^3_+$ with the Lorentzian metric α_L defined below, i.e.,

$$\mathbb{R}^3_+ = \left\{(\tilde{x}^1,\tilde{x}^2,\tilde{x}^3) \in \mathbb{R}^3 : \tilde{x}^3 > 0\right\},$$

$\alpha_L(\tilde{x},\tilde{v}) = \sqrt{(\tilde{v}^1)^2 + (\tilde{v}^2)^2 - (\tilde{v}^3)^2}$ with $\tilde{x} \in \mathbb{R}^3_+$ and $\tilde{v} \in T_{\tilde{x}}\mathbb{R}^3_+ \cong \mathbb{R}^3$.

Now consider the deformation F_L of α_L by $\beta_L = \frac{1}{\tilde{x}^3}d\tilde{x}^3$ in $\mathbb{R}^3_+$ as follows:

$$F_L(\tilde{x},\tilde{v}) = \alpha_L(\tilde{x},\tilde{v}) + \beta_L(\tilde{x},\tilde{v}). \tag{3.3}$$

It should be noted that F_L is *a non-positive definite* Randers metric. Recall that upper half of hyperboloid of two sheets in $\mathbb{R}^3$ is

$$\mathbb{H}_+ = \left\{(\tilde{x}^1,\tilde{x}^2,\tilde{x}^3) \in \mathbb{R}^3 : \tilde{x}^3 = \sqrt{1+(\tilde{x}^1)^2+(\tilde{x}^2)^2}\right\}.$$

We parametrize the upper half portion $\mathbb{H}_+$ of the hyperboloid of two sheets in $\mathbb{R}^3$ as:

$$\eta : \mathbb{D} \subset \mathbb{R}^2 \to \mathbb{H}_+ \subset \mathbb{R}^3_+,$$

$$\eta(x) = \left(\frac{x}{\sqrt{1-|x|^2}}, \frac{1}{\sqrt{1-|x|^2}}\right). \tag{3.4}$$

Note that η is a smooth diffeomorphism between $\mathbb{D}$ and $\mathbb{H}_+$.

PROPOSITION 3.1 ([**49**]). *The pullback of the metric F_L defined as above, on the upper sheet of the hyperboloid of two sheets by the map η is the realization of the Funk metric on the unit disc, i.e., $\eta^* F_L(x, v) = F_F(x, v)$.*

REMARK 4. It is interesting to note that the Randers metric F_L is not positive definite on upper half space, however its pullback on the upper sheet of the hyperboloid of two sheets is the positive definite Funk metric on the unit disk.

3.5. The Teichmüller space. The Teichmüller and the Thurston metrics are natural examples of Finsler metrics on the Teichmüller spaces of a connected oriented surface S (with or without boundary). The Teichmüller space can be defined either as a space of equivalence classes of conformal structures on a connected oriented surface S or as a space of equivalence classes of hyperbolic structures on S. The two metrics considered here are natural from one of these points of view: Teichmüller's metric from the point of view of conformal geometry and Thurston's asymmetric metric from the point of view of hyperbolic geometry. It is interesting to note that the Teichmüller metric does not satisfy the Gromov hyperbolicity (for details see [**55**], [**56**], [**63**], [**65**]).

4. The Finsler metrics of constant/scalar flag/Ricci curvature

As discussed in Definition 2.7, the flag curvature of a Finsler manifold is the analogue to the sectional curvature of a Riemannian manifold. In the two-dimensional case, it reduces to the well-known Gaussian curvature. Finsler manifolds of constant flag curvature have more variety than the Riemannian case. Classifications of those spaces stimulated intensive research. One of the first major contributions was due to Akbar-Zadeh ([**2**]). For simplicity, we state the result in the compact case:

THEOREM 4.1. *Let (M, F) be a compact Finsler manifold of constant flag curvature κ. (i) If $\kappa = 0$, then (M, F) is locally Minkowskian. (ii) If $\kappa < 0$, then (M, F) is Riemannian.*

Although there are rich families of reversible and non-reversible non-compact Finsler manifolds of constant negative flag curvature, for instance, Hilbert and Funk metrics in an open convex subset of $\mathbb{R}^n$. In the positively curved case, we have different phenomena for reversible and non-reversible metrics. On one hand, Bryant ([**13**]) constructed a family of non-reversible Finsler metrics of constant positive flag curvature on the 2-sphere S^2. On the other hand, Kim and Min ([**44**]) showed that any reversible Finsler manifold of constant positive flag curvature is necessarily Riemannian.

In Riemannian geometry, the well known Schur's theorem states that Riemannian manifolds with isotropic sectional curvature have constant curvature. There is also a direct generalization of Schur's theorem in Finsler geometry.

THEOREM 4.2. *(Schur's theorem) ([**6**]) Let (M, F) be an n-dimensional ($n \geq 3$) Finsler manifold. If (M, F) has isotropic flag curvature, i.e., $K = K(x)$, then K must be constant.*

Cheng and Shen [**20**], showed that the Finsler metric with scalar flag and isotropic S- curvature are weakly isotropic flag curvature.

THEOREM 4.3. *Let (M, F) be an n-dimensional Finsler manifold with scalar flag curvature $K = K(x, y)$. If the S-curvature of F is isotropic, then there is a scalar function $\sigma = \sigma(x)$ on M such that*

$$K = 3\frac{c_{x^k} y^k}{F} + \sigma(x).$$

COROLLARY 4.1. *Let (M, F) be an n-dimensional Finsler manifold with scalar flag curvature $K = K(x, y)$. If F has constant S-curvature, then F has isotropic flag curvature $K = K(x)$. In particular, K is constant if $n \geq 3$.*

In Riemannian geometry, the Beltrami theorem states that Riemannian spaces of constant sectional curvature are locally isometric to one of the three model spaces, i.e., Euclidean, hyperbolic and spherical. In Finsler case, it is very far from reach. In particular, Randers spaces of constant flag curvature are classified using the Zermelo navigation description of the Randers metric by Bao, Robles and Shen (see [**5**]), which is discussed in next subsection.

4.1. Zermelo navigation description of Randers metric. In 1931, Zermelo [**96**] posed a problem: Consider a ship is sailing on open, calm water. Suppose a mild wind comes up. How must the ship be steered in order to reach a given destination in the shortest time? Zermelo assumed that the open sea was R^2 with the Euclidean metric. Shen [**80**] generalised this problem to the setting where the sea is an arbitrary differentiable manifold M equipped with a Riemannian metric h. Shen found that when the wind is time-independent, the paths of shortest time are the geodesics of a Randers metric, which is an interesting non Riemannian Finsler metric. Later on, Bao, Robles and Shen [**5**] showed that every strongly convex Randers metric arises as a solution to Zermelo's navigation problem on some Riemannian landscape (M, h), under the influence of an appropriate wind W on M with $|W| < 1$.

PROPOSITION 4.1 ([**5**]). *Let $h = \sqrt{h_{ij}(x)y^i y^j}$ be a Riemannian metric equiped on a manifold M and let $V = V^i(x)(\frac{\partial}{\partial x^i})$ be a vector field on a manifold M with $h(x, -V_x) = \|V\|_x := \sqrt{h_{ij}(x)V^i(x)V^j(x)} < 1$, $x \in M$. Then the corresponding Randers metric F on the manifold M is given by $F(x, y) = \alpha(x, y) + \beta(x, y)$ where $\alpha(x, y) = \sqrt{a_{ij}(x)y^i y^j}$ be a Riemannian metric and $\beta(x, y) = b_i(x)y^i$ be a one form with $a_{ij} = \frac{h_{ij}}{\lambda} + \frac{W_i W_j}{\lambda^2}$, $b_i = \frac{-W_i}{\lambda}$. Here $W_i = h_{ij}W^j$ and $\lambda = 1 - W^i W_i$.*

In particular, there is a canonical Randers metric associated to each Zermelo navigation problem with data (h, W). Conversely let us be given an arbitrary Randers metric F with Riemannian metric $\alpha = \sqrt{a_{ij}(x)y^i y^j}$ and a differential one form $\beta = b_i(x)y^i$ satisfying $\|b\|^2 = a^{ij}b_i b_j < 1$. Set $b^i = a^{ij}b_j$ and $\epsilon = 1 - \|b\|^2$. Construct h and W as

$$h_{ij} = \epsilon(a_{ij} - b_i b_j), W^i = -\frac{b^i}{\epsilon}. \tag{4.1}$$

It can be checked that perturbing h by the stipulated W gives back the Randers metric we started with.

Using the Zermelo navigation description of Randers metric described above, the classification of Randers metric of constant flag curvature was done by Bao, Robles and Shen, (see [**5**]) in 2004. More precisely, they prove that

THEOREM 4.4. *Supppose the Randers metric F solves the Zermelo navigation problem on the Riemannian manifold (M, h) with $h = \sqrt{h_{ij}(x)y^iy^j}$ under the influence of vector field $V = V^i(x)(\frac{\partial}{\partial x^i})$ as in Proposition 4.1. Then F is of constant flag curvature μ if and only if there exists a constant σ such that*

- *h is of constant sectional curvature $(\mu + \sigma^2)$ and*
- *S-curvature is constant, i.e., $S = (n+1)cF$.*

Moreover, σ must vanish whenever h is not flat.

REMARK 5. The constancy of S-curvature in Theorem 4.4 is equivalent to infin itesimal homothety of the vector field V with respect to the Riemannian metric h, i.e., $(\mathcal{L}_V h)_{ij} = V_{i|j} + V_{j|i} = -\sigma h_{ij}$, where $V_{i|j}$ denote the components of covariant derivative of vector field V with respect to Levi-Civita connection of the Riemannian metric h.

In 2005, Mo and Shen, [**58**] proved that if a closed Finsler manifold (M, F) has negative flag curvature with constant S-curvature, then F must be a Randers metric.

In 2014, Yoshikawa and Sabau, [**95**] studied globally defined Kropina metrics as solutions to the Zermelo's navigation problem. Moreover, they characterized the Kropina metrics of constant flag curvature, showing that up to local isometry, there are only two model spaces of them: the Euclidean space and the odd-dimensional spheres.

4.2. Ricci curvature and Einstein metrics.

DEFINITION 4.1 (see section 1.3, [**3**]). Let (M, F) be an n-dimensional Finsler manifold. F is called an Einstein metric if there is a scalar function $K = K(x)$ on M such that the Ricci curvature of F satisfies

$$Ric = (n-1)K(x)F^2. \tag{4.2}$$

The Ricci tensor of F is defined as

$$Ric_{ij} := [\frac{1}{2}Ric]_{y^iy^j} = \frac{1}{2}[R^s_s]_{y^iy^j}. \tag{4.3}$$

Thus F is Einstein if and only if $Ric_{ij} = (n-1)K(x)g_{ij}$. That is, if F is Riemannian, the Finsler-Einstein metric is usual Einstein metric. The above Definition was introduced firstly by Akbar-Zadeh, who wanted to show that Finsler-Einstein metric is a critical point of the Einstein-Hilbert functional as in the Riemannian case. By above Definition, if an n-dimensional Finsler metric F has constant flag curvature K, then its Ricci curvature is $Ric = (n-1)KF^2$, which implies that it is Einstein.

THEOREM 4.5 (see Theorem 9, [**3**]). *Let $F = \alpha + \beta$ be a Randers metric constructed from a Riemannian metric $h = \sqrt{h_{ij}y^iy^j}$ and a vector field $V = V^i\frac{\partial}{\partial x^i}$ by the navigation (see Proposition 4.1). Then F is an Einstein metric satisfying* (4.2) *if and only if there is a constant c such that*

(1) *h is a Riemann-Einstein metric with the Ricci curvature*

$$^hRic = (n-1)(K(x) + c^2),$$

(2) *V is an infinitesimal homothetic vector field with respect to h, i.e.,*

$$(\mathcal{L}_V h)_{ij} = -4ch_{ij}.$$

Furthermore, c must vanish whenever h is not Ricci-flat.

COROLLARY 4.2. *For any Einstein-Randers metric with $n \geq 3$, $K(x)$ in (4.2) is constant.*

REMARK 6. For general Finsler metrics the result of Corollary 4.2 is still an open problem.

PROPOSITION 4.2. *Let F be a three-dimensional Randers metric. Then F is Einstein if and only if F has constant flag curvature.*

THEOREM 4.6. *Let (M, F) be a closed Einstein Randers manifold with constant Ricci curvature Ric:*

- *If $Ric < 0$, then (M, F) is Riemannian.*
- *If $Ric = 0$, then (M, F) is Berwald.*

Chern's question: S.-S. Chern asked the question: *"Does every smooth manifold admit a Finsler Einstein metric"?* Topological obstructions prevent some manifolds, such as $S^2 \times S^1$, from admitting Riemannian Einstein metrics. By the navigation description of Theorem 4.5, any manifold that admits an Einstein Randers metric must also admit a Riemannian Einstein metric. Thus the same topological obstructions prevent Einstein metrics of Randers type. Thus for a Finsler Einstein metric on this 3-manifold, one must looks beyond those of Randers type.

5. Global Riemann-Finsler geometry

5.1. Forward cut locus and forward conjugate locus. Let (M, F) be a forward geodesically complete Finsler manifold, i.e. at each point $x \in M$ the exponential map $\exp_x : T_xM \to (M, F)$ is defined on the whole tangent space. The forward cut locus and forward conjugate locus to a point $x \in (M, F)$, denoted by $C(x)$ and $J(x)$ respectively, are subsets of M and have a significant role in the study of global differential geometry of Finsler manifolds. Let $\gamma_u : [0, a] \to (M, F)$, for a unit vector $u \in \Sigma_x$, be a unit speed geodesic with $\gamma_u(0) := x$, $\dot{\gamma}_u(0) = u$. Define a function $i_x : \Sigma_x \to \mathbb{R}$ by

$$i_x(u) := \sup\{s > 0 \,|\, t = d(x, \gamma_u(t)), \text{ for all } t \in (0, s)\}.$$

The point $\gamma_u(i_x(u))$ is called the *forward cut point to x along γ_u*, $u \in \Sigma_x$. In the case where $i(x) = \infty$, we call x *a forward pole of M*. The forward injectivity radius function at x is defined by

$$i(x) := \inf\{i_x(u) \,|\, u \in \Sigma_x\}.$$

Let $v \in \Sigma_x$ be a unit vector with $g_u(u, v) = 0$. Here g_u is the Riemannian metric defined in Definition 2.4 and will be described in detail in geodesic polar coordinates (see equation(5.2)). We define the Jacobi field $J_{u,v} : [0, a] \to TM$ along γ_u such that $J_{u,v}(0) = 0$, $\nabla_{\frac{\partial}{\partial t}} J_{u,v}(0) = v$. Namely, it is defined by

$$J_{u,v}(t) := d(\exp_x)_{tu} tv, \quad v \in \Sigma_x, \quad g_u(u, v) = 0, \quad t \in [0, a]. \tag{5.1}$$

The forward first conjugate point along γ_u is defined as follows: Let $c_x : \Sigma_x \to (0, \infty)$ be a function defined by

$$c_x(u) := \sup\{\, s > 0 \,|\, \det(d(\exp_x)|_{tu}) \neq 0, \quad \text{for all } t \in (0, s)\}, \quad u \in \Sigma_x.$$

In other words, a non-trivial Jacobi field Y along γ_u, with $J(0) = 0$, exists such that $J(c_x(u)) = 0$ and $J(t) \neq 0$ for all $t \in (0, c_x(u))$. The point $\gamma_u(c_x(u))$ is called the *forward first conjugate point to x along γ_u*.

The forward tangential cut locus $\widetilde{C}(x) \subset T_xM$ and the forward first tangential conjugate locus $\widetilde{J}(x) \subset T_xM$ are defined by

$$\widetilde{C}(x) := \{i_x(u)u \,|\, u \in \Sigma_x\}, \quad \widetilde{J}(x) := \{c_x(u)u \,|\, u \in \Sigma_x\},$$

and their exponential images are the forward cut locus and forward conjugate locus to x respectively, and denoted by

$$C(x) := \exp_x \widetilde{C}(x), \quad J(x) := \exp_x \widetilde{J}(x).$$

The domain containing the origin of T_xM and bounded by $\widetilde{C}(x)$ is denoted by $\widetilde{U}_x$. Clearly we have $\partial\widetilde{U}_x = \widetilde{C}(x)$, and $\widetilde{U}_x \subset T_xM$ is the maximal domain on which $\exp_x$ is an embedding and denote $\exp_x \widetilde{U}_x$ by U_x. We observe from the definition of the cut locus to a point $x \in (M, F)$ that $C(x)$ contains all the information of M. In fact $M \setminus C(x)$ is just an open disk and the identification structure of $\widetilde{C}(x)$ via the exponential map defines the manifold.

Geodesic polar coordinates: We define geodesic polar coordinates around an arbitrary fixed point $x \in M$. Let $\psi : \Sigma_x \times (0, i_x) \to (M, F)$ be defined by

$$\psi(u, t) := \exp_x tu, \quad 0 < t < i_x.$$

The map ψ is a diffeomorphism of $S^{n-1} \times (0, i_x)$ via identification of the indicatrix with the unit sphere $S^{n-1} \subset T_xM$ through the central projection. In view of Definition 2.4 we have a Riemannian metric g_u along γ_u. Let ξ be a radial vector field on U_x, i.e., $\xi(y) := d\exp_{tu}(u)$, $u \in \Sigma_x$, $y = \exp_x(tu)$, $0 < t < i_x$. Thus we have a smooth Riemannian metric g on $U_x \setminus \{x\}$ defined by,

$$g(y) := \bigcup_{y \in U_x} g(y, \xi(y)),\ y \in U_x \setminus \{x\}. \tag{5.2}$$

The polar coordinates centered at x are defined using ψ. All the F-geodesics emanating from x are identified with geodesics as a Riemannian manifold (U_x, g). The well-known first and second variation formulas along a geodesic γ_u with $u \in \Sigma_x$ are valid for (U_x, g). Thus we know that $q := \gamma_u(c_x(u))$ is a conjugate point to x along γ_u if and only if there is a non-trivial Jacobi field J along γ_u such that $J(0) = J(c_x(u)) = 0$. If a unit speed geodesic $\gamma : [0, a] \to (M, F)$ admits a conjugate pair in its interior, then there is a 1-parameter variation $\sigma : (-\epsilon, \epsilon) \times [0, a] \to M$ along σ with $\sigma(t, 0) = x$ and $\sigma(t, a) = \sigma(a)$ for all $t \in (-\epsilon, \epsilon)$, such that all of its variational curves have lengths less than a. This means that

$$c_x(u) \geq i_x(u), \quad \text{for all } u \in \Sigma_x, \quad \text{for all } x \in M. \tag{5.3}$$

We observe that if $\gamma : [0, a] \to (M, F)$ is a minimizing geodesic and if $\gamma(a)$ is conjugate to $\gamma(0)$ along γ, then $\gamma(a)$ is the cut point to $\gamma(0)$ along it.

We now summarize important facts on the properties of cut loci on complete Riemannian and Finsler manifolds [**41**].

PROPOSITION 5.1. *Let (M, F) be a complete Finsler manifold and $x \in M$ an arbitrary fixed point such that $C(x) \neq \emptyset$. If $y \in C(x)$ and if $\gamma : [0, d(x, y)] \to (M, F)$ is a minimizing geodesic with $\gamma(0) = x$, and $\gamma(d(x, y)) = y$, then one of the following holds:*

1. *y is conjugate to x along γ, or*
2. *there exists another minimizing geodesic $\eta : [0, d(x, y)] \to (M, F)$ such that $\eta(0) = x$ and $\eta(d(x, y)) = y$.*

PROPOSITION 5.2. *Let (M, g) be a compact Riemannian manifold and $x \in M$, $y \in C(x) \subset (M, g)$ satisfy (2) in Proposition* 5.1. *If*

$$d(x, y) = i(x) = d(x, C(x)), \tag{5.4}$$

then there are exactly two distinct minimizing geodesics $\gamma, \eta : [0, i(x)] \to (M, g)$ such that $\gamma(0) = \eta(0) = x$, $\gamma(i(x)) = \eta(i(x)) = y$, and $\dot{\gamma}(i(x)) = -\dot{\eta}(i(x))$.

REMARK 7. It turns out that Klingenberg's result in Proposition 5.2 that $\dot{\gamma}(i(x)) = -\dot{\eta}(i(x))$ does not hold for a Finsler manifold, in general. This phenomenon shows an essential difference between Riemannian and Finsler geometry (see [**87**]).

For more results on cut and conjugate loci on complete Riemannian and Finsler manifolds we refer ([**41**, **42**, **87**]).

5.2. Some comparison theorems. As we have seen in subsections 2.3 and 2.5 that the first variation formula is closely related to the geodesic, while second variation formula is related to the flag curvature. As a standard application of the second variation formula, Synge's theorem in Finsler geometry can be shown in the same way as the Riemannian case:

THEOREM 5.1 (Synge's theorem; see Sect. 8.8, [**6**]). *Let (M, F) be a forward complete, oriented, even-dimensional Finsler manifold whose flag curvature is bounded from below by a positive constant. Then M is simply connected.*

The Riemannian characterization of the Finsler curvature has direct applications in comparison geometry. For instance, in Bonnet–Myers and Cartan–Hadamard theorems:

THEOREM 5.2 (Bonnet–Myers theorem; see Sec. 7.7, [**6**]). *Let the Finsler manifold (M, F) is forward or backward complete and satisfies $Ric \geq K$ for some $K > 0$, then $diam(M) = \frac{1}{\pi}\sqrt{\frac{n-1}{K}}$. In particular, M is compact and has finite fundamental group.*

THEOREM 5.3 (Cartan–Hadamard theorem; see Sec. 8.2, [**59**]). *Let the Finsler manifold (M, F) is forward complete and satisfies $K \leq 0$. Then we have the following:*

(1) *Any geodesic does not have conjugate points.*
(2) *For any $x \in M$, the exponential map $exp_x : T_xM \to M$ is a C^1-covering.*
(3) *If M is simply connected, then $exp_x : T_xM \to M$ is a C^1-diffeomorphism for every $x \in M$, and its restriction $exp_x : T_xM \setminus \{0\} \to M \setminus \{x\}$ is a C^∞-diffeomorphism.*

Volume comparison theorems in Riemannian geometry were given by P. Günther and R. L. Bishop in 1960's ([**18**]). Later, they are widely applied in global differential geometry and geometric analysis. In particular, the Bishop-Gromov volume comparison theorem plays a very important role in the global differential geometry of Riemannian manifolds. One of its applications is the Gromov precompactness theorem. Because of their complexity, such comparison theorems in Finsler geometry were open until the end of the last century. The main reason behind the complexity was that in Riemannian manifolds, there is a canonical notion of the volume form. However, the situation is different in Finsler geometry. The Finsler volume form can be defined in various ways, and essentially different results may be obtained. Therefore, it is an interesting and important problem to investigate the relations between the volume forms and the geometric properties of a Finlser manifold. A Finsler volume form used frequently is the Busemann– Hausdorff volume form, with respect to which Z. Shen obtained the following Bishop–Gromov type volume comparison theorem ([**77**]):

THEOREM 5.4. *Let (M, F) be a forward complete Finsler n-manifold satisfying $Ric \geq (n-1)k$ and $S_{BH} \geq (n-1)h$. Then the function*

$$\mathcal{P}(r) = \frac{\mu_{BH}(B^+(p,r))}{V_{k,h,n}(r)}$$

is decreasing and for any $r > 0$, $\mu_{BH}(B^+(p,r)) \leq V_{k,h,n}(r)$, where

$$V_{k,h,n}(r) := c_{n-1} \int_0^r [e^{-ht} \mathscr{S}_k(t)]^{n-1} dt,$$

c_{n-1} is the volume of the standard Euclidean unit $(n-1)$-sphere, $\mathscr{S}_k(t)$ is the unique solution to $y'' + ky = 0$ with $y(0) = 0$ and $y'(0) = 1$, and S_{BH} is the S-curvature with respect to Busemann–Hausdorff volume form.

Using this theorem Shen also obtained some precompactness and finiteness theorems for Finsler manifolds. Using the Laplacian comparison theorem in Finsler geometry, Wu and Xin ([**92**]) investigated a different proof of the above theorem. Besides extension of the volume comparison theorem in Finsler geometry, some other comparison theorems like Berger–Kazdan comparison theorem and Santalo's formula have also been extended in Finsler geometry. For further details on comparison theorem in Finsler geometry see ([**77, 79, 81, 92, 97**]).

5.3. Gauss-Bonnet-Chern theorem. Recall that the Gauss-Bonnet-Chern formula for an oriented closed n-dimensional Riemannian manifold (M, g) is

$$\int_M \Omega = \chi(M) \tag{5.5}$$

where $\chi(M)$ is the Euler characteristic of M and

$$\Omega = \begin{cases} \frac{(-1)^p}{2^{2p}\pi^p p!} \epsilon_{i_1,\ldots,i_{2p}} \Omega^{i_2}_{i_1} \wedge \ldots \Omega^{i_{2p}}_{i_{2p-1}}, & n = 2p. \\ 0, & n = 2p+1. \end{cases} \tag{5.6}$$

here $\epsilon_{i_1,\ldots,i_{2p}}$ is the multi-Kronecker delta and Ω^j_i is the local curvature 2-forms on (M, g). In partcular, when M is a closed surface $\int_M K dA = 2\pi\chi(M)$, where K is the Gaussian curvature of the surface. S.-S. Chern ([**22**]) gave an intrinsic proof of this formula. It is interesting to note here that the Gauss-Bonnet-Chern formula (5.5) is the simplest case of the Atiyah-Singer Index Theorem. In the 1949, Lichnerowicz ([**51**]) and Busemann ([**15**]) tried to generalize (5.5) to Finsler manifolds but could not succeeded in general set up. One of the main difficulties is that there is no connection on non-Riemannian Finsler manifold which is both "torsion-free" and "metric-compatible". In 1996, Bao and Chern ([**4**]) reconsidered this problem and established the GBC theorem for Finselr manifolds with $V(x) =$ constant using Chern connection, here $V(x)$ denotes the Riemannian volume of indicatrix at the point x. In 2002, Lackey ([**50**]) used a trick to deal with $V(x)$ and generalized the Gauss-Bonnet-Chern to any Finsler manifolds. In fact, Lackey established a GBC formula for any torsion-free connection:

THEOREM 5.5 ([**50**]). *Let (M, F) be an n-dimensional oriented closed Finsler manifold, and $X : M \to TM$ be a vector field possibly with isolated zeros $\{x_i\} \subset M$. Let ∇ be any torsion-free Finsler connection with curvature form Ω^∇ defined by*

$$\Omega^\nabla = \begin{cases} \frac{(-1)^p}{2^{2p}\pi^p p!}\epsilon_{i_1,\dots,i_{2p}}\Omega_{i_1}^{i_2} \wedge \dots \Omega_{i_{2p-1}}^{i_{2p}}, & n = 2p. \\ 0, & n = 2p+1. \end{cases} \tag{5.7}$$

where Ω_j^i is the local curvature 2-forms of the connection ∇. Then we have

$$\int_M ([X]^* \frac{\Omega + \mathfrak{D}}{V(x)}) = \frac{\chi(M)}{Vol(\mathbb{S}^{n-1})}, \tag{5.8}$$

where $\chi(M)$ is the Euler characteristic of M , $[X] : M \cup \{x_i\} \to SM$ is the section induced by X, $V(x)$ is the Riemannian volume of the indicatrix S_xM , and $\mathfrak{D}$ is an n-form on SM.

Recently, Zhao and Yuan ([**98**]) also provided a simple proof of the Gauss-Bonnet-Chern formula for any metric-compatible connection from the point of view of index theory.

5.4. The Sphere Theorem. In Riemannian geometry, a metric with constant positive sectional curvature on a compact, simply connected manifold is isometric to the standard sphere of the same curvature. The classical Sphere Theorem states that a compact, simply connected manifold of dimension n with sectional curvature K such that $1/4 < K \leq 1$ everywhere is homeomorphic to the n-sphere ([**1, 47**]). The earlier contributions are due to M. Berger, H. Rauch and V. A. Toponogov ([**7**]). The proof uses an estimate for the injectivity radius and the Toponogov comparison theorem for geodesic triangles. In ([**45**]) it is shown that one can prove the Sphere Theorem without making use of Toponogov's comparison result for geodesic triangles. Instead, one uses Morse theory of the energy functional on the space of curves between two fixed points and on the space of loops. The injectivity radius is bounded from below by π, so geodesic loops have length at least 2π and their Morse index is bounded below by $n-1$. This implies that the loop space is $(n-2)$-connected and therefore the manifold is homotopy equivalent to the n-sphere. Though in Finsler geometry the condition of constant positive flag curvature no longer determines the metric up to isometry, one can show using the exponential map that a simply connected Finsler manifold of constant positive flag curvature is homeomorphic to the n-sphere. P. Dazord ([**26, 27**]) remarked that in the case of a reversible Finsler metric, one can carry over the Morse theoretic proof of the Sphere Theorem found in ([**45**]). The original proof of the Sphere Theorem does not carry over since the triangle comparison result cannot be extended to the Finsler case. In 2004, Rademacher ([**68**]) showed that the estimates for the injectivity radius, the length of a nonminimal geodesic between two fixed points, the length of a nonconstant geodesic loop and the length of a nonconstant closed geodesic can be extended to the case of a nonreversible Finsler metric by introducing the notion of reversibility constant. Rademacher derived from a length estimate the following Sphere Theorem for non-reversible Finsler metrics:

THEOREM 5.6 ([**68**]). *If a simply connected and compact Finsler manifold M of dimension $n \geq 3$ with reversibility λ and with flag curvature K satisfies $0 < (1 - \frac{1}{1+\lambda})^2\delta \leq K < 4\delta$, then M is homotopy equivalent to $\mathbb{S}^n$.*

It is still an open problem whether one can improve the Sphere Theorem in the nonreversible case by choosing the lower curvature bound $1/4$ as in the reversible case. The sphere theorem in Riemannian geometry was extended by Grove and Shiohama [**36**], which states that "For a complete connected Riemannian manifold M satisfying $0 < \delta \leq K$ and $d(M) > \frac{\pi}{2\sqrt{\delta}}$, M is homeomorphic to $\mathbb{S}^n$". This result is further extended in the Finsler setting in 2015 by K. Kondo [**48**], under certain restrictions of radial tangent curvature.

6. Busemann functions and convex functions

6.1. Busemann functions. It is known that Busemann functions play an important role in the investigation of the geometry of noncompact complete Riemannian manifolds with negative sectional curvature and harmonic manifolds ([**83**]). The convexity of Busemann functions is essential for the study of a complete simply connected Riemannian manifold of non positive curvature, what is called *Hadamard Riemannian manifolds* (see section 3.G.2, [**31**]). An effective tool to study various topics in differential geometry, such as the structure of harmonic spaces in Riemannian geometry, is Busemann functions ([**67**]). On the other hand, Busemann functions have been used in the study of reversible Finsler manifolds of negative flag curvature [**28**], the splitting theorems for Finsler manifolds of non-negative Ricci curvature [**61**] and recently, the relation between affine functions and Busemann functions on a complete Finsler manifold are treated in [**40**]. The Busemann function has its origin in §22 of [**17**] by Busemann.

DEFINITION 6.1. A forward Busemann function $F_\gamma : (M, F) \to \mathbb{R}$ on the Finsler manifold (M, F) for a forward ray $\gamma : [0, \infty) \to (M, F)$ is defined as follows:

$$F_\gamma(x) := \lim_{t\to\infty}(t - d(x, \gamma(t))), \quad x \in M.$$

A backward ray $\gamma : (-\infty, 0] \to (M, F)$ and a backward Busemann function for the backward ray γ are similarly defined. A *super Busemann function* F_x : $(M, F) \to \mathbb{R}$ *at* x is defined by

$$F_x(y) := \sup\{\, F_\gamma(y) \mid \gamma \text{ is a forward ray with } \gamma(0) = x\}, \quad y \in M.$$

Clearly, the function $t \to (t - d(x, \gamma(t)))$ is monotone increasing in t and bounded above by $d(x, \gamma(0))$. Thus F_γ is well defined, for $t - d(x, \gamma(t))$ converges uniformly on a compact set and F_γ is locally Lipschitz continuous. A unit speed forward ray $\eta : [0, \infty) \to (M, F)$ is by definition *asymptotic to* γ if there exists a sequence of unit speed minimizing geodesics $\{\eta_j : [0, \ell_j] \to (M, F)\}_{j=1,2,\ldots}$ such that $\lim_{j\to\infty} \dot\eta_j(0) = \dot\eta(0)$, $\eta_j(\ell_j) = \gamma(t_j)$ for a monotone divergent sequence $\{t_j\}$. We denote by $F_\gamma^{-1}(\{a\})$ and $F_\gamma^{-1}(-\infty, a]$ the a-level set and the a-sublevel set of F_γ respectively. The basic properties of Busemann functions are stated in §§22 and 23 of [**17**] and those on complete Riemannian and Finsler manifolds (M, F) are summarized in [**84**] and [**85**] as follows:

PROPOSITION 6.1. *Let* $\gamma : [0, \infty) \to (M, F)$ *be a forward ray and* $F_\gamma : (M, F) \to \mathbb{R}$ *a Busemann function for* γ. *We then have:*

(1) F_γ *is locally Lipschitz.*
(2) *A level set* $F_\gamma^{-1}(\{a\})$ *for* $a \in F_\gamma(M)$ *is obtained by*

$$F_\gamma^{-1}(\{a\}) = \lim_{t\to\infty} S^{-1}(\gamma(t), t - a).$$

(3) *If* $a, b \in F_\gamma(M)$ *satisfies* $a \leq b$, *then*

$$F_\gamma^{-1}(-\infty, a] = \{y \in F_\gamma^{-1}(-\infty, b] \mid d(y, F_\gamma^{-1}(\{b\}) \geq b - a\},$$

and

$$\begin{aligned} F_\gamma^{-1}(\{a\}) &= \{y \in F_\gamma^{-1}(-\infty, b]) \mid d(y, F_\gamma^{-1}(\{b\})) = b - a\} \\ &= S^{-1}(F_\gamma^{-1}(\{b\}), b - a) \cap F_\gamma^{-1}(-\infty, b]. \end{aligned}$$

(4) *A unit speed geodesic* $\eta : [0, \infty) \to (M, F)$ *is a forward ray asymptotic to* γ *if and only if*

$$F_\gamma \circ \eta(t) = t + F_\gamma \circ \eta(0), \quad \textit{for all } t \geq 0.$$

(5) *If* $x \in M$ *and* $a \in F_\gamma(M)$ *satisfy* $a > F_\gamma(x)$, *and if* $\eta : [0, \ell] \to (M, F)$ *is a unit speed minimizing geodesic with* $\eta(0) = x$ *such that* $\eta(\ell)$ *is a foot of* x *on* $F_\gamma^{-1}(\{a\})$, *then the extension* $\eta : [0, \infty) \to (M, F)$ *of* η *is a forward ray asymptotic to* γ.

(6) F_γ *is differentiable at a point* $x \in M$ *if* x *is an interior of some ray asymptotic to* γ.

The detailed proof of Proposition 6.1 on Riemannian manifolds was given in [**84**] and the same proof technique for the Finsler case is seen in [**85**].

In the pioneering works [**35**] and [**19**], the authors have proved that a Busemann function on a complete and non-compact Riemannian manifold (M, g) is strongly convex if its sectional curvature is positive (see [**35**]) and convex if its sectional curvature is non-negative (see [**19**]). In particular, every super Busemann function is a convex exhaustion if its sectional curvature is non-negative. If the minimum set of a super Busemann function has non-empty boundary, then the negative of the distance function on the minimum set to the boundary is convex, and hence attains its minimum. Thus by iterating this, a totally convex compact totally geodesic submanifold without boundary, called a soul of M, is found in the minimum set. The well-known Cheeger-Gromoll structure theorem (see [**19**]) states that a complete non-compact Riemannian manifold is homeomorphic to the normal bundle over the soul in M. If the sectional curvature is positive, the soul is a point, and hence M is diffeomorphic to $\mathbb{R}^n$. However, anologues of the above results stated in [**35**], [**19**] are still open problem on complete Finsler manifolds with positive (or non-negative) flag curvature.

6.2. Convex functions. A set $U \subset M$ is by definition *convex* if any pair of points $x, y \in U$ is joined by a unique minimizing forward geodesic whose image is contained entirely in U(see [**74**]). The existence of a convex ball centered at every point on (M, F) is guaranteed by the Whitehead Convexity Theorem (see Theorem 3.1, [**87**]). Let $B(x, \delta(x))$ be a convex $\delta(x)$-ball around x. A set $U \subset M$ is called *locally convex* if every $x \in U$ has the property that $U \cap B(x, \delta(x))$ is convex. A set $U \subset M$ is called *totally convex* if every forward geodesic joining two points in U is contained entirely in U.

A function $\varphi : (M, F) \to \mathbb{R}$ is said to be *convex* if along every geodesic $\gamma : [a, b] \to (M, F)$, the restriction $\varphi \circ \gamma : [a, b] \to \mathbb{R}$ is convex:

$$\varphi \circ \gamma((1 - \lambda)a + \lambda b) \leq (1 - \lambda)\varphi \circ \gamma(a) + \lambda \varphi \circ \gamma(b), \quad 0 \leq \lambda \leq 1 \tag{6.1}$$

If the inequality in the above (6.1) is strict for all γ and for all $\lambda \in (0, 1)$, φ is called *strictly convex*, and *strongly convex* if the second order difference quotient, namely $\{\varphi o\gamma(h) - \varphi o\gamma(-h) - 2\varphi o\gamma(0)\}/h^2$ is positive for all γ and all $\lambda \in (0, 1)$ (**[74, 87]**). In the special case where equality in (6.1) holds for every γ and for every $\lambda \in [0, 1]$, the function is called an *affine function.* If a non-trivial convex function φ is constant on an open set, then it assumes its minimum on the open set and the number of components of a level set $\varphi^{-1}(\{a\})$, $a > \inf_M \varphi$, is equal to that of the boundary components of the minimum set of φ. A convex function φ is said to be *locally non-constant* if it is not constant on any non-empty open set of M. From now on, *we assume that our convex functions are locally non-constant.*

The slope inequality of a one variable convex function is very much applicable in the analysis of convex functions on Riemannian or Finsler manifolds. Let $f : (\alpha, \beta) \to \mathbb{R}$ be a convex function. Let $\alpha < a < b < c < d < \beta$ and $A := (a, f(a))$, $B := (b, f(b))$, $C := (c, f(c))$ and $D := (d, f(d))$ be points on the graph of f. The slope inequality is expressed as

$$\text{slope}(AB) \leq \text{slope}(AC) \leq \text{slope}(BC) \leq \text{slope}(BD) \leq \text{slope}(CD). \tag{6.2}$$

We observe from (6.2) that the right and left derivatives $f'_+(t)$ and $f'_-(t)$ of f exist and $f'_+(t) \geq f'_-(t)$ for all $t \in (\alpha, \beta)$, and the equality holds if and only if f is differentiable at t.

The topology of Riemannian manifolds admitting locally nonconstant convex functions has been investigated in **[33]** and **[32]**. The topology of complete Alexandrov surfaces admitting locally nonconstant convex functions has been studied in **[54]** and **[52]**. The classification of Busemann G-surfaces admitting convex functions has been obtained in **[37]**. The isometry groups of complete Riemannian manifolds admitting strictly convex functions have been discussed in **[93]**. There are several extensions of convex functions, such as peakless functions introduced by Busemann **[17]**, uniformly locally convex filtrations **[94]**, etc. The splitting theorem for Alexandrov surfaces admitting affine functions has been established in **[53]**. Also in **[54]**, the condition for compact Alexandrov surfaces to admit locally non-constant convex functions has been studied. A detailed discussion of convex sets on Riemannian manifolds of non-negative curvature is carried out in **[14]**. It is emphasized that the notion of convexity makes sense, irrespective of whether the distance function is symmetric or not.

The assumption for a convex function to be locally non-constant is really necessary. As we can construct on every noncompact manifold a complete Riemannian metric and a non-trivial smooth convex function whose minimum set contains a non empty open set. Therefore the existence of such a non-trivial convex function gives no restriction on the topology of a manifold. Without assuming the continuity of a convex function on (M, F), we have the following Proposition. For its proof see (**[33]**, **[74]**):

PROPOSITION 6.2. *Any convex function on (M, F) is locally Lipschitz.*

PROPOSITION 6.3 (compare Proposition 2.3 in **[33]**). *Let $\varphi : (M, F) \to \mathbb{R}$ be a convex function and $a > \inf_M \varphi$. Then the a-level set $\varphi^{-1}(\{a\})$ is a topological submanifold of dimension $n - 1$.*

PROPOSITION 6.4. *Let $C \subset (M, F)$ be a closed locally convex set. Then there exists a totally geodesic submanifold W of M such that $W \subset C$ and its closure is C.*

An elementary observation based on the slope inequality (6.2) gives the following facts on a locally non-constant convex function $\varphi : (M, F) \to \mathbb{R}$ (and $\varphi : (M, g) \to \mathbb{R}$).

LEMMA 6.1 ([**87**]). *If there exists a compact level set $\varphi^{-1}(\{a\})$ of a convex function φ on a Rieamannian manifold (M, g), then so are all the other level sets. However, in Finsler case if $\varphi^{-1}(\{a\}) \subset (M, F)$ is compact, then so are $\varphi^{-1}(\{b\})$ for all $b \geq a$.*

LEMMA 6.2 ([**87**]). *Let $\varphi : (M, F) \to \mathbb{R}$ be a locally non-constant convex function. Let $\varphi^{-1}(\{a\})$ be a compact level set. Then, $\varphi^{-1}[a, b]$ for a fixed $b > a$ is homeomorphic to the product $\varphi^{-1}(\{a\}) \times [a, b]$.*

In particular for $n = 2$, we have the following classification theorem for non-compact surfaces in presence of a locally non-constant convex function on it:

PROPOSITION 6.5 ([**33**]). *Let M is a complete Riemannian manifold of dimension two. If there is a locally non-constant convex function on M, then M is diffeomorphic to either plane $\mathbb{R}^2$ or the cylinder $S^1 \times \mathbb{R}$ or the open Mobius strip.*

For a ray $\gamma : [0, \infty) \to M$ (resp., $\gamma^- : (-\infty, 0] \to M$), a ray $\alpha : [0, \infty) \to M$ (resp., $\alpha^- : (-\infty, 0] \to M$) is called a *co-ray* to γ (resp., γ^-), if there exists a sequence of minimal geodesics $\alpha_k : [0, l_k] \to M$ (resp., $\alpha_k^- : [-l_k, 0] \to M$) such that $\alpha_k(0) \to \alpha(0)$ and $\alpha_k(l_k) \to \gamma(t_k)$ (resp., $\alpha_k^-(0) \to \alpha^-(0)$ and $\alpha_k^-(l_k) \to \gamma^-(t_k)$) with $t_k \to \infty$(resp., $t_k \to -\infty$) as $k \to \infty$. For a straight line $\gamma : (-\infty, \infty) \to M$, a straight line $\alpha : (-\infty, \infty) \to M$ is called parallel to γ if $\alpha|_{[s,\infty)}$ and $\alpha|_{(-\infty,s)}$ are co-rays to $\gamma|_{[0,\infty)}$ and $\gamma|_{(-\infty,0)}$, respectively, for any $s \in (-\infty, \infty)$. Recently, Finsler maniflod with affine function has been classified as follows:

THEOREM 6.1 ([**40**]). *Let φ be an affine function on (M, F) and $p \in M$. If φ is not constant, then there exists a unique straight line $\gamma_p : (-\infty, \infty) \to M$ with $\gamma_p(0) = p$ and a constant $a = a(\gamma_p)$ such that $\varphi = aF_{\gamma_p} + b$, where $b = \varphi(p)$. In particular, (M, F) is topologically a product manifold $[\varphi = 0] \times \mathbb{R}$. Further, those straight lines $\gamma_p, p \in M$, are parallel to each other.*

REMARK 8. In above theorem if (M, F) is Riemannian, then the parallel condition is replaced by "Riemannian product".

Let $\varphi : (M, g) \to \mathbb{R}$ be a locally non-constant convex function on a complete Riemannian manifold. The Sharafutdinov construction of flow curves along $-\operatorname{grad}(\varphi)$ implies that the diameter function $t \mapsto \operatorname{diam}(\varphi^{-1}(\{t\}))$ is monotone non-decreasing. Here

$$\operatorname{diam}(\varphi^{-1}(\{t\})) := \sup\{d(x, y) \mid x, y \in \varphi^{-1}(\{t\}) \subset (M, g)\}. \tag{6.3}$$

The monotone property of $t \mapsto \operatorname{diam}(\varphi^{-1}(\{t\}))$ may be roughly explained as follows:

Let $C \subset (M, g)$ be a closed convex set and $x, y \in M \setminus C$ be taken sufficiently close to C such that there exist unique foot $f(x), f(y)$ of x, y on C respectively. We observe that $d(x, y) \geq d(f(x), f(y))$. This shows that the diameter function has everywhere non-negative derivative (See [**32**], [**33**]). We therefore get that if φ admits a level set that is compact, then so are all the others.

However, it is not known yet, whether the monotone property of the diameter function is valid for Finsler manifolds. Irrespective of the monotonicity of the diameter function, the following result can be derived using Lemma 6.2:

THEOREM 6.2 (see Theorem 1.1-1.2 in [**74**]). *Let (M, F) be a complete Finsler manifold and $\varphi : (M, F) \to \mathbb{R}$ be a locally non-constant convex function whose level sets are all compact. Then we have the following:*

(1) *If $\varphi^{-1}(\{c\})$ is connected for some $c > \inf_M \varphi$ and $\inf_M \varphi$ is not attained, then there exists a homeomorphism $H : \varphi^{-1}(\{c\}) \times (\inf_M \varphi, \infty) \to M$ such that $H(x, t) \in \varphi^{-1}(\{t\})$ for all $(x, t) \in \varphi^{-1}(\{c\}) \times (\inf_M \varphi, \infty)$.*

(2) *If $\varphi^{-1}(\{c\})$ is connected and φ attains its infimum, say $m := \inf_M \varphi$, then M is homeomorphic to the normal bundle over $\varphi^{-1}(\{m\})$ in M.*

(3) *If there is a disconnected level set, then φ attains its minimum $m = \inf_M \varphi$, and $\varphi^{-1}(\{m\})$ is a compact totally geodesic smooth hypersurface with trivial normal bundle. Moreover, M is homeomorphic to $\varphi^{-1}(\{m\}) \times \mathbb{R}$.*

REMARK 9. Without the assumption of the existence of a compact level set of $\varphi : (M, g) \to \mathbb{R}$, all the above statements are still valid in the Riemannian case. However, it is not known yet in the Finsler case.

The number of ends of complete Riemannian manifolds admitting locally non-constant convex functions is estimated by using the slope inequality (6.2), Lemma 6.1 and Theorem 6.2.

THEOREM 6.3 (Ends of (M, g), [**33**]). *Let (M, g) be a connected geodesically complete Riemannian manifold admitting a locally nonconstant convex function φ.*

(1) *If φ has a noncompact level set, then M has one end.*

(2) *If φ assumes its minimum and if it admits a compact level set, then M has one end.*

(3) *If φ has a disconnected compact level set, then M has two ends.*

(4) *If φ has a compact level set and if its infimum is not attained, then M has two ends.*

However the ends of geodesically complete Finsler manifolds admitting locally non-constant convex functions have not been fully understood yet. We do not know any example of a convex function $\varphi : (M, F) \to \mathbb{R}$ with compact and non-compact level sets simultaneously.

6.3. The properness of exponential map and isometry group.

DEFINITION 6.2. A map $f : M \to \mathbb{R}$ is said to be *proper* if $f^{-1}(K)$ is compact for all compact set $K \subset \mathbb{R}$ and f is said to be an *exhaustion* of M if $f^{-1}(-\infty, a]$ is compact for all $a \in f(M)$.

The slope inequality of convex functions along geodesics leads us to the properness of the exponential maps on manifolds with convex functions. Clearly the exponential map $\exp_x : T_xM \to M$ at each point on a complete and simply connected Riemannian manifold M gives a diffeomorphism, and hence it is proper. The properness of exponential map on Finsler manifold has been extended as follows:

THEOREM 6.4 (see [**43**]). *If (M, F) is a geodesically complete non-compact Finsler manifold, and if $\varphi : (M, F) \to \mathbb{R}$ is a strictly convex exhaustion function, then the exponential map at each point of (M, F) is proper.*

Notice that the exhaustion property in Theorem 6.4 is needed for the conclusion to hold. For instance, let $\mathcal{F} \subset \mathbb{R}^3$ be a surface of revolution with profile curve $y = e^x$, $x \in \mathbb{R}$. Then the exponential map is not proper at any point of $\mathcal{F}$.

Let (H,g) be a Hadamard manifold, namely H is a complete and simply connected Riemannian manifold of non-positive sectional curvature. Then the distance function $d(x,.)$ from a fixed point $x \in H$ is convex, with a unique minimum point at x. A well-known classical theorem by Cartan states that if G is a compact subgroup of the isometry group $I(H)$ of H, then it has a common fixed point. In fact, if $x \in H$ is a fixed point, then the G-orbit $G(x)$ of x is compact, and hence there exists a unique smallest ball $B(y,r)$ with $G(x) \subset \bar{B}(y,r)$. Clearly $B(y,r)$ is invariant under the action of G, and hence the center y is fixed under the actions of G.

We finally discuss how the existence of a convex function on (M,g) and (M,F) influences the group of isometries on them. The splitting theorem for Riemannian manifolds admitting affine functions has been discussed in [**39**]. It is proved in [**34**] that if (M,g) is a complete Riemannian manifold with non-compact isometry group and if (M,g) admits a convex function without minimum whose level sets are all compact, then (M,g) is isometric to the Riemannian product $N \times \mathbb{R}$, where N is a compact smooth manifold. In [**19**] Cheeger and Gromoll constructed the compact totally convex filtration obtained by a super Busemann function on a complete non-compact Riemannian manifold of non-negative sectional curvature. They proved:

THEOREM 6.5. *A complete Riemannian manifold (M,g) of non-negative sectional curvature splits off isometrically as the product:*

$$M = \overline{M} \times \mathbb{R}^k,$$

where the isometry group $I(\overline{M})$ of $\overline{M}$ is compact and $I(M) = I(\overline{M}) \times I(\mathbb{R}^k)$.

Without assuming that the sectional curvature is non-negative, there are some results on the relation between the isometry groups and convex functions defined on (M,g) and on (M,F) respectively.

THEOREM 6.6 (see [**93**]). *Let (M,g) be a complete Riemannian manifold admitting a strictly convex function $\psi : (M,g) \to \mathbb{R}$. We then have*

(1) *If ψ admits a minimum, then every compact subgroup G of the isometry group of (M,g) has a common fixed point.*
(2) *If ψ has a compact level set and if it has no minimum, then the group of isometries of (M,g) is compact.*

We know very little about the isometry groups of complete Finsler manifolds admitting convex functions. More precisely, what we have is the following:

THEOREM 6.7 (see [**43**]). *Let $\psi : (M,F) \to \mathbb{R}$ be a strictly convex exhaustion function. Then every compact subgroup of the group of isometries on (M,F) has a common fixed point.*

References

[1] Uwe Abresch and Wolfgang T. Meyer, *Injectivity radius estimates and sphere theorems*, Comparison geometry (Berkeley, CA, 1993), Math. Sci. Res. Inst. Publ., vol. 30, Cambridge Univ. Press, Cambridge, 1997, pp. 1–47. MR1452866

[2] H. Akbar-Zadeh, *Bulletins de l'Académie Royale de Belgique Année*, **74** (1988), 281–322.

[3] David Bao and Colleen Robles, *Ricci and flag curvatures in Finsler geometry*, A sampler of Riemann-Finsler geometry, Math. Sci. Res. Inst. Publ., vol. 50, Cambridge Univ. Press, Cambridge, 2004, pp. 197–259. MR2132660

[4] David Bao and S. S. Chern, *A note on the Gauss-Bonnet theorem for Finsler spaces*, Ann. of Math. (2) **143** (1996), no. 2, 233–252, DOI 10.2307/2118643. MR1381986

[5] David Bao, Colleen Robles, and Zhongmin Shen, *Zermelo navigation on Riemannian manifolds*, J. Differential Geom. **66** (2004), no. 3, 377–435. MR2106471

[6] D. Bao, S.-S. Chern, and Z. Shen, *An introduction to Riemann-Finsler geometry*, Graduate Texts in Mathematics, vol. 200, Springer-Verlag, New York, 2000, DOI 10.1007/978-1-4612-1268-3. MR1747675

[7] M. Berger, *Riemannian geometry during the second half of the twentieth century*, Jahresber. Deutsch. Math.-Verein. **100** (1998), no. 2, 45–208. MR1637246

[8] Marcel Berger, *Sur les variétés à courbure positive de diamètre minimum* (French), Comment. Math. Helv. **35** (1961), 28–34, DOI 10.1007/BF02567002. MR139121

[9] Marcel Berger, *Sur quelques variétés riemanniennes suffisamment pincées* (French), Bull. Soc. Math. France **88** (1960), 57–71. MR133781

[10] L. Berwald, *On Finsler and Cartan geometries. III. Two-dimensional Finsler spaces with rectilinear extremals*, Ann. of Math. (2) **42** (1941), 84–112, DOI 10.2307/1968989. MR3992

[11] L. Berwald, *Ueber Systeme von gewöhnlichen Differentialgeichungen zweiter Ordnung deren Integralkurven mit dem System der geraden Linien topologisch aequivalent sind* (German), Ann. of Math. (2) **48** (1947), 193–215, DOI 10.2307/1969166. MR21178

[12] Arthur L. Besse, *Manifolds all of whose geodesics are closed*, Ergebnisse der Mathematik und ihrer Grenzgebiete [Results in Mathematics and Related Areas], vol. 93, Springer-Verlag, Berlin-New York, 1978. With appendices by D. B. A. Epstein, J.-P. Bourguignon, L. Bérard-Bergery, M. Berger and J. L. Kazdan. MR496885

[13] Robert L. Bryant, *Projectively flat Finsler 2-spheres of constant curvature*, Selecta Math. (N.S.) **3** (1997), no. 2, 161–203, DOI 10.1007/s000290050009. MR1466165

[14] Ju. D. Burago and V. A. Zalgaller, *Convex sets in Riemannian spaces of nonnegative curvature* (Russian), Uspehi Mat. Nauk **32** (1977), no. 3(195), 3–56, 247. MR464119

[15] Herbert Busemann, *Angular measure and integral curvature*, Canad. J. Math. **1** (1949), 279–296, DOI 10.4153/cjm-1949-024-7. MR30808

[16] Herbert Busemann, *Selected works. I*, Springer, Cham, 2018. Edited by Athanase Papadopoulos. MR3822319

[17] Herbert Busemann, *The geometry of geodesics*, Academic Press, Inc., New York, 1955. MR75623

[18] Jeff Cheeger and David G. Ebin, *Comparison theorems in Riemannian geometry*, AMS Chelsea Publishing, Providence, RI, 2008. Revised reprint of the 1975 original, DOI 10.1090/chel/365. MR2394158

[19] Jeff Cheeger and Detlef Gromoll, *On the structure of complete manifolds of nonnegative curvature*, Ann. of Math. (2) **96** (1972), 413–443, DOI 10.2307/1970819. MR309010

[20] Xinyue Chen, Xiaohuan Mo, and Zhongmin Shen, *On the flag curvature of Finsler metrics of scalar curvature*, J. London Math. Soc. (2) **68** (2003), no. 3, 762–780, DOI 10.1112/S0024610703004599. MR2010010

[21] Xinyue Cheng and Zhongmin Shen, *Finsler geometry: An approach via Randers spaces*, Science Press Beijing, Beijing; Springer, Heidelberg, 2012, DOI 10.1007/978-3-642-24888-7. MR3015145

[22] Shiing-shen Chern, *A simple intrinsic proof of the Gauss-Bonnet formula for closed Riemannian manifolds*, Ann. of Math. (2) **45** (1944), 747–752, DOI 10.2307/1969302. MR11027

[23] Shiing-Shen Chern, *Finsler geometry is just Riemannian geometry without the quadratic restriction*, Notices Amer. Math. Soc. **43** (1996), no. 9, 959–963. MR1400859

[24] S. S. Chern, W. H. Chen, and K. S. Lam, *Lectures on differential geometry*, Series on University Mathematics, vol. 1, World Scientific Publishing Co., Inc., River Edge, NJ, 1999, DOI 10.1142/3812. MR1735502

[25] Shiing-Shen Chern and Zhongmin Shen, *Riemann-Finsler geometry*, Nankai Tracts in Mathematics, vol. 6, World Scientific Publishing Co. Pte. Ltd., Hackensack, NJ, 2005, DOI 10.1142/5263. MR2169595

[26] Pierre Dazord, *Variétés finslériennes de dimension paire δ-pincées* (French), C. R. Acad. Sci. Paris Sér. A-B **266** (1968), A496–A498. MR232327

[27] Pierre Dazord, *Variétés finslériennes en forme de sphères* (French), C. R. Acad. Sci. Paris Sér. A-B **267** (1968), A353–A355. MR239547

[28] Daniel Egloff, *Uniform Finsler Hadamard manifolds* (English, with English and French summaries), Ann. Inst. H. Poincaré Phys. Théor. **66** (1997), no. 3, 323–357. MR1456516

[29] P. Finsler, *Über Kurven und Flächen in allgemeinen Räumen* (German), Verlag Birkhäuser, Basel, 1951. MR42195

[30] Paul Funk, *Über Geometrien, bei denen die Geraden die Kürzesten sind* (German), Math. Ann. **101** (1929), no. 1, 226–237, DOI 10.1007/BF01454835. MR1512527

[31] Sylvestre Gallot, Dominique Hulin, and Jacques Lafontaine, *Riemannian geometry*, 3rd ed., Universitext, Springer-Verlag, Berlin, 2004, DOI 10.1007/978-3-642-18855-8. MR2088027

[32] R. E. Greene and K. Shiohama, *Convex functions on complete noncompact manifolds: differentiable structure*, Ann. Sci. École Norm. Sup. (4) **14** (1981), no. 4, 357–367 (1982). MR654202

[33] R. E. Greene and K. Shiohama, *Convex functions on complete noncompact manifolds: topological structure*, Invent. Math. **63** (1981), no. 1, 129–157, DOI 10.1007/BF01389196. MR608531

[34] Robert E. Greene and Katsuhiro Shiohama, *The isometry groups of manifolds admitting nonconstant convex functions*, J. Math. Soc. Japan **39** (1987), no. 1, 1–16, DOI 10.2969/jmsj/03910001. MR867983

[35] Detlef Gromoll and Wolfgang Meyer, *On complete open manifolds of positive curvature*, Ann. of Math. (2) **90** (1969), 75–90, DOI 10.2307/1970682. MR247590

[36] Karsten Grove and Katsuhiro Shiohama, *A generalized sphere theorem*, Ann. of Math. (2) **106** (1977), no. 2, 201–211, DOI 10.2307/1971164. MR500705

[37] Nobuhiro Innami, *A classification of Busemann G-surfaces which possess convex functions*, Acta Math. **148** (1982), 15–29, DOI 10.1007/BF02392724. MR666106

[38] Nobuhiro Innami, *On the terminal points of co-rays and rays*, Arch. Math. (Basel) **45** (1985), no. 5, 468–470, DOI 10.1007/BF01195373. MR815987

[39] Nobuhiro Innami, *Splitting theorems of Riemannian manifolds*, Compositio Math. **47** (1982), no. 3, 237–247. MR681608

[40] Nobuhiro Innami, Yoe Itokawa, Tetsuya Nagano, and Katsuhiro Shiohama, *Affine functions and Busemann functions on complete Finsler manifolds*, Proc. Amer. Math. Soc. **149** (2021), no. 4, 1723–1732, DOI 10.1090/proc/15304. MR4242326

[41] Nobuhiro Innami, Yoe Itokawa, Tetsuya Nagano, and Katsuhiro Shiohama, *Blaschke Finsler manifolds and actions of projective Randers changes on cut loci*, Trans. Amer. Math. Soc. **371** (2019), no. 10, 7433–7450, DOI 10.1090/tran/7603. MR3939582

[42] Nobuhiro Innami, Katsuhiro Shiohama, and Toshiro Soga, *The cut loci, conjugate loci and poles in a complete Riemannian manifold*, Geom. Funct. Anal. **22** (2012), no. 5, 1400–1406, DOI 10.1007/s00039-012-0177-4. MR2989437

[43] Yoe Itokawa, Katsuhiro Shiohama, and Bankteshwar Tiwari, *Strictly convex functions on complete Finsler manifolds*, Proc. Indian Acad. Sci. Math. Sci. **126** (2016), no. 4, 623–627, DOI 10.1007/s12044-016-0307-2. MR3568254

[44] Chang-Wan Kim and Kyungha Min, *Finsler metrics with positive constant flag curvature*, Arch. Math. (Basel) **92** (2009), no. 1, 70–79, DOI 10.1007/s00013-008-2981-5. MR2471989

[45] Wilhelm Klingenberg, *Manifolds with restricted conjugate locus*, Ann. of Math. (2) **78** (1963), 527–547, DOI 10.2307/1970539. MR159289

[46] W. Klingenberg, *Contributions to Riemannian geometry in the large*, Ann. of Math. (2) **69** (1959), 654–666, DOI 10.2307/1970029. MR105709

[47] Wilhelm P. A. Klingenberg, *Riemannian geometry*, 2nd ed., De Gruyter Studies in Mathematics, vol. 1, Walter de Gruyter & Co., Berlin, 1995, DOI 10.1515/9783110905120. MR1330918

[48] Kei Kondo, *Grove-Shiohama type sphere theorem in Finsler geometry*, Osaka J. Math. **52** (2015), no. 4, 1143–1162. MR3426633

[49] Ashok Kumar, Hemangi Madhusudan Shah, and Bankteshwar Tiwari, *Isometric models of the funk disc and the Busemann function*, Results Math. **79** (2024), no. 2, Paper No. 77, 31, DOI 10.1007/s00025-023-02100-4. MR4693336

[50] Brad Lackey, *On the Gauss-Bonnet formula in Riemann-Finsler geometry*, Bull. London Math. Soc. **34** (2002), no. 3, 329–340, DOI 10.1112/S002460930100892X. MR1887705

[51] André Lichnerowicz, *Quelques théorèmes de géométrie diifférentielle globale* (French), Comment. Math. Helv. **22** (1949), 271–301, DOI 10.1007/BF02568060. MR29251

[52] Yukihiro Mashiko, *A splitting theorem for Alexandrov spaces*, Pacific J. Math. **204** (2002), no. 2, 445–458, DOI 10.2140/pjm.2002.204.445. MR1907900

[53] Yukihiro Mashiko, *Affine functions on Alexandrov surfaces*, Osaka J. Math. **36** (1999), no. 4, 853–859. MR1745653

[54] Yukihiro Mashiko, *Convex functions on Alexandrov surfaces*, Trans. Amer. Math. Soc. **351** (1999), no. 9, 3549–3567, DOI 10.1090/S0002-9947-99-02193-5. MR1473452

[55] Howard Masur, *On a class of geodesics in Teichmüller space*, Ann. of Math. (2) **102** (1975), no. 2, 205–221, DOI 10.2307/1971031. MR385173

[56] Howard Masur, *Geometry of Teichmüller space with the Teichmüller metric*, Surveys in differential geometry. Vol. XIV. Geometry of Riemann surfaces and their moduli spaces, Surv. Differ. Geom., vol. 14, Int. Press, Somerville, MA, 2009, pp. 295–313, DOI 10.4310/SDG.2009.v14.n1.a11. MR2655331

[57] Makoto Matsumoto, *Foundations of Finsler geometry and special Finsler spaces*, Kaiseisha Press, Shigaken, 1986. MR858830

[58] Xiaohuan Mo and Zhongmin Shen, *On negatively curved Finsler manifolds of scalar curvature*, Canad. Math. Bull. **48** (2005), no. 1, 112–120, DOI 10.4153/CMB-2005-010-3. MR2118768

[59] Shin-ichi Ohta, *Comparison Finsler geometry*, Springer Monographs in Mathematics, Springer, Cham, [2021] ©2021, DOI 10.1007/978-3-030-80650-7. MR4386101

[60] Shin-ichi Ohta, *Weighted Ricci curvature estimates for Hilbert and Funk geometries*, Pacific J. Math. **265** (2013), no. 1, 185–197, DOI 10.2140/pjm.2013.265.185. MR3095118

[61] Shin-ichi Ohta, *Splitting theorems for Finsler manifolds of nonnegative Ricci curvature*, J. Reine Angew. Math. **700** (2015), 155–174, DOI 10.1515/crelle-2013-0011. MR3318514

[62] Hideki Omori, *A class of Riemannian metrics on a manifold*, J. Differential Geometry **2** (1968), 233–252. MR239625

[63] Athanase Papadopoulos and Guillaume Théret, *On Teichmüller's metric and Thurston's asymmetric metric on Teichmüller space*, Handbook of Teichmüller theory. Vol. I, IRMA Lect. Math. Theor. Phys., vol. 11, Eur. Math. Soc., Zürich, 2007, pp. 111–204, DOI 10.4171/029-1/3. MR2349669

[64] Athanase Papadopoulos and Marc Troyanov (eds.), *Handbook of Hilbert geometry*, IRMA Lectures in Mathematics and Theoretical Physics, vol. 22, European Mathematical Society (EMS), Zürich, 2014. MR3309067

[65] A. Papadopoulos and W. Su, *On the Finsler structure of Teichmüller's metric and Thurston's metric*, Expo. Math. **33** (2015), no. 1, 30–47, DOI 10.1016/j.exmath.2013.12.007. MR3310926

[66] G. Perelman, *Proof of the soul conjecture of Cheeger and Gromoll*, J. Differential Geom. **40** (1994), no. 1, 209–212. MR1285534

[67] Akhil Ranjan and Hemangi Shah, *Busemann functions in a harmonic manifold*, Geom. Dedicata **101** (2003), 167–183, DOI 10.1023/A:1026369930269. MR2017901

[68] Hans-Bert Rademacher, *A sphere theorem for non-reversible Finsler metrics*, Math. Ann. **328** (2004), no. 3, 373–387, DOI 10.1007/s00208-003-0485-y. MR2036326

[69] Hans-Bert Rademacher, *Nonreversible Finsler metrics of positive flag curvature*, A sampler of Riemann-Finsler geometry, Math. Sci. Res. Inst. Publ., vol. 50, Cambridge Univ. Press, Cambridge, 2004, pp. 261–302. MR2132661

[70] Gunnar Randers, *On an asymmetrical metric in the fourspace of general relativity*, Phys. Rev. (2) **59** (1941), 195–199. MR3371

[71] H. Rauch, Geodesics and Curvature in Differential Geometry in the Large, Yeshiva University Press, New York (1959).

[72] Bernhard Riemann, *On the hypotheses which lie at the bases of geometry*, Classic Texts in the Sciences, Birkhäuser/Springer, [Cham], 2016. Edited and with commentary by Jürgen Jost; Expanded English translation of the German original, DOI 10.1007/978-3-319-26042-6. MR3469866

[73] Hanno Rund, *The differential geometry of Finsler spaces*, Die Grundlehren der mathematischen Wissenschaften, Band 101, Springer-Verlag, Berlin-Göttingen-Heidelberg, 1959. MR105726

[74] Sorin V. Sabau and Katsuhiro Shiohama, *Topology of complete Finsler manifolds admitting convex functions*, Pacific J. Math. **276** (2015), no. 2, 459–481, DOI 10.2140/pjm.2015.276.459. MR3374069

[75] Takashi Sakai, *Riemannian geometry*, Translations of Mathematical Monographs, vol. 149, American Mathematical Society, Providence, RI, 1996. Translated from the 1992 Japanese original by the author, DOI 10.1090/mmono/149. MR1390760

[76] V. A. Šarafutdinov, *The Pogorelov-Klingenberg theorem for manifolds that are homeomorphic to* $\mathbf{R}^n$ (Russian), Sibirsk. Mat. Ž. **18** (1977), no. 4, 915–925, 958. MR487896

[77] Zhongmin Shen, *Volume comparison and its applications in Riemann-Finsler geometry*, Adv. Math. **128** (1997), no. 2, 306–328, DOI 10.1006/aima.1997.1630. MR1454401

[78] Zhongmin Shen, *Differential geometry of spray and Finsler spaces*, Kluwer Academic Publishers, Dordrecht, 2001, DOI 10.1007/978-94-015-9727-2. MR1967666

[79] Zhongmin Shen, *Lectures on Finsler geometry*, World Scientific Publishing Co., Singapore, 2001, DOI 10.1142/9789812811622. MR1845637

[80] Zhongmin Shen, *Finsler metrics with* $\mathbf{K} = 0$ *and* $\mathbf{S} = 0$, Canad. J. Math. **55** (2003), no. 1, 112–132, DOI 10.4153/CJM-2003-005-6. MR1952328

[81] Yi-Bing Shen and Zhongmin Shen, *Introduction to modern Finsler geometry*, Higher Education Press, Beijing; World Scientific Publishing Co., Singapore, 2016, DOI 10.1142/9726. MR3526104

[82] Zhongmin Shen and Qiaoling Xia, *A class of Randers metrics of scalar flag curvature*, Internat. J. Math. **24** (2013), no. 7, 1350055, 9, DOI 10.1142/S0129167X13500559. MR3084736

[83] Katsuhiro Shiohama, *Busemann functions and total curvature*, Invent. Math. **53** (1979), no. 3, 281–297, DOI 10.1007/BF01389768. MR549403

[84] Katsuhiro Shiohama, *Topology of complete noncompact manifolds*, Geometry of geodesics and related topics (Tokyo, 1982), Adv. Stud. Pure Math., vol. 3, North-Holland, Amsterdam, 1984, pp. 423–450, DOI 10.2969/aspm/00310423. MR758662

[85] K. Shiohama, *Riemannian and Finsler geometry in the large*, Recent advances in mathematics, Ramanujan Math. Soc. Lect. Notes Ser., vol. 21, Ramanujan Math. Soc., Mysore, 2015, pp. 163–179. MR3380372

[86] Katsuhiro Shiohama and Minoru Tanaka, *Cut loci and distance spheres on Alexandrov surfaces* (English, with English and French summaries), Actes de la Table Ronde de Géométrie Différentielle (Luminy, 1992), Sémin. Congr., vol. 1, Soc. Math. France, Paris, 1996, pp. 531–559. MR1427770

[87] Katsuhiro Shiohama and Bankteshwar Tiwari, *The global study of Riemannian-Finsler geometry*, Geometry in history, Springer, Cham, 2019, pp. 581–621. MR3965775

[88] J. L. Synge, *A generalisation of the Riemannian line-element*, Trans. Amer. Math. Soc. **27** (1925), no. 1, 61–67, DOI 10.2307/1989164. MR1501298

[89] J. L. Synge, *The First and Second Variations of the Length-Integral in Riemannian Space*, Proc. London Math. Soc. (2) **25** (1926), 247–264, DOI 10.1112/plms/s2-25.1.247. MR1575283

[90] James Henry Taylor, *A generalization of Levi-Civita's parallelism and the Frenet formulas*, Trans. Amer. Math. Soc. **27** (1925), no. 2, 246–264, DOI 10.2307/1989064. MR1501309

[91] James Henry Taylor, *Parallelism and transversality in a sub-space of a general (Finsler) space*, Ann. of Math. (2) **28** (1926/27), no. 1-4, 620–628, DOI 10.2307/1968402. MR1502810

[92] B. Y. Wu and Y. L. Xin, *Comparison theorems in Finsler geometry and their applications*, Math. Ann. **337** (2007), no. 1, 177–196, DOI 10.1007/s00208-006-0031-9. MR2262781

[93] Takao Yamaguchi, *The isometry groups of Riemannian manifolds admitting strictly convex functions*, Ann. Sci. École Norm. Sup. (4) **15** (1982), no. 1, 205–212. MR672480

[94] Takao Yamaguchi, *Uniformly locally convex filtrations on complete Riemannian manifolds*, Curvature and topology of Riemannian manifolds (Katata, 1985), Lecture Notes in Math., vol. 1201, Springer, Berlin, 1986, pp. 308–318, DOI 10.1007/BFb0075664. MR859593

[95] Ryozo Yoshikawa and Sorin V. Sabau, *Kropina metrics and Zermelo navigation on Riemannian manifolds*, Geom. Dedicata **171** (2014), 119–148, DOI 10.1007/s10711-013-9892-8. MR3226790

[96] E. Zermelo, *Uber das Navigationsproblem bei ruhender oder veranderlicher Windverteilung*, Z. Angew. Math. Mech. 11 (1931), 114–124.

[97] Wei Zhao and Yibing Shen, *A universal volume comparison theorem for Finsler manifolds and related results*, Canad. J. Math. **65** (2013), no. 6, 1401–1435, DOI 10.4153/CJM-2012-053-4. MR3121676

[98] W. Zhao and L. Yuan, *A simple proof of the Gauss-Bonnet-Chern formula for Finsler manifolds*, Preprint, `arXiv:1303.1918`, 2013.

Centre for Interdisciplinary Mathematical Sciences, Institute of Science, Banaras Hindu University, Varanasi 221 005, India

Contemporary Mathematics
Volume **811**, 2025
https://doi.org/10.1090/conm/811/16250

Revisiting Kulkarni's topological proof of Millington's theorem

Devendra Tiwari and Harshavardhan Reddy

ABSTRACT. In this work, we will revisit Ravi Kulkarni's topological proof [Bull. London Math. Soc. 17 (1985) 458–462] of Millington's theorem [J. London Math. Soc. 1 (1969), 351–357] on subgroups of the modular group $\mathrm{PSL}(2,\mathbb{Z})$. Millington's original proof was algebraic and used permutation representation methods. In particular, we simplify Kulkarni's proof by constructing a combinatorial complex for a given finite index subgroup, which is an analogue of a *dessin d'enfant*. We shall also discuss Millington's question about realizable cusp-widths of a finite index subgroup of $\mathrm{PSL}(2,\mathbb{Z})$ and briefly describe an approach to compute cusp-widths of a given finite index subgroup.

1. Introduction

It is well known that the modular group, $\Gamma = \mathrm{PSL}_2(\mathbb{Z}) \simeq \mathbb{Z}_2 * \mathbb{Z}_3$ is abstractly isomorphic to a free product of $\mathbb{Z}_2$ and $\mathbb{Z}_3$. Let ϕ be a subgroup of Γ of finite index d. By the Kurosh subgroup theorem, (see e.g. [**Kul83**]), it is abstractly isomorphic to a free product of (say) e_2 copies of $\mathbb{Z}_2$, e_3 copies of $\mathbb{Z}_3$ and r copies of $\mathbb{Z}$. e_2, e_3, r and d are the intrinsic algebraic invariants of ϕ as a subgroup of finite index. We also have the genus g and the number of cusps t of the surface $Y_\phi := \mathbb{H}^2/\phi$.

The invariants e_2, e_3, r, d also have equivalent geometric definitions in the sense that they can be defined just in terms of the surface Y_ϕ along with the branched covers $p_\phi : \mathbb{H}^2 \to Y_\phi$ and $m_\phi : Y_\phi \to Y_\Gamma$. To geometrically define e_2 (resp. e_3), we can count the number of branch points on the curve Y_ϕ, whose preimages in $\mathbb{H}^2$ have ramification index 2 (resp. 3). (All preimages will have the same ramification index as ϕ acts transitively on the fibers of the covering $\mathbb{H}^2 \to \mathbb{H}^2/\phi$). We will show that $\pi_1(Y_\Gamma)$ is the free group on r generators. Finally, d is also the index of the branched covering $m_\phi : Y_\phi \to Y_\Gamma$.

We define the *short-signature* of a subgroup $\phi \leq \Gamma$ to be the tuple $(d, r; e_2, e_3)$.

Since the work of Fricke and Klein [**FK1890**], the problem of the existence of subgroups of the modular group $\mathrm{PSL}(2,\mathbb{Z})$ with given signature, has been investigated by many authors [**HP53, HP71, Ran77, Mil69A, Kul83**]. This problem is closely related to the problem of realizing branched coverings of surfaces with prescribed branch data. The data must satisfy the obvious Riemann-Hurwitz conditions, but in general, these are not sufficient to guarantee existence. In the case

2020 *Mathematics Subject Classification.* Primary 11F06.
Key words and phrases. Modular group, modular curves, Riemann surfaces.

of $\mathrm{PSL}(2,\mathbb{Z})$, Millington [**Mil69A**] proved that there is at least one subgroup of every short-signature satisfying the Riemann-Hurwitz relations. Millington's original proof was algebraic and used permutation representation methods. Kulkarni using his work in [**Kul83**] gave a pictorial/topological proof of Millington's theorem in [**Kul85**].

In this work, we will give an exposition of Kulkarni's constructive and topological proof in [**Kul85**]. This work by Kulkarni uses combinatorial diagrams to construct modular curves as finite degree branched covers of Y_Γ. The construction is done by gluing together 4 types of blocks which represent spaces that are covers of two halves of Y_Γ. By covering space theory, these explicitly constructed branched covers will be of the form Y_ϕ for some finite index subgroup $\phi \leq \Gamma = \mathrm{PSL}_2(\mathbb{Z})$.

We will ensure that the cover is constructed so that it has the desired invariants $(d, r; e_2, e_3, g, t)$ by using the geometric definitions of these invariants. To achieve the invariants e_2, e_3, d, it will be enough to check that our diagram has the right number of blocks of the given types. Then, we will repeatedly interchange pairs of handles until the diagram has the required values for g, t. Thus, ϕ itself will have these desired invariants.

The constructed branched cover also provides a way to compute the cusp-splitting. This helps us to study Millington's question (see Page 356 of [**Mil69A**]) to find a ϕ with a given cusp-splitting. We briefly discuss Millington's question and want to tackle this with a topological approach in a forthcoming work.

We would like to emphasize that the work of and Millington can be seen as analogous to the study of Belyi maps and dessins d'enfant, popularized by Grothendieck in his manuscript [**AG85**]. In fact, Millington's prescient definition of 'permutation pairing' (defined up to simultaneous conjugacy as given on page 352, [**Mil69A**]) anticipates the modern definition of a passport of a dessin d'enfant, with the extra condition that the first two elements of the permutation triple have order 2 and 3 (for comparison see [**SV13**]). Similarly, Kulkarni's diagrams in [**Kul83, Kul85**] are examples of dessin d'enfants for finite index subgroups of the Modular group. Therefore, we hope their methods and in particular the combinatorial construction of Y_ϕ elaborated upon in this paper would be useful in important computations related to dessin d'enfants. For the interested reader we would like to mention that the idea of representing subgroups of modular groups by diagrams, under various names, has occurred to many people (see e.g. [**BB93, JS78, Kul91, DT75**] and references therein).

Plan of paper. After setting up our building blocks for branched coverings and obtaining relations between their invariants. In section 3, we apply this method to give a constructive and geometric proof of Millington's theorem [**Mil69A**]. We conclude the paper by briefly discussing Millington's cusp-splitting problem and describing an algorithm to compute the cusp-splitting for a given finite index subgroup of $\mathrm{PSL}(2,\mathbb{Z})$.

2. Preliminaries

In this section, we will briefly recall some basic facts about finite index subgroups of the modular group including their topological invariants and relations between them. We will show how to combinatorially describe branched covers of Y_Γ.

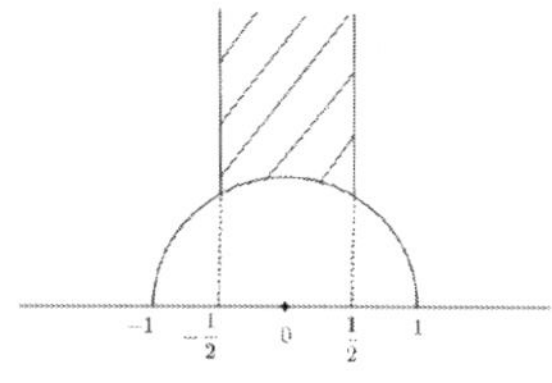

FIGURE 1. Standard Fundamental Domain

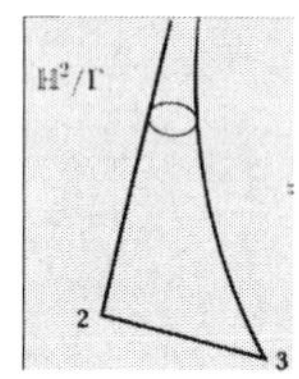

FIGURE 2. Modular Curve

2.1. Modular Curves and Modular Group. Figure 3 shows a picture of the modular curve $Y_\Gamma = \mathbb{H}^2/\Gamma$. This space is homeomorphic to a punctured sphere and hence an open disk. We crunch the open disk in the middle to get a dumbbell-shaped region with the images of the elliptic points of order 2 and 3 at the centers of the left and right disks respectively (which we denote by p_2 and p_3). The dumbbell also has an arm homeomorphic to $[0,1] \times (0,1)$ joining the two disks. The arm is open in the vertical direction and its left and right edges are identified with sectors on the boundaries of the two disks.

This dumbbell model helps us to construct covers for this space by splitting it into two subspaces. The **left half of the** Y_Γ is the union of the left open disk and the left half of the arm which joins the two disks and similarly the **right half of the** Y_Γ. Both these left and right halves include the vertical open interval $(\{1/2\} \times (0,1))$ at the middle of the arm. We will call this the **central interval** of Y_Γ.

Γ is generated by the holomorphic transformations $S : z \to -\frac{1}{z}$, and $T : z \to z+1$. Let $U = ST$. Note that $\langle S \rangle \approx \mathbb{Z}/2\mathbb{Z}$, and $\langle U \rangle \approx \mathbb{Z}/3\mathbb{Z}$, and $\Gamma = \langle S, U \rangle \approx \langle S \rangle * \langle U \rangle \approx \mathbb{Z}/2\mathbb{Z} * \mathbb{Z}/3\mathbb{Z}$.

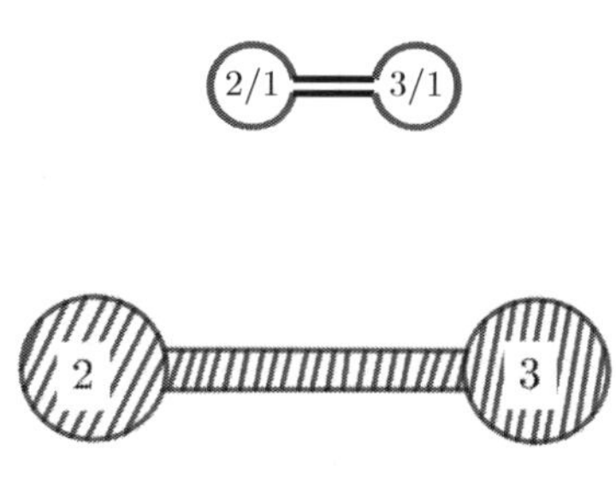

FIGURE 3. Topological Modular Curve

2.2. Building blocks. We will build branched covers of the Y_Γ by gluing the four spaces listed below in various ways.

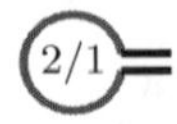

(1) A (2/1)-block is just the left half of Y_Γ as defined above. Hence, it is trivially a cover of the left half of Y_Γ via the identity map. It has a boundary at the right which is homoeomorphic to $(0,1)$. We call this the free edge of the (2/1) block. We will construct coverings of Y_Γ by identifying this free edge with the free edges of (3/1) and (3/3) blocks.

(2) A (2/2)-block is a space consisting of the open disk with two arms in both directions along with the two open boundary intervals on both the right and the left. It represents the space which is a branched double cover of the left half of Y_Γ. The covering map can be taken to be $z \to z^2$ in the circular portion (so, it is ramified over p_2) and the two arms map to the single arm. The two boundary intervals are called the free edges of the (2/2) blocks.

(3) Similarly, a (3/1)-block is an open subset which is the right half of Y_Γ. It covers the right half trivially via the identity map and has one free edge at the right.

(4) Similarly, a (3/3)-block is an open disk with three arms with a boundary consisting of three open intervals. It represents the branched triple cover of the right half of Y_Γ. The covering map can be taken to be $z \to z^3$ in the circular portion (so, it is ramified over p_3) and the three arms map to the single arm.

There is a natural cyclic order on the three arms of a (3/3) block, i.e. every arm has a 'next arm' which appears next in counterclockwise order. In general, there are 2 different possible cyclic orders on a set with 3 elements.

We can glue two blocks along their free edges. As both free edges map homemorphically down into the central interval of Y_Γ, there is a unique way to perform this identification which respects these maps into Y_Γ defined on the two blocks.

Definition 1. A **block diagram** is a finite degree, connected, branched cover of Y_Γ which can be obtained by taking several disjoint copies of the above 4 types of blocks and identifying every free edge with some other free edge in the above unique way. There is a further bipartite restriction that a free edge of a (2/1) or a (2/2) block is identified with a free edge of a (3/1) or a (3/3) block. There is a natural map to Y_Γ from this space as the projections on the individual blocks agree on their shared free edges.

The bipartite condition implies that the projection map on the block diagram is a branched cover. The covering condition is trivially true on the interiors of the left and right halves. For any point p in the central interval, we can take an open disk D centered at p with the two halves of the disk projecting to the left and right halves of Y_Γ. The bipartite condition implies that each connected component of the pullback of D to the block diagram splits into two half disks which project homeomorphically to the above two half disks. Conversely, if the bipartite condition doesn't hold at some free edge or if the free edge is not identified with any free edge, the resulting space fails to be a branched cover as a point on this free edge will have neighbourhoods that only project to one half of Y_Γ.

PROPOSITION 2. *Any finite degree, connected, branched cover $f : E \to Y_\Gamma$, which is a subcover of the universal cover $p_\Gamma : \mathbb{H}^2 \to Y_\Gamma$ is isomorphic to the cover coming from a block diagram.*

PROOF. We have the following factorization of the universal branched cover of Y_Γ, $p_\Gamma : \mathbb{H}^2 \to E \to Y_\Gamma$ (ie $p_\Gamma = f \circ g$). f will be unbranched above all points except at the two special points p_2, p_3 as p_Γ is only branched at these special points and branching indices multiply under composition of maps. For the same reason, at a preimage of p_2(resp. p_3), f can either have branching index 1(unbranched) or 2 (resp. 3).

Let C be one of the connected components of the inverse image of the left half of Y_Γ and q_2 be a preimage of p_2 in C. C will be closed as it is the inverse image of a closed set. We will show that C is a (2/1) or a (2/2) block. Firstly, the preimage q_2 is unique as any other such q_2' can be joined to q_2 via a path α such that $\beta = f(\alpha)$ is in the unbranched locus (missing p_2, p_3) except at its endpoints, where it will be p_2. β can be contracted to the trivial loop via a homotopy which again remains in the unbranched locus except at the left and right edges where it always takes the value p_2. This homotopy lifts over f and since at time 1 it is the constant loop, we get that $q_2 = q_2'$.

If f is unbranched at q_2, C will be a (2/1) block as any connected unbranched cover of a simply connected space is itself. If, on the other hand, the map f ramifies at q_2, then after deleting q_2, f is an unbranched double cover over the left half with p_2 deleted. Such a cover is unique up to isomorphism as there is only one subgroup of index 2 in the fundamental group which is $\mathbb{Z}$. So, we can take this cover to be of a standard type $z \to z^2$ and with the two arms mapping to the single arm. The above map continuously extends to C by mapping q_2 to p_2. Hence, we get that C is a (2/2) block. Similarly, any preimage of the right half will be a (3/1) or a (3/3) block.

This implies that the cover E decomposes into closed sets which are homeomorphic to one of the 4 blocks and which can only intersect along the free edges which are the preimage of the boundary interval. □

DEFINITION 3. An **M-graph** is a connected, finite graph with 4 distinguished classes of vertices which we call E_2, F_2, E_3, F_3 together with a specified cyclic orientation on the edges at each F_3 vertex. The degrees of the E_2, F_2, E_3, F_3 vertices have to be $1, 2, 1, 3$ respectively and the graph has to be bipartite with every edge connecting a vertex of type E_2 or F_2 with a vertex of type E_3 or F_3.

An isomorphism of M-graphs is a graph isomorphism which preserves the type of the vertex and the cyclic order at each F_3 vertex.

COROLLARY 4. *The isomorphism classes of connected branched covers of Y_Γ of finite degree up to isomorphism are in bijective correspondence with isomorphism classes of $M-$graphs. Further, this correspondence can be chosen so that the branched cover deformation retracts to the corresponding graph.*

PROOF. This follows from the above proposition, where for a given branched cover coming from a block diagram b, we associate a graph $M(b)$ whose vertices of type E_2, F_2, E_3, F_3 correspond to blocks of type $(2/1), (2/2), (3/1), (3/3)$ respectively. Edges are drawn exactly when the free edges of the blocks are identified. The vertex degree conditions follow as the corresponding blocks have the required number of free edges and the bipartite condition follows as any covering will connect the free edge of a lift of the left half of X_Γ to a free edge on a lift of the right half of X_Γ. The finite degree of the covering implies finitely many blocks in the covering. This, in turn, implies that the number of vertices is finite. The orientation on a $(3/3)$ block results in a cyclic order around the edges attached to the associated vertex.

$M(b)$ depends only on the isomorphism class of b as a branched cover. If two such covers b, b' are isomorphic via a map $f : b \to b'$ which commutes with the projections to Y_Γ, then both b, b' will have exactly the same number of blocks of each type as these counts can be determined by looking at the number of unbranched and branched preimages of p_2, p_3. If two blocks k, k' are identified at a free edge e, then $f(k)$ and $f(k')$ will also share a free edge $f(e)$. Hence, the graph structure is the same for $M(b)$ and $M(b')$. The cyclic order is the same as f is necessarily orientation-preserving since it commutes with the two orientation-preserving projections.

In the opposite direction, given an M-graph, g, we can define a block diagram, $B(g)$ with $(2/1), (2/2), (3/1), (3/3)$ blocks for each vertex of type E_2, F_2, E_3, F_3 respectively and identify free edges of blocks which correspond to vertices joined by an edge. The cyclic order of the arms on $(3/3)$ blocks is determined by the cyclic order at the $F3$ vertices. It is clear that $B(g)$ depends only on the isomorphism class of g as any isomorphism function on vertices and edges of a graph can be used to define a corresponding homeomorphism on blocks and arms.

These two constructions are inverses of each other so that $B(M(b))$ is isomorphic to b and $M(B(g))$ is isomorphic to g and hence, we have the said correspondence.

By contracting the disks to points and the arms to edges, we can deformation retract the cover to its corresponding graph. □

3. Relations between invariants

In this section, we derive various relations between invariants for block diagrams. In the next section, we show that some of these relations are sufficient conditions for the invariants to be realized by a block diagram.

In the block diagrams for branched covers of Y_Γ, we will denote the number of $(2/1), (2/2), (3/1), (3/3)$ blocks as e_2, f_2, e_3, f_3 respectively. This agrees with our previous definition of e_2, e_3 as the number of elliptic points in the cover of order 2 and 3 as only the $(2/1)$ block has an elliptic point of order 2 and a $(3/1)$ block has an elliptic point of order 3.

The number of arms equals the degree of the covering as the covering is unbranched over the unique arm in Y_Γ. As each arm is attached to exactly one block which is of type (2/1) or (2/2) blocks which have exactly 1 or 2 such arms, we get that $d = e_2 + 2f_2$ and similarly, $d = e_3 + 3f_3$.

The Wall Euler Characteristic $\chi(\Gamma)$ (see [**CW61**]) of the group Γ is $\frac{1}{2} + \frac{1}{3} - 1 = -\frac{1}{6}$ and that of ϕ is

$$\chi(\phi) = \frac{e_2}{2} + \frac{e_3}{3} - (e_2 + e_3 + r - 1)$$

Then the possible index of ϕ in Γ is

$$d = \frac{\chi(\phi)}{\chi(\Gamma)} = 3e_2 + 4e_3 + 6(r-1). \tag{1}$$

The above formula (1) can also be derived from the classical Riemann Hurwitz identity in the context of the modular curve, $m_\phi : Y_\phi \to Y_\Gamma$ which says that $(1-r) = d(1) - f_2(2-1) + f_3(3-1)$ (The Euler characteristic of the contractible space Y_Γ is 1. The Euler characteristic of Y_ϕ is $1-r$ as we will show below that Y_ϕ is homotopically equivalent to a wedge product of r circles. The cover m_ϕ branches only on (2/2) and (3/3) blocks with multiplicities 2 and 3 respectively).

By rearranging the terms of note (1) above we get that $f_2 := \frac{(d-e_2)}{2} = e_2 + 2e_3 + 3(r-1)$ and $f_3 := \frac{(d-e_3)}{3} = e_2 + e_3 + 2(r-1)$.

PROPOSITION 5. *For a given finite index subgroup $\phi \leq \Gamma$ with r factors of $\mathbb{Z}$ in its free product decomposition, $\pi_1(Y_\phi)$ is the free group on r generators. Further, $r = 2g - t + 1$*

PROOF. Consider the block diagram decomposition for the covering $m_\phi : Y_\phi \to Y_\Gamma$. An arm between a (2/1) block and a (3/1) block will lead to these two blocks forming a single connected component and hence will be the whole space. This space is just Y_Γ. So apart from this special case, a (2/1) blocks connect to (3/3) blocks and (3/1) blocks will connect to a (2/2) block. The proposition holds in the special case, as $r = 0$ for $\Gamma = \mathbb{Z}/2\mathbb{Z} * \mathbb{Z}/3\mathbb{Z}$ and the fundamental group of Y_Γ is trivial.

A (2/2) block with both of its arms connected to a (3/1) block will also form a single connected component which will be a degree 2 cover of Y_Γ. The fundamental group of this cover is trivial as the space is contractible. Since the only index 2 subgroup of Γ is isomorphic to $\mathbb{Z}/2\mathbb{Z} * \mathbb{Z}/3\mathbb{Z} * \mathbb{Z}/3\mathbb{Z}$ which doesn't have a $\mathbb{Z}$ factor, the proposition also holds in this special case.

Apart from these two special cases, every arm in our space is of three types. It is either (i) an arm connecting a (2/1) block to a (3/3) block, or (ii) one of the two arms of a (2/2) block which is connected to a (3/1) and a (3/3) block or (iii) one of the two arms of a (2/2) block both of which connect to a (3/3) block.

There are e_2 arms of the first type as they are in correspondence to the (2/1) blocks. Similarly, there are e_3 cases of the second type, giving us $2e_3$ arms of the second type. Hence, the number of arms of the third type is $d - e_2 - 2e_3 = 2e_2 + 2e_3 + 6(r-1)$. The third type of arms comes in pairs connected to single (2/2) blocks. These pairs can be considered as edges on a graph which consists of (3/3) blocks. There are $e_2 + e_3 + 3(r-1)$ such pairs connecting $f_3 = e_2 + e_3 + 2(r-1)$ (3/3) blocks. The pair of arms along with the intervening (2/2) block deformation retracts to an edge and the (3/3) block to a point. So, the space Y_ϕ deformation retracts to a graph with $e_2 + e_3 + 3(r-1)$ edges joining $e_2 + e_3 + 2(r-1)$ vertices. As a connected graph with v vertices and $v + k$ edges deformation retracts to a wedge product of $k+1$ circles, our space will have the free group on $(r-1)+1=r$ generators as its fundamental group.

The surface Y_ϕ is homeomorphic to $S_{g,t}$ - the surface obtained by removing t points (the cusps) from the compact oriented genus g surface with $t \geq 1$. This surface deformation retracts to the wedge product of $2g+t-1$ circles and hence, $r = 2g+t-1$. □

We define the *long-signature* of subgroup ϕ of Γ to be the tuple $(d, r, g, t; e_2, e_3)$. We summarize the following relation between different invariants of ϕ:

NOTE 6.

(1) $d = 3e_2 + 4e_3 + 6(r-1) = 3e_2 + 4e_3 + 12g + 6t - 12$
(2) $r = 2g + t - 1$
(3) $d = d_1 + d_2 + \cdots + d_t$ is the cusp-splitting of the cusp of Y_Γ, and $d_1 \leq d_2 \leq \cdots \leq d_t$.

4. Millington's theorem on subgroups of modular group

Millington's proves the following theorem in [**Mil69A**].

THEOREM. *Given integers $e_2 \geq 0, e_3 \geq 0, g \geq 0, t \geq 1$ and $d \geq 1$, satisfying the Riemann-Hurwitz condition $d = 3e_2 + 4e_3 + 12g + 6t - 12$, there exists a subgroup ϕ of Γ of finite index which realizes these invariants.*

PROOF. To prove this theorem, we construct a block diagram whose number of (2/1) blocks is e_2, the number of (3/1) blocks is e_3 and d edges. Then, we perform certain interchange moves repeatedly. These moves preserve e_2, e_3, d, r but increase g by 1 and decrease t by 2. Then we obtain the group ϕ using the branched cover given by the block diagram.

We first show that $f_2 \geq (f_3 - 1) + e_3$, as this ensures that we have sufficiently many (2/2)-blocks which will be added in Step 1 and Step 3.

> Let $k = 12g + 6t - 12$. Note that $k \geq -6$ as $t \geq 1$ and $g \geq 0$.
> $f_2 = (d - e_2)/2 = e_2 + 2e_3 + k/2$.
> Also, $f_3 - 1 + e_3 = (d - e_3)/3 - 1 + e_3 = e_2 + 2e_3 + k/3 - 1$
> Hence, the required inequality reduces to $k/2 \geq k/3 - 1$ which is equivalent to $k \geq -6$ and hence true.

STEP 1. We first place all the (3/3) blocks (of which there are f_3 in number) along a vertical line in the plane and connect consecutive (3/3) blocks with a (2/2) blocks in between. Here, we have used up an $f_3 - 1$ number of (2/2) blocks out of the $(f_3 - 1) + e_3$ minimum bound given by the inequality.

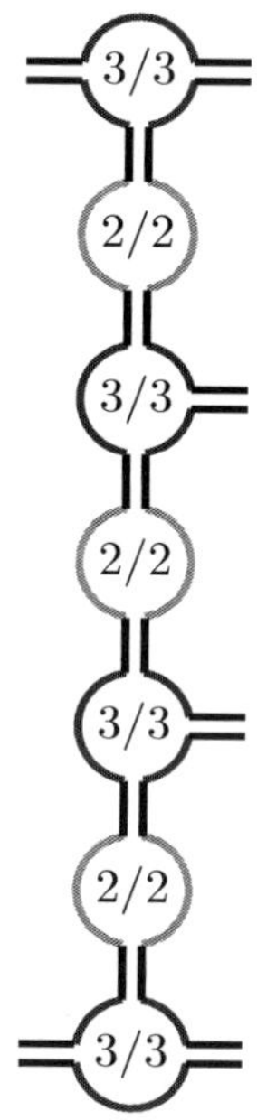

STEP 2. The (3/3) blocks together have $3f_3$ slots out of which $2(f_3 - 1)$ have been used up to connect them into a single component using (2/2) blocks and so, $f_3 + 2$ remain free. Since $f_3 + 2 = e_2 + e_3 + \frac{k}{3} + 2 \geq e_2 + e_3$, the number of free slots in the (3/3) blocks is greater than $e_2 + e_3$ count that we need for step 2 and step 3.

We connect the topmost e_2 free slots to e_2 2 blocks.

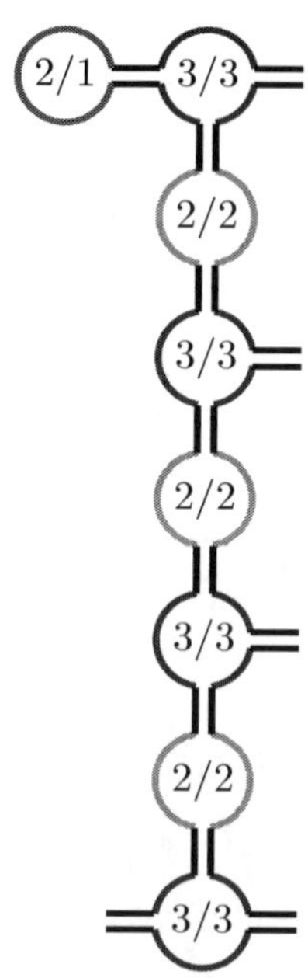

STEP 3. Going counterclockwise, we connect the next e_3 free slots of the (3/3) block to e_3 (3/1) blocks via an intervening (2/2) block. (Here, the e_3 part of the minimum bound for f_2 ensures that we have sufficient (2/2) blocks.)

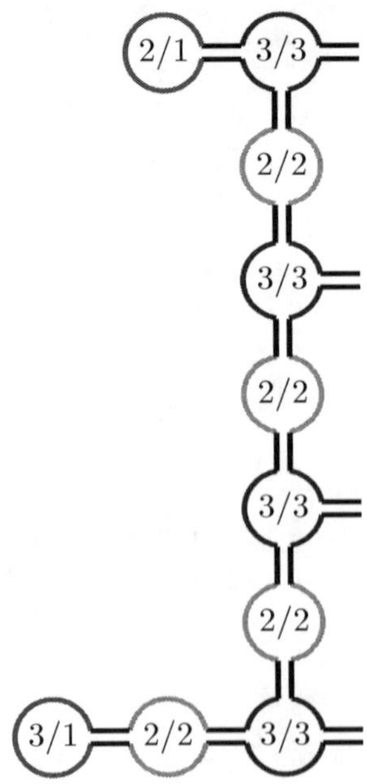

There are now $\frac{k}{3} + 2 = 2(2g + t - 2) + 2 = 2(2g + t - 1)$ free slots on (3/3) blocks. This is exactly the same as the number of free (2/2) slots. This is because at each stage we are connecting a slot on (2/1) or (2/2) block to a slot on a (3/1) or (3/3) block and the initial count ($e_2 + f_2 = e_3 + f_3 = d$) is the same for both types. As all the (2/1) and (3/1) slots have been filled, the number of free (2/2) slots equals the number of free (3/3) slots.

STEP 4. We can successively pair them off via $2g + t - 1$ curved arcs, where a curved arc consists of a (2/2) block along with connections to two successive (3/3) slots.

As we are connecting successive slots in pairs, there are no crossings and the resulting space embeds in the plane as an open set. The genus of this space will be 0. This is because the genus has a characterization as the maximal number of cuts along simple closed curves which still leaves the surface connected. By the Jordan curve theorem, any simple closed curve inside a planar domain breaks it into two pieces.

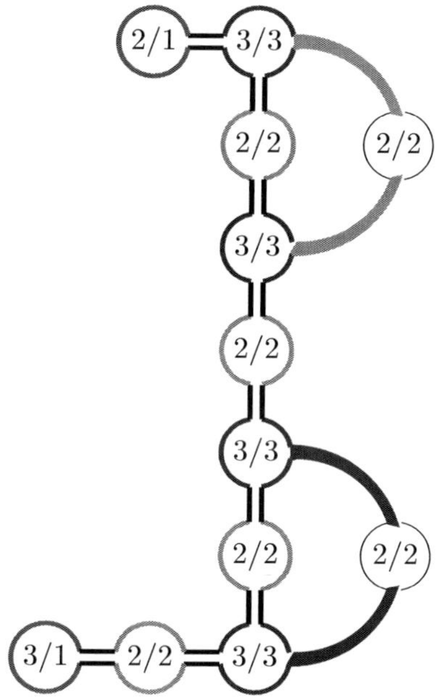

STEP 5. The above steps give us a surface with the required values for e_2, e_3, d. So, the surface is of the form $S_{g',t'}$ with $g' = 0$ initially and $t' = 2g+t$ punctures (as the inner region inside each of the $2g+t-1$ curved arcs contributes one puncture and there is a puncture corresponding to the region external to the diagram). The Euler characteristic for $S_{g',t'}$ is $2-2g'-t'$.

To increase the genus, we will perform a transformation by choosing two successive curved arcs and intertwining their two nearest arms as shown in the picture. We will use two such curved arcs in only one such transformation and new pairs of curved arcs are used for any succeeding transformations.

This transformation leaves e_2, e_3, d unchanged as these invariants can be expressed in terms of the number of blocks of different types ($d = e_2 + 2f_2$) and these numbers are unchanged. We will show below that the Euler characteristic is also unchanged but t' decreases by 2. Hence, the genus g' will increase by 1. We can keep performing this transformation $\lfloor \frac{2g+t-1}{2} \rfloor \geq g$ times by choosing different successive pairs of curved arcs. After g such transformations, $g' = g$ and $t' = (2g+t) - (2g) = t$ which are the desired values.

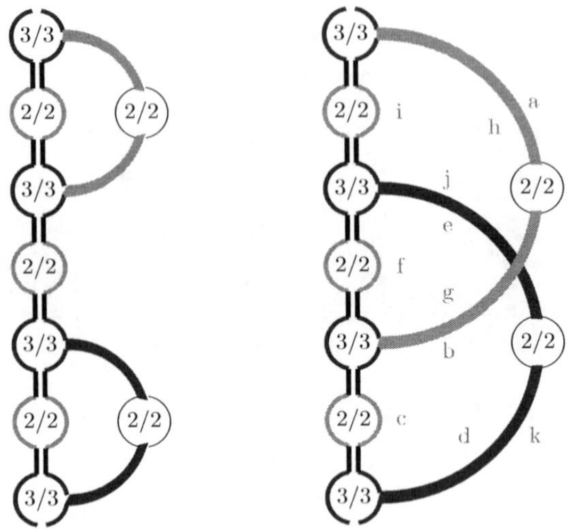

FIGURE 4. Intertwining of two curved arcs. Moving along the boundaries of the arms in alphabetical order shows that the three boundary components collapse into a single component.

We first show that t' decreases by 2. t' can be characterized as the number of boundary components. After the transformation, the number of boundary curves reduces by 2 as can be seen in the figure above, where the alphabetical order flow along the boundary shows that the inner boundaries of the curved arcs are connected to their outer boundaries which is also the outer boundary for the full space. We are working with noncompact surfaces $S_{g',t'}$ which, as such, do not have boundary components. However, we can replace boundary components with the concept of ends. The number of ends is the same as the number of boundary components of the natural compactification. By considering curves in the surface close enough to the boundary, the fact that the three boundary components are connected after interchange implies that the ends also become connected and hence the number of ends decreases by 2.

The reason that the Euler characteristic doesn't change is because we can choose triangulations of the main stem and the curved arcs separately with common edges where they meet. On changing the endpoints of the curved arcs during the transformation, only the identifications change but the counts for vertices, edges, and faces are the same. So, $2 - 2g' - t'$ doesn't change, t' decreases by 2 and so, g' increases by 1.

The above steps show that e_2, e_3, d, g, t are attained by some connected branched covering of Y_Γ. This cover shares the same universal branched cover as Y_Γ and hence is of the form $Y_\phi = \mathbb{H}^2/\phi$ for some subgroup $\phi \leq \Gamma$. The invariants of this subgroup match the geometric invariants of the surface Y_ϕ and hence ϕ is the subgroup with the required e_2, e_3, d, g, t.

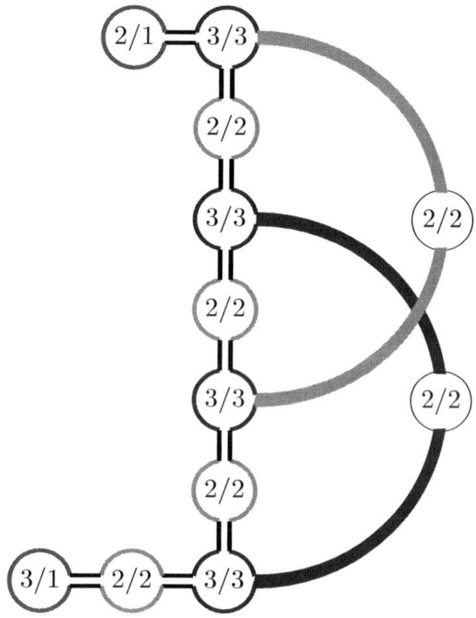

5. Millington's conjecture on cusp-splitting for modular subgroups

In this final section, we briefly discuss Millington's question about the realizability of a given cusp-splitting by a finite-index subgroup $\phi \leq \Gamma$. But first, a definition:

DEFINITION 7. After extending the branched covering $f : Y_\phi \to Y_\Gamma$ to their compactified versions by adding the finitely many cusps, we get a map $f_c : X_\phi \to X_\Gamma$. The ramification of f_c at a cusp (ie a point in X_ϕ which lies above the unique cusp $\infty \in X_\Gamma$) is called the cusp width of that cusp.

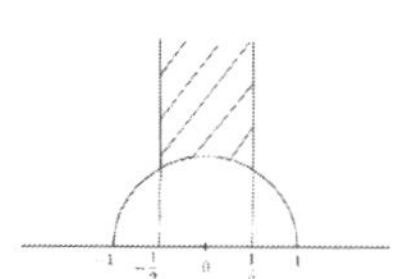

FIGURE 5. Splitting of Cusps in a Branched Covering

REMARK 8. Since $f_c = f$ outside the preimage of ∞, the degree of f_c agrees with the degree of f and hence the sum of the cusp widths which is the total degree of the covering, as the branching indices over any point must sum to the degree of the map.

DEFINITION 9. The **cusp-splitting** of a subgroup $\phi \leq \Gamma$ is a partition of index d into t (number of cusps) parts namely.

$$d = d_1 + d_2 + \cdots + d_t.$$

such that

$$d_1 \leq d_2 \leq d_3 \cdots \leq d_t$$

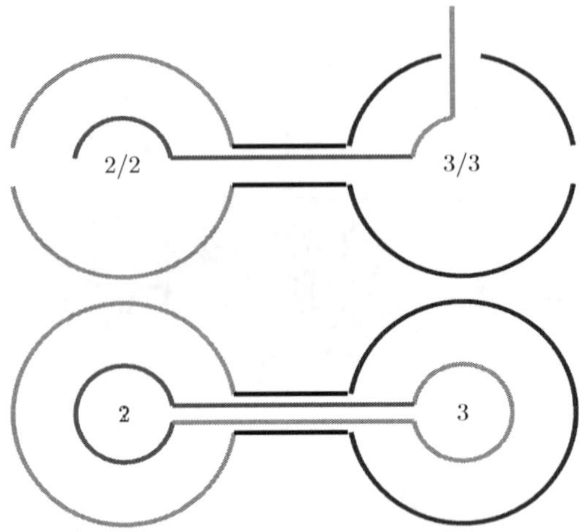

FIGURE 6. The inner curve above partially traverses a single boundary component of Y_ϕ but its projection fully winds along the boundary of Y_Γ. A closed curve along the boundary in Y_ϕ will traverse the projection multiple times.

This cusp-splitting is a geometric invariant of ϕ.

REMARK 10. We can compute the cusp-splitting from our diagrams as follows. Take any boundary component B_i of the diagram for the branched cover, Y_ϕ, and a curve, b_i which is slightly inside this boundary component. Let c_i be its projection to Y_Γ. As each cusp in the compactification X_Γ corresponds to a single boundary component and the neighbourhood of the cusp corresponds to a neighbourhood of this boundary, we have that b_i is the boundary curve around a single cusp C_i in Y_ϕ whereas c_i is the boundary curve winding around the unique cusp in Y_Γ which we denote by C_Γ.

We claim that the winding count of b_i around the cusp C_i will be the ramification at this cusp. This can be seen by choosing local coordinates centered around the cusp C_i in the compactification X_ϕ such that map f_c takes the form $z \to z^k$ in these coordinates. f_c maps C_i to C_Γ. Also, b_i is homotopic to a circle around C_i in these coordinates. The image of this circle will be will also be a circle, but winding k times around C_Γ. Hence, the winding number around C_Γ of c_i, which is the image of b_i, recovers the ramification index k.

When the curve b_i traverses almost one half of a (2/2) block, c_i traverses almost one full sector of the (2/1) block due to (2/2) block being a double cover. Similarly, a one-third sector of (3/3) block in Y_ϕ corresponds to a full sector in Y_Γ. The picture above shows a portion of b_i along the boundaries of a (2/2) and a (3/3) block. For a (2/1) or a (3/1) block, the traversed sector size is the same for b_i and c_i.

b_i alternately visits either a (2/2) or a (2/1) block and then either a (3/3) or a (3/1) block while its projection c_i will traverse around the (2/1) and then the (3/1) blocks downstairs. So, c_i winds once around the cusp in Y_Γ when b_i traverses two consecutive blocks in Y_ϕ.

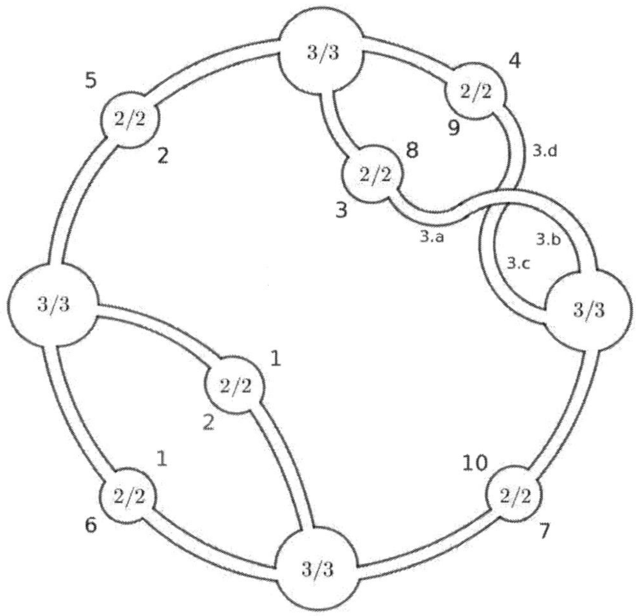

$\Phi = \mathbb{Z} * \mathbb{Z} * \mathbb{Z}$
$d = 2 + 10$

FIGURE 7. Cusp-Splitting for a curve with two boundary components, $12 = 2 + 10$. The numbers indicate the (2/2) blocks along the paths. At the crossing of the arms, the flow of a segment of the path has been marked alphabetically.

Let d_i be the number of (2/2) blocks and (2/1) blocks visited by b_i along its path. Note that a (2/2) block can be visited twice by b_i in which case it is counted twice. Alternatively, each of the two sectors in a (2/2) block might be adjacent to different boundary components and hence the (2/2) is counted once in the counts for two different boundary components. The count d_i will be the number of c_i traversals of the unique (2/1) block in Y_Γ. As c_i visits the (2/1) and (3/1) blocks alternatively, d_i is the same as the number of times b_i winds around along the boundary or the winding number around the cusp C_i.

Each boundary component B_i contributes d_i to the cusp-splitting and these d_i can be directly read from the diagram of the branched cover. As an example, the cover whose diagram is depicted in Figure 7 has 2 boundary components, 6 (2/2) blocks, and no (2/1) blocks. $d_1 = 2, d_2 = 10$ and the cusp-splitting is $12 = 2 + 10$.

DEFINITION 11. A subgroup of the modular group of finite index is cycloidal iff it has exactly one cusp. A subgroup of the modular group of finite index is semi-cycloidal iff it has exactly two inequivalent cusps.

5.1. Millington's question on realizing cusp widths. Systematic study of cycloidal and semi-cycloidal subgroups of the modular group was started by H. Petersson and carried out in his works [**HP53, HP71, HP74**]. M. H. Millington in her work [**Mil69B**] proved the general existence result for cycloidal subgroups. In her proof, Millington introduced a more refined classification of the subgroups in terms of their cusp-splitting. It was shown, however, that certain arithmetically possible combinations of signature and cusp-splitting do not occur for any subgroup. This means that there exist solutions of the Riemann-Hurwitz relation such that for some partitions of d into t parts there is no corresponding subgroup.

In [**Mil69A**] an open question was posed i.e. the existence of an integer N such that, for any allowable group type (d, g, t, e_2, e_3) with d greater than N and any partition of d into t parts, there exists a group of this type and with this cusp-splitting. Later, Stothers [**WS74**] gave a necessary condition on the index and number of cusps such that given a partition of d into t parts, there exists a subgroup G with corresponding cusp-splitting. Note that the congruence subgroups have $t \geq 2$ and are therefore not cycloidal.

5.2. Future work. Using the combinatorial complex constructed for the proof of Millington's theorem, we hope to give a topological proof of Millington's conjecture about the realizability of a cusp-splitting in the special case of semi-cycloidal subgroups.

More diagrams of combinatorial complexes which have been described in this paper can be found at [**DT23**].

Acknowledgments

Both authors are grateful to Professor Ravi Kulkarni for drawing their attention to his works [**Kul83**, **Kul85**] and also for helpful discussions. They are also grateful to organisers of the 80th birthday conference in honour of Professor Ravi Kulkarni for giving an opportunity to present this work. During the work, the first named author was supported by NBHM-PDF. The authors also wish to thank the referee whose suggestions greatly improved this paper.

References

[BB93] Bryan Birch, *Noncongruence subgroups, covers and drawings*, The Grothendieck theory of dessins d'enfants (Luminy, 1993), London Math. Soc. Lecture Note Ser., vol. 200, Cambridge Univ. Press, Cambridge, 1994, pp. 25–46. MR1305392

[FK1890] Felix Klein, *Vorlesungen über die Theorie der elliptischen Modulfunktionen.* (German), Bibliotheca Mathematica Teubneriana, Band 10-11, Johnson Reprint Corp., New York; B. G. Teubner Verlagsgesellschaft, Stuttgart, 1966. Ausgearbeitet und vervollständigt von Robert Fricke; Nachdruck der ersten Auflage. MR247996 and MR247997

[AG85] Alexandre Grothendieck, *Esquisse d'un programme* (French, with French summary), Geometric Galois actions, 1, London Math. Soc. Lecture Note Ser., vol. 242, Cambridge Univ. Press, Cambridge, 1997, pp. 5–48. With an English translation on pp. 243–283. MR1483107

[JS78] Gareth A. Jones and David Singerman, *Theory of maps on orientable surfaces*, Proc. London Math. Soc. (3) **37** (1978), no. 2, 273–307, DOI 10.1112/plms/s3-37.2.273. MR505721

[Kul83] Ravi S. Kulkarni, *An extension of a theorem of Kurosh and applications to Fuchsian groups*, Michigan Math. J. **30** (1983), no. 3, 259–272. MR725780

[Kul85] Ravi S. Kulkarni, *A new proof and an extension of a theorem of Millington on the modular group*, Bull. London Math. Soc. **17** (1985), no. 5, 458–462, DOI 10.1112/blms/17.5.458. MR806012

[Kul91] Ravi S. Kulkarni, *An arithmetic-geometric method in the study of the subgroups of the modular group*, Amer. J. Math. **113** (1991), no. 6, 1053–1133, DOI 10.2307/2374900. MR1137534

[Mil69A] M. H. Millington, *Subgroups of the classical modular group*, J. London Math. Soc. (2) **1** (1969), 351–357, DOI 10.1112/jlms/s2-1.1.351. MR244160

[Mil69B] M. H. Millington, *On cycloidal subgroups of the modular group*, Proc. London Math. Soc. (3) **19** (1969), 164–176, DOI 10.1112/plms/s3-19.1.164. MR248232

[HP53] Hans Petersson, *Über einen einfachen Typus von Untergruppen der Modulgruppe* (German), Arch. Math. **4** (1953), 308–315, DOI 10.1007/BF01899894. MR57910

[HP71] Hans Petersson, *Über die Konstruktion zykloider Kongruenzgruppen in der rationalen Modulgruppe* (German), J. Reine Angew. Math. **250** (1971), 182–212, DOI 10.1515/crll.1971.250.182. MR294255

[HP74] Hans Petersson, *Konstruktionsprinzipien für Untergruppen der Modulgruppe mit einer oder zwei Spitzenklassen* (German), J. Reine Angew. Math. **268(269)** (1974), 94–109, DOI 10.1515/crll.1974.268-269.94. MR354557

[Ran77] Robert A. Rankin, *The modular group and its subgroups*, Ramanujan Institute, Madras, 1969. MR265289

[SV13] J. Sijsling and J. Voight, *On computing Belyi maps* (English, with English and French summaries), Numéro consacré au trimestre "Méthodes arithmétiques et applications", automne 2013, Publ. Math. Besançon Algèbre Théorie Nr., vol. 2014/1, Presses Univ. Franche-Comté, Besançon, 2014, pp. 73–131. MR3362631

[WS74] W. W. Stothers, *Subgroups of the modular group*, Proc. Cambridge Philos. Soc. **75** (1974), 139–153, DOI 10.1017/s0305004100048362. MR332662

[WS78] W. W. Stothers, *The number of subgroups of given index in the modular group*, Proc. Roy. Soc. Edinburgh Sect. A **78** (1977/78), no. 1-2, 105–112, DOI 10.1017/S0308210500009860. MR480341

[DT75] D. J. Tingley, *Elliptic Curves and Modular Functions*, Oxford D.Phil thesis 1975.

[DT23] Devendra Tiwari, *Link for examples combinatorial complex for various ϕ*, `https://www.dropbox.com/sh/2fttsgf9ul9jtjo/AADoyHn5YazDbHgFkuaoGvKVa?dl=0`, 2023.

[CW61] C. T. C. Wall, *Rational Euler characteristics*, Proc. Cambridge Philos. Soc. **57** (1961), 182–184, DOI 10.1017/s030500410003499x. MR122853

Bhaskaracharya Pratishthana, Erandwane, Damle Path, Off Law College Road, Pune Maharashtra 411004, India
Email address: `devendra9.dev@gmail.com`

Tata Institute of Fundamental Research, Homi Bhabha Road, Colaba Mumbai 400005, India
Email address: `harsha.v.r@gmail.com`

Contemporary Mathematics
Volume **811**, 2025
https://doi.org/10.1090/conm/811/16251

Kulkarni-Nomizu tensor fields

Mukut Mani Tripathi

Dedicated to Professor Ravindra Shripad Kulkarni on his 80th birthday

ABSTRACT. The phrases "Kulkarni-Nomizu tensor" and "Kulkarni-Nomizu tensor field" are introduced. Some basic properties have been presented. Schur's theorems for Kulkarni-Nomizu tensors and for proper Kulkarni-Nomizu tensor fields are presented. Some applications are also pointed out.

CONTENTS

1. Introduction
2. Riemannian manifold
3. Kulkarni-Nomizu product
4. Symmetric product
5. Kulkarni-Nomizu tensor
6. Kulkarni-Nomizu tensor field
7. Schur's theorem for proper Kulkarni-Nomizu tensor field
8. Some applications
Acknowledgment
References

1. Introduction

Motivated by the terminology of Kulkarni-Nomizu product, we introduce the phrases Kulkarni-Nomizu tensor and Kulkarni-Nomizu tensor field. The presentation is organized as follows. In Section 2, we collect some basic material about Riemann curvature tensor field and Riemann-Christoffel curvature tensor field. In Section 3, we revisit Kulkarni-Nomizu product and give some basic results. Section 4

2020 *Mathematics Subject Classification.* Primary 53-01, 53-02; Secondary 53B20, 53C15, 53C25, 53C40.

Key words and phrases. Riemann curvature tensor field, Riemann-Christoffel curvature tensor field, Kulkarni-Nomizu product, symmetric product, Kulkarni-Nomizu tensor, algebraic Bianchi identity, algebraic curvature tensor, Kulkarni-Nomizu tensor field; curvature structure, curvature-like tensor, Riemannian tensor field; generalized curvature tensor field, proper generalized curvature tensor field, Ricci contraction, Ricci curvature structure, conformal curvature structure, real space form, complex space form, quaternion space form, Sasakian space form, algebraic Gauss equation.

The research was partially funded by the Professional Development Fund (BHU) and the 'Incentive Grant' under IoE Scheme (BHU) No. R/Dev/D/IoE/Incentive (Phase-IV)/2024-25/82489.

contains concept of symmetric product along with some basic results. In Section 5, the phrase "Kulkarni-Nomizu tensor" is introduced. Some basic results concerning Kulkarni-Nomizu tensors are also given. In Section 6, the phrase "Kulkarni-Nomizu tensor field" is introduced. Some examples and properties are presented. The main goal of presenting Schur's theorem for proper Kulkarni-Nomizu tensor fields is achieved in Section 7. In the last section 8, some applications are pointed out.

2. Riemannian manifold

Let (M, g) be a Riemannian manifold, and ∇ the Levi-Civita connection of (M, g). The *Riemann curvature operator* is defined by

$$R(X,Y) = \nabla_X \nabla_Y - \nabla_Y \nabla_X - \nabla_{[X,Y]}, \tag{2.1}$$

for all vector fields X, Y on M, where $[\ ,\ ]$ is the Lie bracket operation. The *Riemann curvature tensor field* R of type $(1,3)$ is given by the following formula

$$R(X,Y)Z = \nabla_X \nabla_Y Z - \nabla_Y \nabla_X Z - \nabla_{[X,Y]} Z \tag{2.2}$$

for all vector fields X, Y, Z on M. The Riemann curvature tensor field is a local isometry invariant and satisfies

$$R(X,Y)Z = -R(Y,X)Z \tag{2.3}$$

$$g(R(X,Y)Z,W) = -g(R(X,Y)W,Z), \tag{2.4}$$

$$g(R(X,Y)Z,W) = g(R(Z,W)X,Y), \tag{2.5}$$

$$R(X,Y)Z + R(Y,Z)X + R(Z,X)Y = 0, \tag{2.6}$$

$$(\nabla_X R)(Y,Z) + (\nabla_Y R)(Z,X) + (\nabla_Z R)(X,Y) = 0 \tag{2.7}$$

for all vector fields X, Y, Z, W in M. The *Riemann-Christoffel curvature tensor field* R of type $(0,4)$ is defined by

$$R(X,Y,Z,W) = g(R(X,Y)Z,W) \tag{2.8}$$

for all vector fields X, Y, Z, W on M, and satisfies

$$R(X,Y,Z,W) = -R(Y,X,Z,W), \tag{2.9}$$

$$R(X,Y,Z,W) = -R(X,Y,W,Z), \tag{2.10}$$

$$R(X,Y,Z,W) = R(Z,W,X,Y), \tag{2.11}$$

$$R(X,Y,Z,W) + R(Y,Z,X,W) + R(Z,X,Y,W) = 0, \tag{2.12}$$

$$R(X,Y,Z,W) + R(X,Z,W,Y) + R(X,W,Y,Z) = 0, \tag{2.13}$$

$$(\nabla_X R)(Y,Z,U,V) + (\nabla_Y R)(Z,X,U,V) + (\nabla_Z R)(X,Y,U,V) = 0, \tag{2.14}$$

$$(\nabla_X R)(U,V,Y,Z) + (\nabla_Y R)(U,V,Z,X) + (\nabla_Z R)(U,V,X,Y) = 0 \tag{2.15}$$

for all vector fields X, Y, Z, W, U, V in M. For details about Riemannian manifolds, we refer to [**9**, Gallot et al. 1987], [**18**, Kuhnel 2015], [**24**, Lee 2018], [**29**, Petersen 2016].

3. Kulkarni-Nomizu product

We begin this section with the following Definition.

DEFINITION 3.1. Let V be a vector space. Let Θ_1 and Θ_2 be any two second order covariant tensors on V. Then *Kulkarni-Nomizu product* (cf. [**2**, Besse 1987, Definition 1.110, p. 47], [**9**, Gallot-Hulin-Lafontaine 1987, Definition 3.122, p. 151], [**18**, Kuhnel 2015, Definition and Lemma 8.20, p. 335]) $\Theta_1 \circledast \Theta_2$ of Θ_1 and Θ_2 is a $(0,4)$-tensor in V defined by

$$(\Theta_1 \circledast \Theta_2)(X,Y,Z,W) = \Theta_1(Y,Z)\Theta_2(X,W) - \Theta_1(X,Z)\Theta_2(Y,W) + \Theta_2(Y,Z)\Theta_1(X,W) - \Theta_2(X,Z)\Theta_1(Y,W) \tag{3.1}$$

for all $X, Y, Z, W \in V$.

Similarly, Kulkarni-Nomizu product of two $(0,2)$-tensor fields Θ_1 and Θ_2 in a smooth manifold M is a $(0,4)$-tensor field in M also defined by (3.1) for all vector fields X, Y, Z, W on M. More precisely, we have the following:

DEFINITION 3.2. The *Kulkarni-Nomizu product* $\Theta_1 \circledast \Theta_2$ of $(0,2)$-tensor fields Θ_1 and Θ_2 in a smooth manifold M is a $(0,4)$-tensor field defined by (cf. [**2**, Besse 1987, Definition 1.110, p. 47], [**9**, Definition 3.122, Gallot-Hulin-Lafontaine 1987, p. 151])

$$(\Theta_1 \circledast \Theta_2)(X,Y,Z,W) = \Theta_1(Y,Z)\Theta_2(X,W) - \Theta_1(X,Z)\Theta_2(Y,W) + \Theta_2(Y,Z)\Theta_1(X,W) - \Theta_2(X,Z)\Theta_1(Y,W) \tag{3.2}$$

for all vector fields X, Y, Z, W on M.

For any $(0,2)$-tensors $\Theta_1, \Theta_2, \Theta_3$, it follows that

$$\Theta_1 \circledast \Theta_2 = \Theta_2 \circledast \Theta_1, \tag{3.3}$$

$$(\Theta_1 + \Theta_2) \circledast \Theta_3 = \Theta_1 \circledast \Theta_3 + \Theta_2 \circledast \Theta_3. \tag{3.4}$$

REMARK 3.3. In 1972, Kulkarni [**23**] introduced a product, which is more general than Kulkarni-Nomizu product, in the work on double forms. Another form of product was introduced by Nomizu in [**26**, Nomizu 1972, Fourth equation on page 336]. In ([**2**, Besse 1987, Definition 1.110, p. 47] and [**9**, Gallot-Hulin-Lafontaine 1987, Definition 3.122, p. 151]), Kulkarni-Nomizu product is defined for only symmetric $(0,2)$-tensors in a vector space, and with the opposite sign to that in the right hand side of (3.1). Also, in contrast to (2.2), in [**2**, Besse 1987, Definition 1.11, p. 24] and [**9**, Gallot-Hulin-Lafontaine 1987, Definition 3.3, p. 102] the Riemann curvature $(1,3)$-tensor field R in a Riemannian manifold (M,g) is defined by

$$R(X,Y)Z = -\nabla_X \nabla_Y Z + \nabla_Y \nabla_X Z + \nabla_{[X,Y]} Z \tag{3.5}$$

for all vector fields X, Y, Z on M, where ∇ is the Levi-Civita connection of the Riemannian metric g. In Ricci calculus, Kulkarni-Nomizu product was traditionally referred to as the *double transvection* [**18**, Kuhnel 2015, p. 336]. For etymology of *Kulkarni-Nomizu product*, we refer to [**3**, Blaickner 2022].

EXAMPLE 3.4. If u_1, u_2, u_3, u_4 are covariant vectors in a vector space V, then Kulkarni-Nomizu product of $u_1 \otimes u_2$ and $u_3 \otimes u_4$ satisfies

$$\begin{aligned}&((u_1 \otimes u_2) \circledast (u_3 \otimes u_4))(X,Y,Z,W)\\&\quad = \{u_3(X)u_1(Y) - u_1(X)u_3(Y)\}\{u_2(Z)u_4(W) - u_4(Z)u_2(W)\}\end{aligned}$$

for all $X, Y, Z, W \in V$, that is,

$$(u_1 \otimes u_2) \circledast (u_3 \otimes u_4) = -(u_1 \wedge u_3) \otimes (u_2 \wedge u_4).$$

Here $u_1 \otimes u_2$ or $u_3 \otimes u_4$ need not be symmetric.

LEMMA 3.5. *Let Θ_1 and Θ_2 be two $(0,2)$-tensors in a vector space V. Let $X, Y, Z, W \in V$. Then, for Kulkarni-Nomizu product of Θ_1 and Θ_2, the following statements are true.*

(1) *We have*

$$(\Theta_1 \circledast \Theta_2)(X,Y,Z,W) = -(\Theta_1 \circledast \Theta_2)(Y,X,Z,W), \tag{3.6}$$

$$(\Theta_1 \circledast \Theta_2)(X,Y,Z,W) = -(\Theta_1 \circledast \Theta_2)(X,Y,W,Z). \tag{3.7}$$

(2) *If both Θ_1 and Θ_2 are either symmetric or skew-symmetric, then*

$$(\Theta_1 \circledast \Theta_2)(X,Y,Z,W) = (\Theta_1 \circledast \Theta_2)(Z,W,X,Y). \tag{3.8}$$

(3) *If one of Θ_1 and Θ_2 is symmetric and the other is skew-symmetric, then*

$$(\Theta_1 \circledast \Theta_2)(X,Y,Z,W) = -(\Theta_1 \circledast \Theta_2)(Z,W,X,Y). \tag{3.9}$$

(4) *If both Θ_1 and Θ_2 are symmetric, then*

$$(\Theta_1 \circledast \Theta_2)(X,Y,Z,W) + (\Theta_1 \circledast \Theta_2)(Y,Z,X,W) + (\Theta_1 \circledast \Theta_2)(Z,X,Y,W) = 0. \tag{3.10}$$

(5) *If Θ_1 is symmetric and Θ_2 is skew-symmetric, then*

$$\begin{aligned}(\Theta_1 \circledast \Theta_2)(X,Y,Z,W) &+ (\Theta_1 \circledast \Theta_2)(Y,Z,X,W) + (\Theta_1 \circledast \Theta_2)(Z,X,Y,W) \\ &= 2\Theta_1(\Theta_2(X,Y)Z + \Theta_2(Y,Z)X + \Theta_2(Z,X)Y, W).\end{aligned} \tag{3.11}$$

(6) *If both Θ_1 and Θ_2 are skew-symmetric, then*

$$\begin{aligned}(\Theta_1 \circledast \Theta_2)(X,Y,Z,W) &+ (\Theta_1 \circledast \Theta_2)(Y,Z,X,W) + (\Theta_1 \circledast \Theta_2)(Z,X,Y,W) \\ &= 2\Theta_1(\Theta_2(X,Y)Z + \Theta_2(Y,Z)X + \Theta_2(Z,X)Y, W) \\ &\quad + 2\Theta_2(\Theta_1(X,Y)Z + \Theta_1(Y,Z)X + \Theta_1(Z,X)Y, W).\end{aligned} \tag{3.12}$$

(7) *If both Θ_1 and Θ_2 are symmetric, then*

$$(\Theta_1 \circledast \Theta_2)(X,Y,Z,W) + (\Theta_1 \circledast \Theta_2)(X,Z,W,Y) + (\Theta_1 \circledast \Theta_2)(X,W,Y,Z) = 0. \tag{3.13}$$

(8) *If Θ_1 is symmetric and Θ_2 is skew-symmetric, then*

$$\begin{aligned}(\Theta_1 \circledast \Theta_2)(X,Y,Z,W) &+ (\Theta_1 \circledast \Theta_2)(X,Z,W,Y) + (\Theta_1 \circledast \Theta_2)(X,W,Y,Z) \\ &= 2\Theta_1(X, \Theta_2(Y,Z)W + \Theta_2(Z,W)Y + \Theta_2(W,Y)Z).\end{aligned} \tag{3.14}$$

(9) *If both Θ_1 and Θ_2 are skew-symmetric, then*

$$\begin{aligned}(\Theta_1 \circledast \Theta_2)(X,Y,Z,W) &+ (\Theta_1 \circledast \Theta_2)(X,Z,W,Y) + (\Theta_1 \circledast \Theta_2)(X,W,Y,Z) \\ &= 2\Theta_1(X, \Theta_2(Y,Z)W + \Theta_2(Z,W)Y + \Theta_2(W,Y)Z) \\ &\quad + 2\Theta_2(X, \Theta_1(Y,Z)W + \Theta_1(Z,W)Y + \Theta_1(W,Y)Z).\end{aligned} \tag{3.15}$$

PROOF. Omitted. □

In view of Lemma 3.5, we have Propositions 3.6–3.8.

PROPOSITION 3.6. *Let Θ_1 and Θ_2 be any two symmetric $(0,2)$-tensors in a vector space V. Then Kulkarni-Nomizu product of Θ_1 and Θ_2 satisfies*

$$(\Theta_1 \circledast \Theta_2)(X,Y,Z,W) = -(\Theta_1 \circledast \Theta_2)(Y,X,Z,W), \tag{3.16}$$

$$(\Theta_1 \circledast \Theta_2)(X,Y,Z,W) = -(\Theta_1 \circledast \Theta_2)(X,Y,W,Z), \tag{3.17}$$

$$(\Theta_1 \circledast \Theta_2)(X,Y,Z,W) = (\Theta_1 \circledast \Theta_2)(Z,W,X,Y), \tag{3.18}$$

$$(\Theta_1 \circledast \Theta_2)(X,Y,Z,W) + (\Theta_1 \circledast \Theta_2)(Y,Z,X,W) + (\Theta_1 \circledast \Theta_2)(Z,X,Y,W) = 0, \tag{3.19}$$

$$(\Theta_1 \circledast \Theta_2)(X,Y,Z,W) + (\Theta_1 \circledast \Theta_2)(X,Z,W,Y) + (\Theta_1 \circledast \Theta_2)(X,W,Y,Z) = 0 \tag{3.20}$$

for all $X,Y,Z,W \in V$.

PROPOSITION 3.7. *In a vector space V, suppose Θ_1 is a symmetric $(0,2)$-tensor and Θ_2 a skew-symmetric $(0,2)$-tensor. Then Kulkarni-Nomizu product of Θ_1 and Θ_2 satisfies*

$$(\Theta_1 \circledast \Theta_2)(X,Y,Z,W) = -(\Theta_1 \circledast \Theta_2)(Y,X,Z,W), \tag{3.21}$$

$$(\Theta_1 \circledast \Theta_2)(X,Y,Z,W) = -(\Theta_1 \circledast \Theta_2)(X,Y,W,Z), \tag{3.22}$$

$$(\Theta_1 \circledast \Theta_2)(X,Y,Z,W) = -(\Theta_1 \circledast \Theta_2)(Z,W,X,Y), \tag{3.23}$$

$$\begin{aligned}(\Theta_1 \circledast \Theta_2)(X,Y,Z,W) &+ (\Theta_1 \circledast \Theta_2)(Y,Z,X,W) + (\Theta_1 \circledast \Theta_2)(Z,X,Y,W)\\ &= 2\Theta_1(\Theta_2(X,Y)Z + \Theta_2(Y,Z)X + \Theta_2(Z,X)Y, W),\end{aligned} \tag{3.24}$$

$$\begin{aligned}(\Theta_1 \circledast \Theta_2)(X,Y,Z,W) &+ (\Theta_1 \circledast \Theta_2)(X,Z,W,Y) + (\Theta_1 \circledast \Theta_2)(X,W,Y,Z)\\ &= 2\Theta_1(X, \Theta_2(Y,Z)W + \Theta_2(Z,W)Y + \Theta_2(W,Y)Z)\end{aligned} \tag{3.25}$$

for all $X,Y,Z,W \in V$.

PROPOSITION 3.8. *In a vector space V, suppose Θ_1 and Θ_2 are two skew-symmetric $(0,2)$-tensors. Then Kulkarni-Nomizu product of Θ_1 and Θ_2 satisfies*

$$(\Theta_1 \circledast \Theta_2)(X,Y,Z,W) = -(\Theta_1 \circledast \Theta_2)(Y,X,Z,W), \tag{3.26}$$

$$(\Theta_1 \circledast \Theta_2)(X,Y,Z,W) = -(\Theta_1 \circledast \Theta_2)(X,Y,W,Z), \tag{3.27}$$

$$(\Theta_1 \circledast \Theta_2)(X,Y,Z,W) = (\Theta_1 \circledast \Theta_2)(Z,W,X,Y), \tag{3.28}$$

$$\begin{aligned}(\Theta_1 \circledast \Theta_2)(X,Y,Z,W) &+ (\Theta_1 \circledast \Theta_2)(Y,Z,X,W) + (\Theta_1 \circledast \Theta_2)(Z,X,Y,W)\\ &= 2\Theta_1(\Theta_2(X,Y)Z + \Theta_2(Y,Z)X + \Theta_2(Z,X)Y, W)\\ &\quad +2\Theta_2(\Theta_1(X,Y)Z + \Theta_1(Y,Z)X + \Theta_1(Z,X)Y, W),\end{aligned} \tag{3.29}$$

$$\begin{aligned}(\Theta_1 \circledast \Theta_2)(X,Y,Z,W) &+ (\Theta_1 \circledast \Theta_2)(X,Z,W,Y) + (\Theta_1 \circledast \Theta_2)(X,W,Y,Z)\\ &= 2\Theta_1(X, \Theta_2(Y,Z)W + \Theta_2(Z,W)Y + \Theta_2(W,Y)Z)\\ &\quad +2\Theta_2(X, \Theta_1(Y,Z)W + \Theta_1(Z,W)Y + \Theta_1(W,Y)Z)\end{aligned} \tag{3.30}$$

for all $X,Y,Z,W \in V$.

Finally, we give the following:

LEMMA 3.9. (cf. [**18**, Kuhnel 2015, Definition and Lemma 8.20, p. 335]) *Let* ∇ *be a Koszul connection and* Θ_1, Θ_2 *be two* $(0,2)$*-tensor fields in a smooth manifold* M. *Then,*

$$\nabla_X(\Theta_1 \circledast \Theta_2) = (\nabla_X\Theta_1) \circledast \Theta_2 + \Theta_1 \circledast (\nabla_X\Theta_2) \tag{3.31}$$

for every vector field X *in* M.

4. Symmetric product

DEFINITION 4.1. The *symmetric product* of any two $(0,2)$-tensors Θ_1 and Θ_2 is a $(0,4)$-tensor $\Theta_1 \circledcirc \Theta_2$ defined by (see [**35**, Viaclovsky 2011, Eq. (7.17)], [**36**, Z. Wang 2016, Eq. (1.3)])

$$(\Theta_1 \circledcirc \Theta_2)(X,Y,Z,W) = \Theta_1(X,Y)\Theta_2(Z,W) + \Theta_1(Z,W)\Theta_2(X,Y) \tag{4.1}$$

for all $X,Y,Z,W \in V$. Using a basis $(e_1,\ldots,e_n)$ of V, we can write

$$(\Theta_1 \circledcirc \Theta_2)_{ijk\ell} = (\Theta_1)_{ij}(\Theta_2)_{k\ell} + (\Theta_1)_{k\ell}(\Theta_2)_{ij}.$$

Similarly, the *symmetric product* of any two $(0,2)$-tensor fields Θ_1 and Θ_2 in a smooth manifold M is a $(0,4)$-tensor field $\Theta_1 \circledcirc \Theta_2$ defined by (4.1) for all vector fields X,Y,Z,W on M.

For any $(0,2)$-tensors $\Theta_1, \Theta_2, \Theta_3$, it follows that

$$\Theta_1 \circledcirc \Theta_2 = \Theta_2 \circledcirc \Theta_1, \tag{4.2}$$

$$(\Theta_1 + \Theta_2) \circledcirc \Theta_3 = \Theta_1 \circledcirc \Theta_3 + \Theta_2 \circledcirc \Theta_3. \tag{4.3}$$

LEMMA 4.2. *Let* Θ_1 *and* Θ_2 *be two* $(0,2)$*-tensors in a vector space* V. *Let* $X,Y,Z,W \in V$. *Then, for the symmetric product of* Θ_1 *and* Θ_2, *the following statements are true.*

(1) *If both* Θ_1 *and* Θ_2 *are symmetric, then*

$$(\Theta_1 \circledcirc \Theta_2)(X,Y,Z,W) = (\Theta_1 \circledcirc \Theta_2)(Y,X,Z,W), \tag{4.4}$$

$$(\Theta_1 \circledcirc \Theta_2)(X,Y,Z,W) = (\Theta_1 \circledcirc \Theta_2)(X,Y,W,Z). \tag{4.5}$$

(2) *If both* Θ_1 *and* Θ_2 *are skew-symmetric, then*

$$(\Theta_1 \circledcirc \Theta_2)(X,Y,Z,W) = -(\Theta_1 \circledcirc \Theta_2)(Y,X,Z,W), \tag{4.6}$$

$$(\Theta_1 \circledcirc \Theta_2)(X,Y,Z,W) = -(\Theta_1 \circledcirc \Theta_2)(X,Y,W,Z). \tag{4.7}$$

(3) *For arbitrary* $(0,2)$*-tensors* Θ_1 *and* Θ_2,

$$(\Theta_1 \circledcirc \Theta_2)(X,Y,Z,W) = (\Theta_1 \circledcirc \Theta_2)(Z,W,X,Y). \tag{4.8}$$

PROOF. Omitted. □

Finally, we give two results without proof.

PROPOSITION 4.3. *Let* Θ *be a* $(0,2)$*-tensor in a vector space* V. *Then,*

$$(1/2)(\Theta \circledast \Theta)(X,Y,Z,W) = \Theta(Y,Z)\Theta(X,W) - \Theta(X,Z)\Theta(Y,W),$$
$$(\Theta \circledcirc \Theta)(X,Y,Z,W) = 2\Theta(X,Y)\Theta(Z,W).$$

Moreover, if Θ *is skew-symmetric and* $T = (1/2)\Theta \circledast \Theta - \Theta \circledcirc \Theta$, *then*

$$T(X,Y,Z,W) = -T(Y,X,Z,W),$$
$$T(X,Y,Z,W) = -T(X,Y,W,Z),$$
$$T(X,Y,Z,W) = T(Z,W,X,Y),$$
$$T(X,Y,Z,W) + T(Y,Z,X,W) + T(Z,X,Y,W) = 0,$$
$$T(X,Y,Z,W) + T(X,Z,W,Y) + T(X,W,Y,Z) = 0.$$

for all vectors X, Y, Z, W *in* V.

LEMMA 4.4. *Let* ∇ *be a Koszul connection and* Θ_1, Θ_2 *be two* $(0,2)$*-tensor fields in a smooth manifold* M. *Then,*

$$\nabla_X(\Theta_1 \circledcirc \Theta_2) = (\nabla_X\Theta_1) \circledcirc \Theta_2 + \Theta_1 \circledcirc (\nabla_U\Theta_2) \tag{4.9}$$

for every vector field X *in* M.

5. Kulkarni-Nomizu tensor

We begin with the following Lemma.

LEMMA 5.1. *Let* V *be a vector space. For a* $(0,4)$*-tensor* T *consider the following five conditions*

$$T(X,Y,Z,W) = -T(Y,X,Z,W), \tag{5.1}$$
$$T(X,Y,Z,W) = -T(X,Y,W,Z), \tag{5.2}$$
$$T(X,Y,Z,W) = T(Z,W,X,Y), \tag{5.3}$$
$$T(X,Y,Z,W) + T(Y,Z,X,W) + T(Z,X,Y,W) = 0, \tag{5.4}$$
$$T(X,Y,Z,W) + T(X,Z,W,Y) + T(X,W,Y,Z) = 0 \tag{5.5}$$

for all $X, Y, Z, W \in V$. *Then the following six statements are true.*

- **(a)** *If* T *satisfies* (5.1), (5.2), (5.3) *and* (5.4), *then it also satisfies* (5.5).
- **(b)** *If* T *satisfies* (5.1), (5.2), (5.3) *and* (5.5), *then it also satisfies* (5.4).
- **(c)** *If* T *satisfies* (5.1), (5.2) *and* (5.4), *then it also satisfies* (5.3).
- **(d)** *If* T *satisfies* (5.1), (5.2) *and* (5.5), *then it also satisfies* (5.3).
- **(e)** *If* T *satisfies* (5.1) *and* (5.3), *then it also satisfies* (5.2).
- **(f)** *If* T *satisfies* (5.2) *and* (5.3), *then it also satisfies* (5.1).

Consequently, if T *satisfies any two of the three conditions* $\{(5.1), (5.2), (5.3)\}$ *and any one of the two conditions* $\{(5.4), (5.5)\}$, *then it also satisfies the remaining two relations.*

PROOF. The proof is given as follows.

(a) In view of (5.1), (5.2), (5.3) and (5.4), we have

$$\begin{aligned} &T(X,Y,Z,W) + T(X,Z,W,Y) + T(X,W,Y,Z) \\ &\quad = T(X,Y,Z,W) + T(Z,X,Y,W) + T(Y,Z,X,W) = 0, \end{aligned}$$

which gives (5.5).

(b) In view of (5.1), (5.2), (5.3) and (5.5), we have

$$\begin{aligned} &T(X,Y,Z,W) + T(Y,Z,X,W) + T(Z,X,Y,W) \\ &\quad = T(X,Y,Z,W) + T(X,W,Y,Z) + T(X,Z,W,Y) = 0, \end{aligned}$$

which gives (5.4).

(c) (See [**24**, Lee 2018, Proposition 7.12, p. 204]) From (5.4), we have

$$\begin{aligned}T(X,Y,Z,W)+T(Y,Z,X,W)+T(Z,X,Y,W)&=0,\\T(Y,Z,W,X)+T(Z,W,Y,X)+T(W,Y,Z,X)&=0,\\T(Z,W,X,Y)+T(W,X,Z,Y)+T(X,Z,W,Y)&=0,\\T(W,X,Y,Z)+T(X,Y,W,Z)+T(Y,W,X,Z)&=0.\end{aligned}$$

We add all these four equations. Applying (5.2) four times makes all the terms in the first and second columns cancel. Then applying (5.1) and (5.2) in the third column, we get

$$-2T(X,Z,Y,W)+2T(Y,W,X,Z)=0,$$

which is equivalent to (5.3).

(d) From (5.5), we have

$$\begin{aligned}T(X,Y,Z,W)+T(X,Z,W,Y)+T(X,W,Y,Z)&=0,\\T(Y,Z,W,X)+T(Y,W,X,Z)+T(Y,X,Z,W)&=0,\\T(Z,W,X,Y)+T(Z,X,Y,W)+T(Z,Y,W,X)&=0,\\T(W,X,Y,Z)+T(W,Y,Z,X)+T(W,Z,X,Y)&=0.\end{aligned}$$

We add all these four equations. Applying (5.1) four times makes all the terms in the first and third columns cancel. Then applying (5.1) and (5.2) in the second column, we get

$$-2T(X,Z,Y,W)+2T(Y,W,X,Z)=0,$$

which is equivalent to (5.3). A proof of the statement **(d)** is given in [**14**, Kobayashi and Nomizu 1963, Proposition 1.1, p. 198].

(e) Using (5.3) and (5.1) we have

$$T(X,Y,Z,W)=T(Z,W,X,Y)=-T(W,Z,X,Y)=-T(X,Y,W,Z),$$

which gives (5.2).

(f) Using (5.3) and (5.2) we have

$$T(X,Y,Z,W)=T(Z,W,X,Y)=-T(Z,W,Y,X)=-T(Y,X,Z,W),$$

which gives (5.1).

□

In view of the treatments given in [**14**, Kobayashi and Nomizu 1963, p. 198], [**19**, Kulkarni 1967, Section 8 of Chapter 1], [**20**, Kulkarni 1969], [**21**, Kulkarni 1970], [**22**, Kulkarni 1970], [**28**, Ornea and Romani 1993], and [**10**, Gilkey 2001, p. 23], we give the following Definition.

Definition 5.2. A tensor T of type $(0,4)$ in a vector space V is said to be a *Kulkarni-Nomizu tensor* if it satisfies any two among the three conditions $\{(5.1), (5.2), (5.3)\}$ and any one between the two conditions $\{(5.4), (5.5)\}$. The conditions (5.1), (5.2) and (5.3) are known to be *symmetry conditions* of Kulkarni-Nomizu tensor T, while the condition (5.4) or (5.5) is called the *algebraic Bianchi identity*. The equivalent $(1,3)$-tensor of a Kulkarni-Nomizu tensor T in an inner product space (V,g), given by

$$g(T(X,Y)Z,W)=T(X,Y,Z,W),$$

will also be called a *Kulkarni-Nomizu tensor*.

For a Kulkarni-Nomizu (1, 3)-tensor T in an inner product space (V, g), we have

$$T(X,Y)Z = -T(Y,X)Z, \tag{5.6}$$

$$g(T(X,Y)Z,W) = -g(T(X,Y)W,Z), \tag{5.7}$$

$$g(T(X,Y)Z,W) = g(T(Z,W)X,Y), \tag{5.8}$$

$$T(X,Y)Z + T(Y,Z)X + T(Z,X)Y = 0. \tag{5.9}$$

REMARK 5.3. In view of Lemma 5.1 and Definition 5.2, it follows that a Kulkarni-Nomizu tensor in a vector space V satisfies all the five conditions (5.1)–(5.5).

REMARK 5.4. For a "Kulkarni-Nomizu tensor", the phrase "*algebraic curvature tensor*" was used in [**2**, Besse 1987, Definition 1.108, p. 46], [**28**, Ornea and Romani 1993] and [**10**, Gilkey 2001, p. 23].

REMARK 5.5. For a "Kulkarni-Nomizu (1, 3)-tensor", the phrase "*curvature tensor*" (with respect to g) was used in [**19**, Kulkarni 1967, p. 4] and [**21**, Kulkarni 1970]. In [**15**, Kowalski 1972], the phrase "*curvature structure*" was used. For a more general treatment, we refer to [**16**, Kowalski 1973] and [**17**, Kowalski 1973].

EXAMPLE 5.6. In a Riemannian manifold (M, g), at each tangent space T_pM, the Riemann-Christoffel curvature tensor R of type $(0, 4)$ is obviously a Kulkarni-Nomizu tensor.

LEMMA 5.7. *The Kulkarni-Nomizu product $\Theta_1 \circledast \Theta_2$ of two symmetric $(0, 2)$-tensors Θ_1 and Θ_2 in a vector space V is a Kulkarni-Nomizu tensor.*

PROOF. The proof follows from Proposition 3.6. □

EXAMPLE 5.8. Let (V, g) be an inner product space. Let a $(0, 4)$-tensor R_0 in V be defined by

$$R_0(X,Y,Z,W) = g(Y,Z)g(X,W) - g(X,Z)g(Y,W) \tag{5.10}$$

for all vectors $X, Y, Z, W \in V$. Then R_0 is a Kulkarni-Nomizu tensor in (V, g). In fact,

$$R_0 = \frac{1}{2} g \circledast g, \tag{5.11}$$

and Lemma 5.7 is applicable for R_0.

LEMMA 5.9. *The Kulkarni-Nomizu product $\Theta_1 \circledast \Theta_2$ of two $(0, 2)$-tensors Θ_1 and Θ_2 in a vector space V, in which at least one is skew-symmetric, is not a Kulkarni-Nomizu tensor.*

PROOF. The proof follows from Proposition 3.7 and Proposition 3.8. □

EXAMPLE 5.10. Let V be a $2n$-dimensional vector space equipped with a complex structure J and a Hermitian inner product g, so that

$$J^2X = -X, \quad g(JX, JY) = g(X,Y), \qquad X, Y \in V.$$

Let Ω be the Kaehler form defined by

$$\Omega(X,Y) = g(X, JY), \qquad X, Y \in V.$$

Then the Kulkarni-Nomizu product $g \circledast \Omega$ satisfies the symmetry conditions (5.1), (5.2), but does not satisfy the symmetry condition (5.3) and the algebraic Bianchi identities (5.4) and (5.5). Since g is symmetric and Ω is skew-symmetric, therefore Proposition 3.7 is applicable here.

An n-dimensional Riemannian manifold (M, g) with Riemann-Christoffel curvature $(0, 4)$-tensor field R is said to be a *geometric realization* of a Kulkarni-Nomizu tensor T of an n-dimensional inner product space (V, g_V) at a point p of M if there exists an isometry ϕ from the tangent space T_pM to the vector space V so that $\phi^*T = R_p$:

$$g_V(\phi X, \phi Y) = g(X, Y),$$
$$T(\phi X, \phi Y, \phi Z, \phi W) = R_p(X, Y, Z, W)$$

for all $X, Y, Z, W \in T_pM$. It is known that every Kulkarni-Nomizu tensor is geometrically realizable by the germ of a Riemannian metric; so that, according to Gilkey, Kulkarni-Nomizu tensors become important in differential geometry (cf. [**10**, Gilkey 2001, Theorem 1.12.2, p. 23, p. 72]).

LEMMA 5.11. (see [**12**, Helgason 1962, Lemma 12.4, p. 68], [**14**, Kobayashi and Nomizu 1963, Proposition 1.2, p. 198]) *Suppose that T_1 and T_2 are Kulkarni-Nomizu tensors on a vector space V. If*

$$T_1(X, Y, Y, X) = T_2(X, Y, Y, X)$$

for every pair of vectors $X, Y \in V$, then

$$T_1 = T_2.$$

PROOF. Setting $T = T_1 - T_2$, and assuming

$$T(X, Y, Y, X) = 0, \qquad X, Y \in V,$$

it is sufficient to prove that $T = 0$. For any vectors $X, Y, Z \in V$, we have

$$0 = T(X + Y, Z, Z, X + Y) = 2T(X, Z, Z, Y).$$

From this it follows that

$$0 = T(X, Z + W, Z + W, Y) = T(X, Z, W, Y) + T(X, W, Z, Y).$$

Therefore, T is anti-symmetric in any adjacent pair of arguments. Now the algebraic Bianchi identity (5.4) yields

$$\begin{aligned} 0 &= T(X, Y, Z, W) + T(Y, Z, X, W) + T(Z, X, Y, W) \\ &= T(X, Y, Z, W) - T(Y, X, Z, W) - T(X, Z, Y, W) \\ &= T(X, Y, Z, W) + T(X, Y, Z, W) + T(X, Y, Z, W) = 3T(X, Y, Z, W). \end{aligned}$$

Alternatively, the algebraic Bianchi identity (5.5) yields

$$\begin{aligned} 0 &= T(X, Y, Z, W) + T(X, Z, W, Y) + T(X, W, Y, Z) \\ &= T(X, Y, Z, W) - T(X, Z, Y, W) - T(X, Y, W, Z) \\ &= T(X, Y, Z, W) + T(X, Y, Z, W) + T(X, Y, Z, W) = 3T(X, Y, Z, W). \end{aligned}$$

This completes the conclusion. □

LEMMA 5.12. (cf. [**18**, Kuhnel 2015, Theorem 6.5, p. 247], [**29**, Petersen 2016, Exercise 3.4.29, pp. 112-113]) *Suppose that T is a Kulkarni-Nomizu tensor in a vector space V. Suppose that*

$$\hat{T}(X, Y) = T(X, Y, Y, X) \tag{5.12}$$

for all vectors $X, Y \in V$*, so that* $\hat{T}(X,Y) = \hat{T}(Y,X)$*. Then* T *can be constructed from* $\hat{T}$ *by the following formula:*

(5.13)

$$\begin{aligned}6T(X,Y,Z,W) = \{&\hat{T}(X+W,Y+Z) - \hat{T}(X,Y+Z) - \hat{T}(W,Y+Z)\\ &- \hat{T}(X+W,Y) - \hat{T}(X+W,Z) + \hat{T}(X,Z) + \hat{T}(Y,W)\}\\ &- \{\hat{T}(Y+W,X+Z) - \hat{T}(Y,X+Z) - \hat{T}(W,X+Z)\\ &- \hat{T}(Y+W,X) - \hat{T}(Y+W,Z) + \hat{T}(Y,Z) + \hat{T}(X,W)\}\end{aligned}$$

for all vectors $X, Y, Z, W \in V$*.*

PROOF. See [**18**, Kuhnel 2015, Theorem 6.5, p. 247] and [**29**, Petersen 2016, Exercise 3.4.29, pp. 112-113]. □

6. Kulkarni-Nomizu tensor field

Because of several important works of Kulkarni and Nomizu, including [**19**, Kulkarni 1967], [**20**, Kulkarni 1969], [**21**, Kulkarni 1970], [**22**, Kulkarni 1970], [**25**, Nomizu 1971], and [**26**, Nomizu 1972], we propose the following:

DEFINITION 6.1. A field T of Kulkarni-Nomizu tensors in a smooth manifold M is called a *Kulkarni-Nomizu tensor field* in M. The equivalent $(1,3)$-tensor field of a Kulkarni-Nomizu tensor field T in a Riemannian manifold (M,g), given by

$$g(T(X,Y)Z,W) = T(X,Y,Z,W),$$

will also be called a *Kulkarni-Nomizu tensor field.*

Let (M,g) be a Riemannian manifold. A Kulkarni-Nomizu tensor field in (M,g) is also known as a *curvature-like tensor* (see [**7**, Chen et al. 2005], [**5**, Bolton et al. 2009]). A Kulkarni-Nomizu $(1,3)$-tensor field in (M,g) is also known as a *curvature structure* [**22**, Kulkarni 1970] or a *generalized curvature tensor field* [**25**, Nomizu 1971]. A Kulkarni-Nomizu $(1,3)$-tensor field T in (M,g) will be called a *proper Kulkarni-Nomizu tensor field* if it satisfies the differential Bianchi identity

$$(\nabla_X T)(Y,Z) + (\nabla_Y T)(Z,X) + (\nabla_Z T)(X,Y) = 0 \tag{6.1}$$

for all vector fields X, Y, Z on M, where ∇ the Levi-Civita connection. In a Riemannian manifold (M,g), a tensor field T of type $(0,4)$ will also be called a *proper Kulkarni-Nomizu tensor field* if its equivalent $(1,3)$-tensor field T given by

$$g(T(X,Y)Z,W) = T(X,Y,Z,W)$$

is a proper Kulkarni-Nomizu tensor field. It is clear that proper Kulkarni-Nomizu tensor fields are closed under addition and multiplication by real numbers. A proper Kulkarni-Nomizu tensor field is also known as a *Riemannian tensor field* [**11**, Gray 1966] or a *proper generalized curvature tensor field* [**25**, Nomizu 1971]. Also, we refer to [**30**, Roter 1977] and [**31**, Roter 1978].

REMARK 6.2. There is another concept of a *generalized curvature tensor field* in a Riemannian manifold with respect to its Riemannian submanifolds given in [**1**, Agashe 1987] as follows. Let M be a Riemannian submanifold of a Riemannian manifold $(\overline{M},g)$. Let $\overline{\nabla}$ be the Levi-Civita connection of $(\overline{M},g)$. Let Z be an

$\overline{M}$-vector field defined on M. Then the *generalized Riemann curvature tensor field* of $(\overline{M}, g)$ associated with Z is defined by

$$R^1(X,Y)Z = \overline{\nabla}_X\overline{\nabla}_Y Z - \overline{\nabla}_Y\overline{\nabla}_X Z - \overline{\nabla}_{[X,Y]}Z,$$

where X, Y are vector fields on M.

Let (M, g) be a Riemannian manifold. The *Ricci contraction* of a symmetric $(0,2)$-tensor field B in (M, g) is defined by ([**2**, Besse 1987, p. 46-47], [**35**, Viaclovsky 2011], [**36**, Z. Wang 2016])

$$\mathfrak{c}(B) = \sum_{i=1}^{\dim M} B(e_i, e_i), \tag{6.2}$$

where (e_i) is any orthonormal basis. The *Ricci contraction* of a Kulkarni-Nomizu tensor field T in (M, g) is a symmetric $(0,2)$-tensor field defined by ([**9**, Gallot-Hulin-Lafontaine 1987, p. 151], [**36**, Z. Wang 2016])

$$\mathfrak{c}(T)(Y,Z) = \sum_{i=1}^{\dim M} T(e_i, Y, Z, e_i), \tag{6.3}$$

where (e_i) is any orthonormal basis.

The T-*Ricci tensor* of a Kulkarni-Nomizu tensor field T is the symmetric $(0,2)$-tensor field S^T given by $S^T = \mathfrak{c}(T)$. The Ricci contraction of the Riemann-Christoffel curvature tensor field is the well-known Ricci tensor field, denoted by S. The Ricci contraction of the Kulkarni-Nomizu product $g \circledast B$ of a Riemannian metric g and a symmetric $(0,2)$-tensor field B in an n-dimensional Riemannian manifold (M, g) satisfies

$$\mathfrak{c}(g \circledast B) = (\dim M - 2)B + \mathfrak{c}(B)g. \tag{6.4}$$

Let (M, g) be an n-dimensional Riemannian manifold, and T a Kulkarni-Nomizu tensor field in M. Let Π^k be a k-plane section of T_pM and $(e_1, \ldots, e_k)$ an orthonormal basis of Π^k. Denoting

$$K_{ij}^T = T(e_i, e_j, e_j, e_i), \tag{6.5}$$

the T-k-*scalar curvature* $\tau^T(\Pi^k)$ of the k-plane section Π^k is defined by

$$\tau^T(\Pi^k) = \sum_{1 \le i < j \le k} K_{ij}^T. \tag{6.6}$$

The *normalized* T-k-*scalar curvature* of the k-plane section Π^k is defined as

$$\tau_{nor}^T(\Pi^k) = \frac{2\tau^T(\Pi^k)}{k(k-1)}. \tag{6.7}$$

For a plane section Π^2, the T-2-scalar curvature $\tau^T(\Pi^2)$ of Π^2 is exactly the T-*sectional curvature* $K^T(\Pi^2)$ of Π^2 (see [**7**, Chen et al. 2005], [**5**, Bolton et al. 2009]).

Let $\{e_1, \ldots, e_n\}$ be any orthonormal basis for the tangent space T_pM at $p \in M$. Then the T-n-scalar curvature $\tau^T(T_pM)$ of T_pM is the T-*scalar curvature* $\tau^T(p)$ of M at p, given by [**5**, Bolton et al. 2009]

$$\tau^T(p) = \sum_{1 \le i < j \le n} K_{ij}^T. \tag{6.8}$$

The *T-normalized scalar curvature* at p is defined as

$$\tau^T_{nor}(p) = \frac{2}{n(n-1)}\tau^T(p). \tag{6.9}$$

As in [**7**, Chen et al. 2005], *T-Chen invariant* is defined by

$$\delta^T(p) = \tau^T(p) - \inf\{K^T(\Pi^2) : \Pi^2 \subset T_pM\}. \tag{6.10}$$

For a plane section Π^2 of T_pM, we note that $K^T(\Pi^2) = \tau^T(\Pi^2) = \tau^T_{nor}(\Pi^2)$.

If T is replaced by the Riemann-Christoffel curvature tensor R, then T-sectional curvature K^T, T-scalar curvature τ^T, normalized T-scalar curvature τ^T_{nor}, T-k-scalar curvature $\tau^T(\Pi^k)$, normalized T-k-scalar curvature $\tau^T_{nor}(\Pi^k)$ and T-Chen invariant δ^T reduce to the *sectional curvature* K, the *scalar curvature* τ, the *normalized scalar curvature* τ_{nor}, *k-scalar curvature* $\tau(\Pi^k)$, *normalized k-scalar curvature* $\tau_{nor}(\Pi^k)$ and *Chen invariant* δ, respectively.

EXAMPLE 6.3. Let M be a smooth manifold equipped with a symmetric $(0,2)$-tensor field Θ and two skew-symmetric $(0,2)$-tensor fields Ω_1 and Ω_2. Then, the $(0,4)$-tensor field

$$\frac{1}{2}f_1\Theta \circledast \Theta + f_2\left(\frac{1}{2}\Omega_1 \circledast \Omega_2 - \Omega_1 \circledcirc \Omega_2\right) \tag{6.11}$$

is a Kulkarni-Nomizu tensor field, where f_1 and f_2 are some smooth functions on M. For $f_1 = 1, f_2 = 0$, one may compare (6.11) with the right hand side of Eq. (21) of [**8**, Chern 1955]; and for $f_1 = 0, f_2 = 1$ and $\Omega = \Omega_1 = \Omega_2$, one may compare (6.11) with the right hand side of Eq. (23) of [**8**, Chern 1955].

EXAMPLE 6.4. (cf. [**22**, Kulkarni 1970, Example (a)]) Let (M,g) be a Riemannian manifold. Then the $(0,4)$-tensor field R_0 defined by

$$R_0 = \frac{1}{2}g \circledast g$$

is a proper Kulkarni-Nomizu tensor field. We call R_0 the *trivial Kulkarni-Nomizu tensor field*. The R_0-sectional curvature K^{R_0} of R_0 is always 1. The equivalent proper Kulkarni-Nomizu $(1,3)$-tensor field R_0 is given by

$$R_0(X,Y)Z = g(Y,Z)X - g(X,Z)Y$$

for vector fields X, Y, Z in M.

EXAMPLE 6.5. (cf. [**11**, Gray 1966], [**22**, Kulkarni 1970, Example (b)]) Let (M,g) be a Riemannian manifold. Then the Riemann-Christoffel curvature tensor field R is a proper Kulkarni-Nomizu tensor field.

EXAMPLE 6.6. (cf. [**22**, Kulkarni 1970, Example (c)]) Let (M,g) be a Riemannian manifold. Let R be the Riemann-Christoffel curvature tensor field of (M,g), S its Ricci tensor field and Q its Ricci operator. Then the $(0,4)$-tensor field defined by

$$\mathcal{R}ic = g \circledast S$$

is a Kulkarni-Nomizu tensor field, called the *Ricci curvature structure*. The $\mathcal{R}ic$-sectional curvature $K^{\mathcal{R}ic}(\Pi^2) = \tau^{\mathcal{R}ic}(\Pi^2)$. The equivalent $(1,3)$-tensor field of $\mathcal{R}ic$ is given by

$$\mathcal{R}ic(X,Y)Z = g(Y,Z)QX - g(X,Z)QY + S(Y,Z)X - S(X,Z)Y$$

for all vector fields X, Y, Z in M. In general, every Kulkarni-Nomizu tensor field T gives rise to the corresponding Ricci-curvature structure $\mathcal{R}ic_T$.

EXAMPLE 6.7. Let (M, g) be a Riemannian manifold of dimension $n > 2$. Let T be a Kulkarni-Nomizu tensor field in (M, g). Then, the $(0,4)$-tensor field defined by

$$\mathfrak{W}^T = T - \frac{1}{n-2} g \circledast S^T + \frac{\tau^T}{(n-1)(n-2)} g \circledast g, \tag{6.12}$$

is a Kulkarni-Nomizu tensor field, called the *conformal curvature structure* (see [**22**, Kulkarni 1970, Example (d)]). In particular, Weyl's well known conformal curvature tensor field, given by

$$\mathfrak{W} = R - \frac{1}{n-2} g \circledast S + \frac{\tau}{(n-1)(n-2)} g \circledast g, \tag{6.13}$$

is a Kulkarni-Nomizu tensor field.

EXAMPLE 6.8. (cf. [**11**, Gray 1966]) In a complete connected Riemannian manifold (M, g), a $(0,4)$-tensor field T_{Real} defined by

$$T_{Real} = R - \frac{c}{2} g \circledast g,$$

where c is a constant, is a proper Kulkarni-Nomizu tensor field. It is well-known that $T_{Real} = 0$ if and only if M has constant sectional curvature c. A complete connected Riemannian manifold of constant section curvature is well known as a *real space form*. To the best knowledge of the author, the phrase 'real space form' was used first time by Takahashi in [**32**, Takahashi 1972].

EXAMPLE 6.9. (cf. [**11**, Gray 1966]) In a complete connected almost Hermitian manifold (M, J, g), let Ω be the Kaehler form given by $\Omega(X, Y) = g(X, JY)$ for all vector fields X, Y on M. If (M, J, g) is Kaehler, then the $(0,4)$-tensor field T_{Comp} given by

$$T_{Comp} = R - \frac{c}{8} g \circledast g - \frac{c}{4}\left\{\frac{1}{2}\Omega \circledast \Omega - \Omega \circledcirc \Omega\right\},$$

where c is a real constant, is a proper Kulkarni-Nomizu tensor field. In a Kaehler manifold (M, J, g), $T_{Comp} = 0$ if and only if (M, J, g) is of constant holomorphic sectional curvature c, that is, (M, J, g) is a *complex space form*. A Kaehler manifold of constant holomorphic sectional curvature was appeared first in [**37**, Yano and Mogi 1953].

EXAMPLE 6.10. In a complete connected quaternion Kaehler manifold (M, J_1, J_2, J_3, g), let Ω_i $(i = 1, 2, 3)$ be the quaternion-Kaehler forms given by $\Omega_i(X, Y) = g(X, J_i Y)$ for all vector fields X, Y on M. Suppose that a $(0,4)$-tensor field T_{Quat} is defined by

$$T_{Quat} = R - \frac{c}{8} g \circledast g - \frac{c}{4}\sum_{i=1}^{3}\left\{\frac{1}{2}\Omega_i \circledast \Omega_i - \Omega_i \circledcirc \Omega_i\right\},$$

where c is a real constant. Then T_{Quat} is a Kulkarni-Nomizu tensor field. In a quaternion Kaehler manifold (M, J_1, J_2, J_3, g), the Kulkarni-Nomizu tensor field T_{Quat} vanishes if and only if (M, J_1, J_2, J_3, g) is of constant quaternion sectional curvature c, that is, (M, J_1, J_2, J_3, g) is a *quaternion space form*. It may be interesting to check whether T_{Quat} is a proper Kulkarni-Nomizun tensor field. It is well-known that quaternion space forms are locally symmetric [**13**, Ishihara 1974]. For some details of quaternion space form, we refer to [**13**, Ishihara 1974] and [**6**, Chen 1978].

EXAMPLE 6.11. In a complete connected almost contact metric manifold $(M, \varphi, \xi, \eta, g)$, let T_{Sas} be a $(0,4)$-tensor field defined by

$$T_{Sas} = R - \frac{c+3}{8} g \circledast g - \frac{c-1}{4}\left\{\frac{1}{2}\Phi \circledast \Phi - \Phi \circledcirc \Phi\right\} + \frac{c-1}{4} g \circledast (\eta \otimes \eta),$$

where c is a real constant. If $(M, \varphi, \xi, \eta, g)$ is a Sasakian manifold; then it can be proved that T_{Sas} is a Kulkarni-Nomizu tensor field. In a Sasakian manifold of dimension ≥ 5, $T_{Sas} = 0$ if and only if the Sasakian manifold is of constant φ-sectional sectional curvature c (see [**27**, Ogiue 1964, Theorem 2]). It may be interesting to check whether T_{Sas} is a proper Kulkarni-Nomizu tensor field. For details of almost contact metric manifolds and Sasakian manifolds we refer to [**4**, Blair 2010].

EXAMPLE 6.12. (cf. [**7**, Chen et al. 2005], [**33**, Tripathi 2011]) Let (M, g) be an n-dimensional Riemannian manifold and (B, g_B) a Riemannian vector bundle over M. If ζ is a B-valued symmetric $(1,2)$-tensor field and T a $(0,4)$-tensor field on M such that

$$T(X,Y,Z,W) = g_B(\zeta(X,W), \zeta(Y,Z)) - g_B(\zeta(X,Z), \zeta(Y,W)) \tag{6.14}$$

for all vector fields X, Y, Z, W on M, then the equation (6.14) is said to be an *algebraic Gauss equation.* If T is a $(0,4)$-tensor field on M which satisfies (6.14), then T is Kulkarni-Nomizu tensor field. A typical example of an algebraic Gauss equation is given for a submanifold M of an Euclidean space, if B is the normal bundle, ζ the second fundamental form and T the Riemann-Christoffel curvature tensor field of M. Some nice situations, in which such T and ζ satisfy an algebraic Gauss equation exist, are Lagrangian and Kaehlerian slant submanifolds of complex space forms, C-totally real submanifolds of Sasakian space forms, and positive definite centroaffine hypersurfaces.

7. Schur's theorem for proper Kulkarni-Nomizu tensor field

Taking into account the treatment given in [**14**, Kobayashi and Nomizu 1963, pp. 199-200], we give the definition of T-sectional curvature as follows.

Let (V, g) be an inner product space of dimension $n \geq 3$, and T a Kulkarni-Nomizu tensor in (V, g). Let Π^2 be a 2-dimensional subspace of V. Such a subspace Π^2 is called a 2-plane section of V. Let (X, Y) be a basis for the plane section Π^2. We define the *T-sectional curvature* $K^T(\Pi^2)$ of Π^2 by

$$K^T(\Pi^2) = \frac{T(X,Y,Y,X)}{g(X,X)g(Y,Y) - g(X,Y)^2}. \tag{7.1}$$

Let (X', Y') be another basis for the plane section Π^2. Then

$$X' = aX + bY, \qquad Y' = cX + dY,$$

such that

$$ad - bc \neq 0.$$

Using the symmetry properties of Kulkarni-Nomizu tensor T, it is well-known that

$$T(X', Y', Y', X') = (ad - bc)^2 T(X, Y, Y, X),$$

$$g(X',X')g(Y',Y') - g(X',Y')^2 = (ad-bc)^2(g(X,X)g(Y,Y) - g(X,Y)^2).$$

In view of the above two relations, it follows that $K^T(\Pi^2)$ does not depend on the choice of the basis for the plane section Π^2, and this makes the definition of $K^T(\Pi^2)$ meaningful. Similarly, for a Kulkarni-Nomizu $(1,3)$-tensor T with respect to g, the *T-sectional curvature* $K^T(\Pi^2)$ of a 2-plane Π^2 spanned by $\{X,Y\}$ is defined by [**19**, Kulkarni 1967, p. 4]

$$K^T(\Pi^2) = K^T(X,Y) = \frac{g(T(X,Y)Y,X)}{g(X,X)g(Y,Y) - g(X,Y)^2}.$$

EXAMPLE 7.1. Let (V,g) be an n-dimensional inner product space. Let Π^2 be any plane section of V. Then the R_0-sectional curvatures $K^{R_0}(\Pi^2)$, of the trivial Kulkarni-Nomizu tensor R_0 given by (5.10), satisfies

$$K^{R_0}(\Pi^2) = 1.$$

One important class of quantitative information provided by a Kulkarni-Nomizu tensor T is the T-sectional curvature of all plane sections. It turns out that this is the only information contained in Kulkarni-Nomizu tensor T. In fact, T-sectional curvatures $K^T(\Pi^2)$ completely determine Kulkarni-Nomizu tensor T as shown in the following Proposition.

PROPOSITION 7.2. *Suppose that T_1 and T_2 are Kulkarni-Nomizu tensors on an n-dimensional inner product space (V,g). If*

$$\frac{T_1(X,Y,Y,X)}{g(X,X)g(Y,Y) - g(X,Y)^2} = \frac{T_2(X,Y,Y,X)}{g(X,X)g(Y,Y) - g(X,Y)^2}$$

for every pair of independent vectors $X, Y \in V$, then

$$T_1 = T_2.$$

PROOF. The proof follows from Lemma 5.11. □

THEOREM 7.3. **(Schur's theorem for Kulkarni-Nomizu tensor)** (cf. [**14**, Kobayashi and Nomizu 1963, Proposition 1.4, p. 200], [**19**, Kulkarni 1967, Lemma 1, p. 5]) *Suppose that T is a Kulkarni-Nomizu tensor in an n-dimensional inner product space (V,g). The T-sectional curvature*

$$K^T(\Pi^2) = c$$

for all plane sections Π^2 of V if and only if

$$T = cR_0 = \frac{c}{2} g \circledast g. \tag{7.2}$$

PROOF. From

$$c = K^T(\Pi^2) = \frac{T(X,Y,Y,X)}{g(X,X)g(Y,Y) - g(X,Y)^2},$$

we have $T(X,Y,Y,X) = cR_0(X,Y,Y,X)$ for all $X, Y \in V$. Applying Lemma 5.11 to T and cR_0, we get (7.2). The converse is straightforward. □

An orthonormal basis of an inner product space (V,g) is called a *frame*. An orthonormal set of vectors (e_i) is said to be an *acceptable frame* if for every triple of distinct indices i,j,k, the values $K^T(e_i,e_j)$, $K^T(e_j,e_k)$, $K^T(e_k,e_i)$ are all distinct (see [**19**, Kulkarni 1967, p. 5]).

LEMMA 7.4. (cf. [**19**, Kulkarni 1967, Lemma 2, p. 5]) *Let (V,g) be an inner product space of dimension $n \geq 3$, and T a Kulkarni-Nomizu tensor in (V,g). If $K^T \neq$ constant, then (V,g) admits an acceptable frame.*

REMARK 7.5. (cf. [**19**, Kulkarni 1967, Remark, p. 8]) Let (V,g) be an inner product space of dimension $n > 3$, and T a Kulkarni-Nomizu tensor in (V,g). Suppose that $K^T \neq$ constant. Then for a frame $(e_1,\ldots,e_n)$, all $K^T(e_i,e_j)$, $i,j \in \{1,\ldots,n\}$, $i \neq j$, need not be different. A qualitative reason is that the number of independent components $T_{ijk\ell}$ is $n^2(n^2-1)/12$ (but only $n(n-1)^2/2$ with repeated indices). For $n > 3$, the condition $K^T \neq$ constant sets up too few restrictions. As an explicit counterexample, we observe that, in an Einstein manifold of dimension 4 (not of constant sectional curvature), mutually orthogonal 2-plane sections always yield the same sectional curvature.

THEOREM 7.6. (cf. [**19**, Kulkarni 1967, Theorem 1, pp. 4-5]) *Let (V_1,g_1) and (V_2,g_2) be two inner product spaces of dimension $n \geq 3$. Let T_1 and T_2 be two Kulkarni-Nomizu tensors, with respect to g_1 and g_2, respectively; and let K^{T_1}, K^{T_2} be corresponding sectional curvatures. Suppose that $K^{T_1} \neq$ constant, and $f : V_1 \to V_2$ a sectional curvature preserving vector space isomorphism, then f is a homothety.*

Now, we turn our attention to Kulkarni-Nomizu tensor fields in Riemannian manifolds.

DEFINITION 7.7. Let (M,g) be a Riemannian manifold, and T a Kulkarni-Nomizu tensor field in (M,g). If the T-sectional curvatures at a point $p \in (M,g)$ all have the same value c_p^T, then (M,g) is said to be *T-isotropic* at the point $p \in M$, and p is called a *T-isotropic point* of (M,g). Moreover, (M,g) is said to be *T-isotropic*, if every point $p \in M$ is a T-isotropic point of (M,g).

LEMMA 7.8. *A point $p \in (M,g)$ is a T-isotropic point if and only if*

$$T(X,Y,Z,W) = c_p^T\left(g(Y,Z)g(X,W) - g(X,Z)g(Y,W)\right) \tag{7.3}$$

for all vectors $X,Y,Z,W \in T_pM$, where c_p^T is a real constant.

PROOF. Assume that $K^T\left(\Pi^2\right) = c_p^T$ for all sections $\Pi^2 \subset T_pM$. Then, we get

$$T(X,Y,Y,X) = c_p^T R_0\left(X,Y,Y,X\right)$$

for all vectors $X,Y \in T_pM$, where R_0 is defined by (5.10). Hence, by Lemma 5.11, we have $T = c_p^T R_0$. The converse is obvious. □

THEOREM 7.9. **(Schur's theorem for proper Kulkarni-Nomizu tensor field)** *Let (M,g) be a Riemannian manifold of dimension $n \geq 3$, ∇ the Levi-Civita connection of g, and T a proper Kulkarni-Nomizu tensor field in (M,g). If (M,g) is of isotropic T-sectional curvature, then it is of constant T-sectional curvature.*

PROOF. Suppose that M is T-isotropic at each point $p \in M$, that is,

$$T = c^T R_0, \tag{7.4}$$

where c^T is a smooth function on M. Since $\nabla g = 0$, we have $\nabla R_0 = 0$. Hence, we have

$$\left(\nabla_W T\right) = \left(\nabla_W c^T\right) R_0,$$

from which we get

$$((\nabla_W T)(X,Y))Z = (W(c^T))(g(Y,Z)X - g(X,Z)Y). \tag{7.5}$$

Taking the cyclic sum of the equation (7.5) with respect to W, X, Y and applying the differential Bianchi identity (6.1), we get

$$\begin{aligned} 0 &= (W(c^T))(g(Y,Z)X - g(X,Z)Y) \\ &\quad + (X(c^T))(g(W,Z)Y - g(Y,Z)W) \\ &\quad + (Y(c^T))(g(X,Z)W - g(W,Z)X). \end{aligned} \tag{7.6}$$

Since $\dim(M) \geq 3$, we can choose X, Y, Z, W in such a manner that X, Y, $Z = W$ are mutually orthogonal and $g(Z,Z) = 1$. Then from (7.6) we obtain

$$\left(X(c^T)\right)Y - \left(Y(c^T)\right)X = 0.$$

Since X and Y are linearly independent, therefore it follows that

$$X(c^T) = Y(c^T) = 0.$$

Since X can be chosen as an arbitrary vector field, therefore the function c^T is a constant. □

PROBLEM 7.10. *To prove or disprove that, for a non-proper Kulkarni-Nomizu tensor field in a Riemannian manifold, Schur's theorem does not hold.*

8. Some applications

If a $(0,4)$-tensor field satisfies the algebraic Gauss equation for a Riemannian vector bundle of a Riemannian manifold, then it becomes a Kulkarni-Nomizu tensor field. A series of sharp inequalities (like for δ-invariant, Chen invariant, different Casorati curvatures, Wintgen curvature etc.) can be/have been obtained in this situation, and this theory can be applied in different situations, like Riemannian submanifolds of real space forms, different kinds of submanifolds of complex space forms and their generalizations (see [**34**, Tripathi 2021]), different kinds of submanifolds of quaternion space forms and their generalizations, different kinds of submanifolds of different kinds of space forms in contact geometry, different kinds of submanifolds of different kinds of space forms in the geometry of Yano f-structure, different kinds of submanifolds of complex two-plane Grassmannians, and in hypersurfaces in affine spaces as well.

Acknowledgment

The author is grateful to the honoured referee for several striking comments towards the improvement of the paper.

References

[1] Nirmala S. Agashe, *Generalized curvature tensors of a Riemannian manifold $\bar{M}$*, Tensor (N.S.) **46** (1987), 92–94. Zbl 0685.53021

[2] Arthur L. Besse, *Einstein manifolds*, Classics in Mathematics, Springer-Verlag, Berlin, 2008. Reprint of the 1987 edition. MR2371700

[3] G. Blaickner, *Etymology "Kulkarni-Nomizu product"*, December 30, 2022. `https://mathoverflow.net/questions/437531/etymology-kulkarni-nomizu-product`

[4] David E. Blair, *Riemannian geometry of contact and symplectic manifolds*, 2nd ed., Progress in Mathematics, vol. 203, Birkhäuser Boston, Ltd., Boston, MA, 2010, DOI 10.1007/978-0-8176-4959-3. MR2682326

[5] John Bolton, Franki Dillen, Johan Fastenakels, and Luc Vrancken, *A best possible inequality for curvature-like tensor fields*, Math. Inequal. Appl. **12** (2009), no. 3, 663–681, DOI 10.7153/mia-12-50. MR2540985

[6] Bang-yen Chen, *Totally umbilical submanifolds of quaternion-space-forms*, J. Austral. Math. Soc. Ser. A **26** (1978), no. 2, 154–162. MR511599

[7] Bang-Yen Chen, Franki Dillen, and Leopold Verstraelen, *δ-invariants and their applications to centroaffine geometry*, Differential Geom. Appl. **22** (2005), no. 3, 341–354, DOI 10.1016/j.difgeo.2005.01.008. MR2166127

[8] Shiing-shen Chern, *On curvature and characteristic classes of a Riemann manifold*, Abh. Math. Sem. Univ. Hamburg **20** (1955), 117–126, DOI 10.1007/BF02960745. MR75647

[9] Sylvestre Gallot, Dominique Hulin, and Jacques Lafontaine, *Riemannian geometry*, Universitext, Springer-Verlag, Berlin, 1987, DOI 10.1007/978-3-642-97026-9. MR909697

[10] Peter B. Gilkey, *Geometric properties of natural operators defined by the Riemann curvature tensor*, World Scientific Publishing Co., Inc., River Edge, NJ, 2001, DOI 10.1142/9789812799692. MR1877530

[11] Alfred Gray, *Spaces of constancy of curvature operators*, Proc. Amer. Math. Soc. **17** (1966), 897–902, DOI 10.2307/2036279. MR198392

[12] Sigurđur Helgason, *Differential geometry and symmetric spaces*, Pure and Applied Mathematics, Vol. XII, Academic Press, New York-London, 1962. MR145455

[13] Shigeru Ishihara, *Quaternion Kählerian manifolds*, J. Differential Geometry **9** (1974), 483–500. MR348687

[14] Shoshichi Kobayashi and Katsumi Nomizu, *Foundations of differential geometry. Vol I* (1963), xi+329. MR0152974

[15] Oldřich Kowalski, *On regular curvature structures*, Math. Z. **125** (1972), 129–138, DOI 10.1007/BF01110924. MR295250

[16] Oldřich Kowalski, *Partial curvature structures and conformal geometry of submanifolds*, J. Differential Geometry **8** (1973), 53–70. MR358608

[17] Oldřich Kowalski, *On the Gauss-Kronecker curvature tensors*, Math. Ann. **203** (1973), 335–343, DOI 10.1007/BF01351912. MR353201

[18] Wolfgang Kühnel, *Differential geometry: Curves-surfaces-manifolds*, Student Mathematical Library, vol. 77, American Mathematical Society, Providence, RI, 2015. Third edition [of MR1882174]; Translated from the 2013 German edition by Bruce Hunt, with corrections and additions by the author, DOI 10.1090/stml/077. MR3443721

[19] Ravindra Shripad Kulkarni, *Curvature and metric*, ProQuest LLC, Ann Arbor, MI, 1968. Thesis (Ph.D.)–Harvard University. MR2939822

[20] Ravindra Shripad Kulkarni, *Curvature structures and conformal transformations*, Bull. Amer. Math. Soc. **75** (1969), 91–94, DOI 10.1090/S0002-9904-1969-12155-5. MR233306

[21] Ravindra Shripad Kulkarni, *Curvature and metric*, Ann. of Math. (2) **91** (1970), 311–331, DOI 10.2307/1970580. MR257932

[22] Ravindra Shripad Kulkarni, *Curvature structures and conformal transformations*, J. Differential Geometry **4** (1970), 425–451. MR284949

[23] Ravindra Shripad Kulkarni, *On the Bianchi Identities*, Math. Ann. **199** (1972), 175–204, DOI 10.1007/BF01429873. MR339004

[24] John M. Lee, *Introduction to Riemannian manifolds*, Graduate Texts in Mathematics, vol. 176, Springer, Cham, 2018. Second edition of [MR1468735]. MR3887684

[25] Katsumi Nomizu, *On the spaces of generalized curvature tensor fields and second fundamental forms*, Osaka Math. J. **8** (1971), 21–28. MR298587

[26] Katsumi Nomizu, *On the decomposition of generalized curvature tensor fields. Codazzi, Ricci, Bianchi and Weyl revisited*, Differential geometry (in honor of Kentaro Yano), Kinokuniya Book Store, Tokyo, 1972, pp. 335–345. MR383296

[27] Koichi Ogiue, *On almost contact manifolds admitting axiom of planes or axiom of free mobility*, Kōdai Math. Sem. Rep. **16** (1964), 223–232. MR172223

[28] Liviu Ornea and Giuliano Romani, *The fundamental equations of conformal submersions*, Beiträge Algebra Geom. **34** (1993), no. 2, 233–243. MR1264292

[29] Peter Petersen, *Riemannian geometry*, 3rd ed., Graduate Texts in Mathematics, vol. 171, Springer, Cham, 2016, DOI 10.1007/978-3-319-26654-1. MR3469435

[30] Witold Kazimierz Roter, *On generalized curvature tensors on some Riemannian manifolds*, Colloq. Math. **37** (1977), no. 2, 233–240, DOI 10.4064/cm-37-2-233-240. MR458325

[31] Witold Kazimierz Roter, *Generalized curvature tensors on conformally symmetric manifolds*, Colloq. Math. **39** (1978), no. 1, 45–54, DOI 10.4064/cm-39-1-45-54. MR500645

[32] Toshio Takahashi, *A note on Kählerian hypersurfaces of spaces of constant curvature*, Kumamoto J. Sci. (Math.) **9** (1972), 21–24. MR301657

[33] Mukut Mani Tripathi, *Improved Chen-Ricci inequality for curvature-like tensors and its applications*, Differential Geom. Appl. **29** (2011), no. 5, 685–698, DOI 10.1016/j.difgeo.2011.07.008. MR2831825

[34] Mukut Mani Tripathi, *Different kinds of submanifolds of almost Hermitian manifolds*, Proceedings of The 23rd International Differential Geometry Workshop on Submanifolds in Homogeneous Spaces and Related Topics (Kyungpook National University, Korea) **23** (2021), 59-90.

[35] Jeffrey Alan Viaclovsky, *Topics in Riemannian Geometry* (Math 865), Fall 2011. `https://www.math.uci.edu/~jviaclov/courses/865_Fall_2011.pdf`

[36] Zouqin Wang, *Riemannian geometry, Lecture* 07: *The Weyl curvature tensor*, `http://staff.ustc.edu.cn/~wangzuoq/Courses/16S-RiemGeom/Notes/Lec07.pdf`

[37] Kentaro Yano and Isamu Mogi, *Sur les variétés pseudokählериennes à courbure holomorphique constante* (German), C. R. Acad. Sci. Paris **237** (1953), 962–964. MR58290

Department of Mathematics, Institute of Science, Banaras Hindu University, Varanasi 221005, India

Email address: `mmtripathi66@yahoo.com`

Selected Published Titles in This Series

811 **Krishnendu Gongopadhyay and Shashikant A. Katre, Editors,** Geometry, Groups and Mathematical Philosophy, 2025

810 **Robert J. Berman and Yanir A. Rubinstein, Editors,** Convex and Complex: Perspectives on Positivity in Geometry, 2025

809 **Fabrice Baudoin and Luca Rizzi, Editors,** New Trends in Sub-Riemannian Geometry, 2025

808 **Loring W. Tu, Editor,** Group Actions and Equivariant Cohomology, 2024

807 **Galina Filipuk, Editor,** Recent Progress in Special Functions, 2024

806 **Maximilian Engel, Hildeberto Jardón-Kojakhmetov, and Cinzia Soresina, Editors,** Topics in Multiple Time Scale Dynamics, 2024

805 **Anthony Iarrobino, Pedro Macias Marques, Maria Evelina Rossi, and Jean Vallès, Editors,** Deformation of Artinian Algebras and Jordan Type, 2024

804 **Mahir Bilen Can and Jörg Feldvoss, Editors,** A Glimpse into Geometric Representation Theory, 2024

803 **Peter Gothen, Margarida Melo, and Montserrat Teixidor i Bigas, Editors,** Moduli Spaces and Vector Bundles—New Trends, 2024

802 **Ralph M. Kaufmann, Martin Markl, and Alexander A. Voronov, Editors,** Higher Structures in Topology, Geometry, and Physics, 2024

801 **K. A. Brown, T. J. Hodges, M. Vancliff, and J. J. Zhang, Editors,** Recent Advances in Noncommutative Algebra and Geometry, 2024

800 **Avinoam Mann, Louis H. Rowen, David J. Saltman, Aner Shalev, Lance W. Small, and Uzi Vishne, Editors,** Amitsur Centennial Symposium, 2024

799 **Alberto A. Condori, Elodie Pozzi, William T. Ross, and Alan A. Sola, Editors,** Recent Progress in Function Theory and Operator Theory, 2024

798 **Marat V. Markin, Igor V. Nikolaev, and Carsten Trunk, Editors,** Advances in Functional Analysis and Operator Theory, 2024

797 **Sangita Jha, Mrinal Kanti Roychowdhury, and Saurabh Verma, Editors,** Recent Developments in Fractal Geometry and Dynamical Systems, 2024

796 **John Cremona, John Jones, Jennifer Paulhus, Andrew V. Sutherland, and John Voight, Editors,** LuCaNT: LMFDB, Computation, and Number Theory, 2024

795 **Michael A. Jones, David McCune, and Jennifer M. Wilson, Editors,** Mathematical Analyses of Decisions, Voting and Games, 2024

794 **Jean-Pierre Magnot, Editor,** Recent Advances in Diffeologies and Their Applications, 2024

793 **Abba Gumel, Editor,** Mathematical and Computational Modeling of Phenomena Arising in Population Biology and Nonlinear Oscillations, 2024

792 **Shaoming Guo, Zane Kun Li, and Brian Street, Editors,** Recent Developments in Harmonic Analysis and its Applications, 2024

791 **Mee Seong Im, Bach Nguyen, and Arik Wilbert, Editors,** Algebraic and Topological Aspects of Representation Theory, 2024

790 **Peter Crooks and Alexandru I. Suciu, Editors,** Compactifications, Configurations, and Cohomology, 2023

789 **I. S. Krasil′shchik, A. B. Sossinsky, and A. M. Verbovetsky, Editors,** The Diverse World of PDEs, 2023

788 **I. S. Krasil′shchik, A. B. Sossinsky, and A. M. Verbovetsky, Editors,** The Diverse World of PDEs, 2023

787 **Hemen Dutta, Editor,** Mathematical Modelling, 2023

786 **Hemen Dutta, Editor,** Mathematical Modelling, 2023

For a complete list of titles in this series, visit the
AMS Bookstore at **www.ams.org/bookstore/conmseries/**.